BIOCHEMISTRY RESEARCH TRENDS

CAFFEIC ACID

BIOLOGICAL PROPERTIES, STRUCTURE AND HEALTH EFFECTS

Biochemistry Research Trends

Additional books in this series can be found on Nova's website
under the Series tab.

Additional e-books in this series can be found on Nova's website
under the e-book tab.

BIOCHEMISTRY RESEARCH TRENDS

CAFFEIC ACID

BIOLOGICAL PROPERTIES, STRUCTURE AND HEALTH EFFECTS

LEANNA VAUGHN
EDITOR

nova publishers
New York

NOTICE TO THE READER

Library of Congress Cataloging-in-Publication Data

ISBN: 978-1-63483-121-5
Library of Congress Control Number: 2015942063

Published by Nova Science Publishers, Inc. † New York

CONTENTS

PREFACE

This book addresses research advances on caffeic acid, a naturally occurring hydroxycinnamic acid with various biological and pharmacological properties. Some of the topics discusses in this book include the derivatives of caffeic acid; the role it plays in winemaking; the comprehensive understanding of anti-diabetic and anti-obesity effects of coffe intake via nutrigenomics; the antimicrobial properties; and other useful properties of caffeic acid.

Chapter 1 - With rapid socio-ecomonic advancement and compulsion for cosmopolitan human society, the pressure of overwork is bending the society to its toe. Therefore, the result is different neurological and neurovascular diseases, cardiac diseases and other brain diseases are growingly increased day by day. Human food habits are strikingly different from one country to other and even with the difference of their religions, culture, society structures and economic conditions. Indian "Vedas" is first text to describe the deep rooted relation between foods & mind. The type of foods and therefore its importance for health have an interim connection with the change of climate and play a great role to build mind and health of the individual and the society as a whole. Upanishads tell that 'Annam Brahma' (i.e., Food is God) and state the quality of the mind is influenced by the quality of the food intake. So in this above context this article tries to critically describe an eternal relation of food and healthy mind in terms of antioxidant property. Here through this article the authors mainly want to highlight the role of caffeic acid because caffeic acid is a globally available antioxidant compound and since the last more than a decade researchers have found that caffeic acid and its derivatives have lots of effects on human mind and thus on health. Neurological studies have shown that caffeic acid can create protective effects on neurodegeneration, neurotoxicity and has potential cytoprotective effects against a wide range of injuries in different tissues, including the brain.

Chapter 2 - Caffeic acid is a hydroxycinnamic acid that is involved in a large number of chemical interactions with other grape and wine components. Caffeoylglucoside anthocyanins have shown increased stability compared with monoglucoside anthocyanins because their acylated structures allows for intramolecular copigmentation. During wine aging, the phenyl pyranonathocyanin Pinotin A is formed from the direct reaction of caffeic acid with malvidin3-glucoside. This compound shows specific red color characteristics (i.e., hypsochromic shift from 520 nm) and a higher resistance to pH changes and SO_2 addition (i.e., increased stability over time) than malvidin-3-glucoside. In addition, intermolecular copigmentation interactions between anthocyanins and caffeic acid through weak hydrophobic forces exhibit both a hyperchromic (increased maximum absorbance) and a bathochromic shift towards bluish hues. These interactions increase the anthocyanin extraction during the early stages of winemaking and are thought to be the first step in the stabilization of wine color. On the other hand, caffeic acid also plays an important role in white wine browning reactions. Furthermore, caffeic acid elicits moderate puckering astringency and bitterness; however, its impact on overall wine taste remains unclear. Finally, caffeic acid could also be used as a potential inhibitor of aroma loss during aging with a consequent reduction in the SO_2 addition during winemaking. The participation of caffeic acid in a large number of mechanisms strongly verifies the important role of this molecule in wine chemistry.

Chapter 3 - In recent years, there has been rapid progress in high-throughput technologies and platforms to simultaneously assay global gene expression, mRNA, proteins, metabolites and gut microbiota leading to the promotion of "nutrigenomics" in the field of food and nutrition science. Nutrigenomics, which comprises of genomics, transcriptomics, proteomics, metabolomics, metagenomics and other omics technologies, is a discipline in which all available information about the genome and other biological molecules is effectively utilized to unveil every detail of the interactions between diets and the human body. Coffee, one of the most popular beverages in the world, contains major anti-oxidative phenolic compounds: caffeic acid (CA) and chlorogenic acids (CGA) which is an ester of CA and (-) quinic acid. Epidemiologic findings have associated coffee drinking to numerous health benefits ranging from protecting against metabolic diseases, and non-metabolic diseases like depression and skin cancer. In this chapter, the authors describe in detail the elucidation of the molecular mechanisms of coffee and its constituents: CA and CGA; dietary supplementation on metabolic diseases like obesity and Type 2 Diabetes (T2D) by utilizing the nutrigenomic approach. The authors believe that it would be able to provide evaluation of the effects of coffee and its constituents on host physiology, which is expected to be useful and informative, shaping the future of coffee and the other nutraceutical's research.

Chapter 4 - The resistance of microorganisms to drugs in pharmaceutical industry and the food contamination and spoilage by microorganisms in food industry are serious problems. Control of these microorganisms is of importance not only for industry but also for government health agencies. As a result of this problematic, all these years an intensive search for new compounds able to replace currently used drugs was performed. The utilization of phenolic compounds as antimicrobial agents or a possible pharmaceutical drug would provide additional benefits due to their health beneficial properties. Caffeic acid is one of the most abundant hydroxycinnamic acids in the human diet and chlorogenic acid (5-O-caffeoyl-quinic acid), an ester of caffeic acid with quinic acid, has been received considerable attentions for their potential biological effects. Chlorogenic acid has antioxidant property and probable roles in diseases such as cancer, parkinson and cardiovascular as well as antimicrobial activity. In this chapter, the authors present the last researches that indicate that chlorogenic acid of different plant sources or derived products could be a possible candidate to be utilized in different industries. The advantages in the usefulness of chlorogenic acid as antimicrobial agent are discussed in relation with its antimicrobial effectivity and its health benefits.

Chapter 5 - Caffeic acid (CA) is one of the most abundant hydroxycinnamic acids in plants, fruits, coffee, and propolis. CA and its derivative CAPE (CA phenethyl ester) have been well characterized on their aspects of chemistry, biochemistry, immunnopharmacology, molecular biology, and cell biology. During the past two decades, remarkable progress in the potential benefits of CA and CAPE has been documented. CA and its natural derivatives are multi-functional and non-toxic stabilizing compounds, which can thus be salubrious for food additives, cosmetic, drugs, pharmaceutics, and overall well-being. Chemically, they easily chelate cations such as Zn^{++}, Mn^{++}, and Fe^{++} and modulate metal enzyme activities such as matrix metalloproteinases. This specific property has been attributed to the anti-angiogenic and tissue remodeling capacities of the compounds. In intracellular signaling pathway, they effectively inhibit the phosphorylation of kinases such as mitogen-activated protein kinase (MAPK) and transcription factors such as nuclear factor-κB (NF-κB), thus contributing to the down-regulation of specific genes and cellular phenotype changes. The cellular regulatory functions have stimulated the researchers to study the pharmacological activities of CA and CAPE. However, there has been no definitive conclusion about how to regulate the cell functions when they initially meet the cells, more specifically how to interact with the cellular receptors or counter-parted molecules on the cell surface.

CA, as a leader compound, can potentially be applied for new functional and chemotherapeutic agents using fused chemi-biological synthesis in oxidative stress-related diseases such as inflammation, cell transformation, and degenerative pathology. It is also a candidate for cooperative adjuvant therapy in various human diseases, such as degenerative diseases, inflammatory diseases, and transformed cancer. It is assumed that in contrast to conventional medicines, CA does not have side effects. Understanding the mechanism of the interaction of CA with biological targets at the molecular level will allow researchers to explain its positive effects on human. Chemobiological fusion science, using CA as the chemical scaffold, will construct new drugs for human diseases because the creative drug design of small functional compounds is timely and requisite for various human diseases. Medical scientists know that the advent of Aspirin as an active ingredient of acetylsalicylic acid enhanced human health and welfare since the twentieth century. As acetylsalicylic acid is also a small leader compound, CA should be classified into that category. That is the reason why a naturally occurring CA is useful for drug design by scaffold structural variation and chemobiological fusion. CA should be used as a medium and substrate for the creation and development of new chemotherapeutic agents for many human diseases, several of which are fundamentally based on different origins. One should note that the naturally occurring CA derivative, CAPE, is relatively more active against specific inhibition to signaling pathways, such as intracellular phosphorylation and transcription, than the original CA. To date, the chemical syntheses of ester and amide derivatives has progressed. The present-day studies on CA and CAPE have thus, been analyzed for a future direction in their application to health fields. At present, a comprehensive series of scholarly monographs, which describes CA and related derivatives, are needed for drug design and pharmaceutical application. The purpose of this chapter is to provide a comprehensive review, based on recent reports, on the structural basis and pharmacological activities of CA and its natural derivatives. This chapter, “Molecular, pharmacological and chemical biology of Caffeic acid and derivatives,” is composed of two topics: General and biochemical basis of CA; and Biological and pharmacological functions in the health sciences.

Chapter 6 - The lack of treatment options for neurodegenerative diseases challenges neuroscientists to investigate not only new mechanisms but also to explore the properties of promising non-toxic compounds for a targeted therapy. The plant-originated, naturally occurring flavonoid caffeic acid phenethyl ester (CAPE) has attracted considerable interest due its substantiated efficacy, low toxicity and multi-mechanistic activity in a number of animal models of neurodegenerative diseases. CAPE is commonly stored as the active component of propolis of honeybee hives and has been used traditionally in folklore medicine. It shows redox-balancing activity through its antioxidant property and confers neuroprotective effects through its transcription factor-modulating activity. Major mediators of CAPE activity are stimulation of Nrf2 and inhibition of both NF-κB and STAT3. The lipophilic nature of CAPE renders it suitable to cross the blood-brain barrier and to induce a neurorestorative process in neurodegenerative diseases, including stroke, traumatic brain injury, spinal cord injury, adrenoleukodystrophy, multiple sclerosis, amyotrophic lateral sclerosis, and Alzheimer's disease.

Chapter 7 - Hydroxycinnamic acids including caffeic acid are major phenolic secondary metabolites in plants involved in growth and metabolism regulation as well as response to pathogen challenges. Caffeic acid is broadly distributed in several species of plant kingdom and is widely consumed in human diet. Caffeic acid and derivatives have been extensively studied revealing a broad spectrum of biological activities in different species. In addition to the antioxidant activity of caffeic acid, this molecule has been indicated as an inhibitor of number of proteins conserved across distinct species. This chapter will focus on the effect of caffeic acid and several targeted derivatives on lipoxygenases. Lipoxygenases (LOXs) are enzymes involved in peroxidation of lipids to bioactive lipid mediators such as leukotriene. LOXs are significant and highly conserved enzymes in eukaryotes with known function in metabolism of plant, fungal and mammalian cells. LOX inhibitory potential of caffeic acid has therefore major relevance to public health as well as agriculture and environment.

This chapter is aimed at summarizing the current knowledge of the inhibition of LOXs by caffeic acid and its application as well as potential in human health and agriculture. In addition to reviewing literature on LOX inhibition the authors will provide analysis of the LOX protein conservation in highly distinct species and present computational illustration of pan-eukaryotic LOX inhibition by caffeic acid and its derivatives.

Chapter 8 - Phenolic acids, one of the major classes of polyphenols, are widely distributed in the human diet, particularly in fruit, vegetables and beverages. Caffeic acid is a phenolic acid naturally occurring in many plant foods such as carrots, cabbage, tomatoes and several berries. It is also present in multiple beverages such as coffee, fruit juices and wine. Moreover, caffeic acid is one of the major active components in many traditional Chinese medicines. As a natural product, it possesses several biological and physiological activities. These activities include antioxidant, anti-mutagenic, anti-tumor and anti-obesity effects. For these reasons, caffeic acid is also being developed into drug for sale, for instance for its use in the development of asthmatic and allergenic drugs.

Although caffeic acid as well as other polyphenols are not essential nutrients, they are secondary metabolites synthesized by plants with bioactive properties, also known as phytochemicals, that may have an important effect on human health. For this reason, the interest in the analysis of this kind of compounds for the characterization of plant extracts is raising in the last years. Liquid chromatography coupled to mass spectrometry or to tandem mass spectrometry (LC-MS(/MS)) is one of the most frequently used techniques for that purpose, and the number of LC-MS(/MS) methods published in the literature dealing with the analysis of polyphenols, including caffeic acid, not only for their quantitative determination in food products and beverages but also for the characterization of foodstuffs and pharmaceutical preparations as a way to guarantee authentication regarding the natural product of origin is increasing. Additionally, LC-MS(/MS) techniques has also been used for the analysis of caffeic acid and its active metabolites in biological fluids to carry out pharmacological studies.

Recently, high resolution mass spectrometry (HRMS) is acquiring also an important role in the analysis of caffeic acid, and polyphenols in general, and today several LC-HRMS methods either using time-of-flight (TOF) or Orbitrap mass analyzers are described for the quantitative determination of these compounds as well as a fingerprint analysis for characterization, for instance in the case of herbal medicines.

In this chapter the role of liquid chromatography-mass spectrometry in the analysis of caffeic acid will be discussed. LC-MS(/MS) chromatographic conditions, ionization sources, and MS and HRMS analyzers frequently used, as well as strategies for the structural characterization and the qualitative and quantitative analysis of caffeic acid, its active metabolites, and other phytochemicals will be discussed by means of relevant applications. Coverage of all kind of applications is beyond the scope of the present contribution, so the authors will focus on the most relevant applications published in the last years.

In: Caffeic Acid
Editor: Leanna Vaughn

ISBN: 978-1-63483-121-5

Chapter 1

FOOD AND HEALTHY MIND: CAFFEIC ACID AND ITS DERIVATIVES

Satyabrata Ghosh
Department of Ingeniería Rural, ETSI Agrónomos,
Universidad Politécnica de Madrid (UPM), Madrid, Spain

ABSTRACT

With rapid socio-ecomonic advancement and compulsion for cosmopolitan human society, the pressure of overwork is bending the society to its toe. Therefore, the result is different neurological and neurovascular diseases, cardiac diseases and other brain diseases are growingly increased day by day. Human food habits are strikingly different from one country to other and even with the difference of their religions, culture, society structures and economic conditions. Indian "Vedas" is first text to describe the deep rooted relation between foods & mind. The type of foods and therefore its importance for health have an interim connection with the change of climate and play a great role to build mind and health of the individual and the society as a whole. Upanishads tell that 'Annam Brahma' (i.e., Food is God) and state the quality of the mind is influenced by the quality of the food intake. So in this above context this article tries to critically describe an eternal relation of food and healthy mind in terms of antioxidant property. Here through this article we mainly want to highlight the role of caffeic acid because caffeic acid is a globally available antioxidant compound and since the last more than a decade researchers have found that caffeic acid and its derivatives have lots of effects on human mind and thus on health. Neurological studies have shown that caffeic acid can create protective effects on neurodegeneration, neurotoxicity and has potential cytoprotective effects against a wide range of injuries in different tissues, including the brain.

Keywords: Antioxidant, caffeic acid, nutraceutical, health protection, food habit

INTRODUCTION

For the biological systems food is like a fuel which is essential for our body to produce energy by metabolism. Human belongs to omnivorous class and can change their food habit with changing time and places. Therefore, it is a continuous process of development changes from green leaves, tree barks, and raw meat to modern cooked foods. Every common man has to be aware of their food habit, because foods are the only thing that is directly connected to the body and mind. Researchers have found that getting a good health good food habit is necessary.

Not only in the present but from ancient time, in Vedic period, the most ancient civilisations in the world, people knew about relation between food, health and mind. Dharmashastra is a Sanskrit text which described about 'Bhojana' (or diet) pertaining to Hinduism. It was given great importance for the kind of foods, the direction of preparation and season in which they were offered. The sacrificial offerings of foods also had played a great role in the psychology and health of the individual and in society as a whole. There were other ancient civilizations like Mesopotamian civilization, they are said to make the rice dough and were an inspiration to bakers to add herbs and sesame seeds to enable the bread to rise. According to Greek mythology cultivation of grapes are said to give strength and health nourishment. In Hinduism Goddess Annapurna is worshiped for providing us food abundance in all seasons. The Egyptian God Set was said to have lettuce as his favourite vegetable. Charak Samhita is the oldest text on Ayurvedas (Indian traditional medicine) and it describes good diet is a main factor that promotes the healthy growth of humans and protect from disease. Food helps to modulate the hormonal secretion of the body to retain their state of equilibrium. "Charak" also states: "Even light, easily digested and nutritious food should not be taken in excess of bodily requirements, or after the appetite has been satisfied. Food difficult to digest should not be taken habitually. Even if used, the quantity should not exceed a fraction of a full meal." "An excess or surfeit of food is markedly harmful unless the gastric fire is increased by hard exercise." Shushruta, who described diabetes (madhumeha) for the first time in the history of medicine, advised the proper food and physical exercise are the best medicine. Ayurveda also describes three kinds of "ahar" or diet – "satwik," "rajasik," and "tamasik". The "Rishis" took "Satwik ahar" consisting of kanda, moola, phala" (vegetables and fruits) and lived for hundred years. Epidemiologic studies have clearly shown that diets rich in plant foods protect humans against degenerative diseases such as cancer and cardiovascular diseases (Mishra et al., 2011).

Plant foods contain fiber, vitamins, phytosterols, sulfur compounds, carotenoids, and organic acids, and they also contain a variety of polyphenols (antioxidant compound) which are increasingly regarded as effective protective agents. Polyphenols represent a wide variety of compounds, which are divided into several classes, i.e., hydroxybenzoic acids, hydroxycinnamic acids, anthocyanins, proanthocyanidins, flavonols, flavones, flavanols, flavanones, isoflavones, stilbenes, and lignans. Therefore this review has only heighted importance of Caffeic acid, and it's derivate in human health. It is present in honey, as well as in coffee, fruits and vegetables. It has high antioxidant potentiality and it has effect on suppression of neuroinflammatory and aluminum-induced over expression of 5- lipoxygenase (5-LOX) in brain tissues. It protects neuronal death in the hippocampus and with learning and memory deficits (Yang et al., 2008). In vitro and in-vivo treatment with caffeic acid at several different concentrations has been reported to increase the acetylcholinesterase activity in the cerebral cortex, cerebellum, and hypothalamus. All of these findings strongly support the proposition that caffeic acid improves memory by interfering with cholinergic (derivative of caffeic acid) signaling, in addition to its neuroprotective effects (Anwar et al., 2012).

Table 1. List of common food material containing caffeic acid

Source		Name
Fruit*	Dried drupes	Plum (1.11), prune (1.11)
	Dried fruits	Date (2.52)
	Berries	American cranberry (2.31), Black chokeberry (141.1), Cloudberry (1.0), Kiwi, cherry
	Citrus	Grape (0.16)
	Drupes	Peach (0.63),
	Other fruits	Date (3.37)
	Pomes	Apple Dessert (0.33), Pear (0.14)
Fruit Juice**		Pomegranate (0.07), grape (0.16), plum (5.1), pear (0.74), apple (0.68)
Cereals*		Wheat (0.04), rice parboiled (0.34), maize (0.04), rye (0.77), oat (0.16)
Vegetable*		Artichoke, potato (1.62), cider, tomato (0.45), Eggplant purple (0.38), Carrot (0.02), chicory (2.61), Cauliflower
Oil**		Olive oil (0.02), Soy oil
Seed*		Sunflower (8.17)
Beverage		Coffee (0.03), tea, beer (0.03), red wine (1.88), white wine(0.33),
Spices*		Caraway (16.4), Ceylan cinnamon (24.2), Cumin (16.6), Ginger (15.5), Nutmeg (16.3), Star anise (20.2)
Nuts*		Walnut (0.24)

(*) mg / 100 ml and (**) mg / 100 g (Manach etal., 2004; Svobodova et al., 2003)

PHYSICAL & CHEMICAL NATURE OF CAFFEIC ACID

Caffeic acid is (CA) also called 3,4-Dihydroxycinnamic acid, it is a wide spread hydroxycinnamates phenolic compound that is derived biosynthetically from phenylalanine in plants. It appears as Yellow prisms or plates from water (Lide, 1991). It is sparingly soluble in cold water; very soluble in hot water and cold ethanol (Budavari et al., 1989.). It occurs naturally in many agricultural products such as fruits, vegetables, wine, olive oil, and coffee (Clifford, 1999). The molecular structure of CA containing a catechol group with unsaturated carboxylic acid chain is responsible for its efficient interaction with several types of oxidant radicals (Bors et al., 1995). Caffeic acid exists in *cis* and *trans* forms, *trans* being the predominant naturally occurring form (Janssen, 1991). The o-dihydroxy structure is the natural radical site, producing the o-semiquinone after one-electron donation. The lateral double bond conjugated with the catechol group leads to an extensive electron delocalisation, increasing the stability of the o-semiquinone radical and, the antioxidant activity (Laranjinha et al., 2001). Solutions of caffeic acid and its derivatives (chlorogenic and isochlorogenic acids) are unstable in sunlight and ultraviolet light (Rartley et al., 1975). When solutions of caffeic acid is exposed to sunlight or ultraviolet light, the *trans* form of caffeic acid is partially converted to the *cis* form, which may be converted to the lactone and aesculetin (Grodzinska-Zachwieja et al., 1973.).

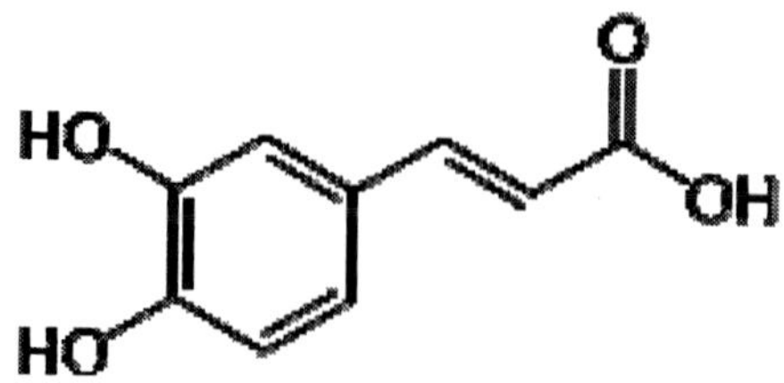

Some Common Sources of Caffeic Acid

Honey

Honey and its constituents can ameliorate oxidative stress and oxidative stress-related effects. The neuroprotective effects of honey are exerted at different stages of neurodegeneration and play prominent roles in early events. The most commonly reported polyphenol and flavonoid compounds in honey caffeic acid and its derevatives one of them (Rahman et al., 2014). One established nootropic property about honey is that it assists the building and development of the entire central nervous system, particularly among newborn babies and preschool age children, which leads to the improvement of memory and growth, a reduction of anxiety, and the enhancement of intellectual performance later in life (Cantarelli et al., 2008). It has been reported that eating honey of postmenopausal women; they improvements in their immediate memory but not in immediate memory after interference or in delayed recall (Othman et al., 2011). Over a one year period of study of honey in the rat model, it showed significantly less anxiety and better spatial memory throughout all stages compared with the control group of rats. More importantly, the spatial memory of honey-fed rats, as assessed by object recognition tasks, was significantly greater during later months (i.e., 9 and12) (Chepulis et al., 2009). Honey can (250mg/kg body weight) significantly decreased the lipid peroxidation in brain tissue and decreased the number of degenerated neuronal cells in the hippocampal CA1 region (Cai et al., 2011). Different neural system such as excitatory neural and inhibitory neural system control exploratory and anxious behaviours. The cholinergic and dopaminergic systems act as xcitatory neural systems and γ-aminobutyric acid (GABA) act as inhibitory neural system, specifically (Patel et al., 2012). In addition to neural effects, glial cells may also respond to honey therapy because honey shows a neuroprotective effect in the cerebral focal-induced ischemia model in rats (Rahman et al., 2014). Moreover, honey attenuated ischemia-induced neuroinflammation by activating microglia, and neuroinflammatory processes in the brain are believed to play a crucial role in the development of neurodegenerative diseases as well as in neuronal injury associated with stroke (Frank-Cannon et al., 2009). Interestingly, ischemia-induced cognitive impairments that result from microglia- and/or astrocyte-mediated neuroinflammation were also significantly attenuated by honey therapy (Frank-Cannon et al., 2009).

Coffee

Coffee, in different preparations, is widely consumed throughout the world, and contains high levels of phenolic compounds. A single serving provides between 20 and 675 mg of chlorogenic acids depending on the type of roast and the volume consumed and regular consumers an easily have an intake in excess of 1 g per day (Yashin et al., 2013). Coffee is a health elixir of the highest magnitude, offering protection to the heart, lowering the risk of several forms of cancer and reducing the risk of Parkinson's disease. Natural properties of coffee shows that the daily brew is a potent protective antioxidant potion. Coffee is super-concentrated with antioxidants Coffee's antioxidant compounds include caffeic acid, caffeine, the chlorogenic acids, eugenol, gamma-tocopherol, isoeugenol, p-coumaric acid, scopoletin and tannic acid. As a mood enhancer, Coffee's greatest effects are exerted upon the brain and mind, since coffee is a bracing awakener. As a delivery system for caffeine, coffee acts upon the central nervous system, stimulating the brain and all the major nerves, and facilitating cognitive function overall. Coffee enhances the flow of blood in the brain and invigorates the mind. It enhances alertness and motivation, facilitates thought formation and concentration, and decreases mental fatigue (Yashin et al., 2013).

Different Caffeic Acid Derivatives and Biological Activity

Caffeic Acid Phenethyl Ester (CAPE)

CAPE is one of the main active components of propolis (a honey bee hive extract). The molecular weight and empirical formula of CAPE are 284.3 g/mol and C17H16O4. CAPE possesses a number of important biological activities, including anti-bacterial, antiviral, anti-fungal, anti-oxidant, anti-inflammatory, and anti-cancer properties (Dhungyal et al., 2014). It is also a specific inhibitor of the nuclear transcription factor nuclear factor (NF)-κB. CAPE has been shown to significantly suppress the lipoxygenase pathway of arachidonic acid metabolism during inflammation. It has also been shown to inhibit HIV-1 integrase 8 and the proliferation of transformed cells and induces apoptosis in transformed fibroblasts (Akcam et al., 2013). Protective effects of CAPE in the nervous system have been reported in cerebral ischemic reperfusion injury, ischemic damage of the spinal cord, Parkinson's disease, convulsions, multiple sclerosis, brain tumors, hepatic encephalopathy, and against toxic effects of therapeutic agents in anticancer therapy (Tolba et al., 2013; Santos et al., 2014).

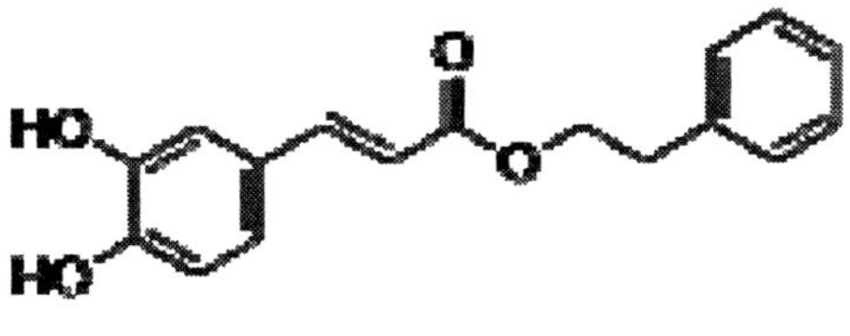

Chlorogenic Acid (5-Caffeoylquinic Acid)

Chlorogenic acids are a group of compounds comprising hydroxycinnamates, such as caffeic acid, ferulic acid, and p-coumaric acid, linked to quinic acid to form a range of conjugated structures known as caffeoylquinic acids (CQA), feruloylquinic acids (FQA), and p-coumaroylquinic acids all of which exist in several isomeric forms (Clifford M.N., 1999). These compounds are the major dietary polyphenols present in various foods and beverages. The similarities in the metabolic patterns observed for caffeic, chlorogenic, and caftaric acids suggest that esterification does not influence the metabolism of caffeic acid by the gut microbiota (Gonthier et al., 2006). Chlorogenic was demonstrated to be more powerful antioxidants in a number of different systems (Koriem et al., 2014). Chlorogenic is substrates of polyphenol oxidases, and under certain conditions they may undergo oxidation in plant tissues or products of plant origin (Bassil et al., 2005). It is one of the major phenolic compounds identified in peach and in prunes. It has protective effect on antihypertensive, neuroinflammatory and dopaminergic neurons etc (Shen et al., 2012; Zhao et al., 2012).

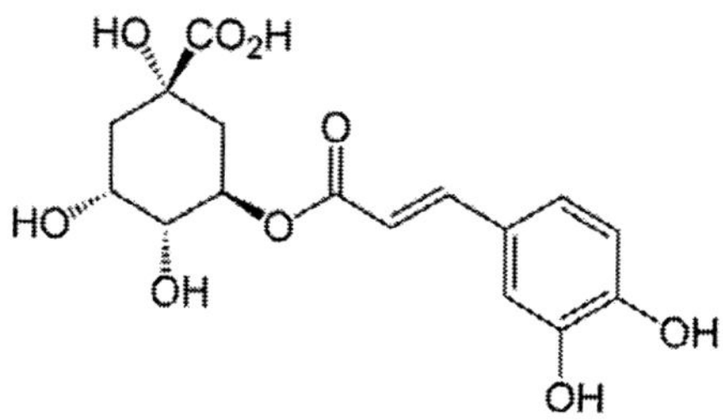

Chlorogenic acid (CGA) is a novel insulin sensitizer that potentiates insulin action similar to the therapeutic action of metformin (McCarty et al., 2005). CGA is capable of improving glucose tolerance and insulin resistance in obese rats, suggesting that CGA may be a promising candidate for the development of antidiabetic agents (Meng et al., 2013). It is reported that CGA promote a significant reduction in the plasma glucose level and attenuating intestinal glucose absorption. It consider as a glycaemic index lowering agent and reducing the risk of developing T2DM (Bassoli et al., 2008). Clinical trials have also testified that CGA in coffee is able to modulate glucose uptake and gastrointestinal hormone andinsulin secretion in humans (Meng et al., 2013). CGA is hypoglycemic agents and may affect lipid metabolism. CGA is possibly effective against weight gain and fat accumulation by inhibition of fat absorption and activation of fat metabolism in the liver (Shimoda et al. 2006). It protects cardiovascular risk status by modestly reducing LDL oxidation susceptibility and decreasing LDL-cholesterol and malondialdehyde (MDA) levels (Yukawa et al., 2004). It also significantly inhibited fatty acid synthase, 3-hydroxy-3-methylglutaryl CoA reductase, and acyl-CoA cholesterol acyltransferase activities, while they increased fatty acid beta-oxidation activity and peroxisome proliferator-activated receptors alpha expression in the liver compared to the high-fat group.

Chicoric Acid

In 1958, while working on the leaves of chicory (*Cichorium intybus L.*) plants, isolated and identified a phenolic compound that was a tartaric acid ester of two caffeic acids (a hydroxycinnamic acid) they proposed naming it chicoric acid. (Scarpati et al., 1958). Reports have indicated that chicoric acid helps a plant protect itself from insects and infection from viruses, bacteria, fungi, and nematodes; and that it aids in wound healing in plants after mechanical damage (Lee et al., 2013; Nishimura et al., 2006).

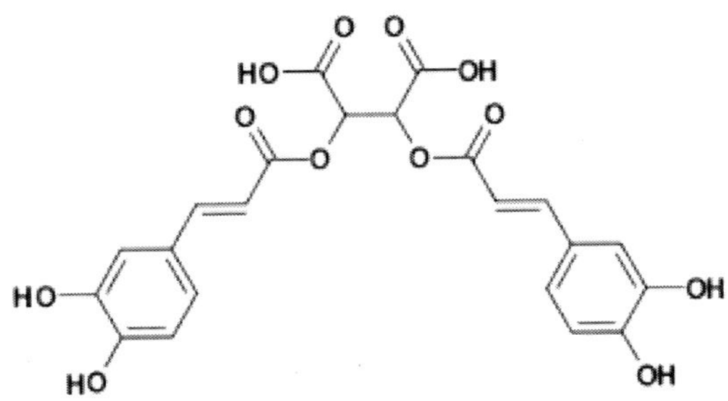

Chicoric acid can be obtained from isolated and purified plant materials orsynthesized (Scarpati et al., 1958). Pure chicoric acid is a white powder now available as a purified standard Lee J 2013. Identification of L-chicoric acid and meso-chicoric acid in plant samples has not been consistent. L-chicoric acid and an isomer (unknown form) were report edinchicory leaves (Heimler et al., 2009). Both L-chicoric acid and meso-chicoric acid were found in ice berg lettuce (Luna et al.,2012). To date, chicoric acid has been found in plant so fat least 13 orders, 25 families, and 63 genera and species. It has been most often reported in either the family Asteraceae (Asterfamily)—20genera and species, orthe family Dryopteridaceae (Wood fern family)—8 genera and Species. Production of chicoric acid does not appear to be ubiquitous in taxon within a plant family or sub-family. Chicoric acid has been detected in leaves of many genera in the Asteraceae, however, some genera within this family may not produce chicoric acid in leaves (e.g., Achillea, Arnica, Cnicus, Echinops, Inula, Petasites, Solidago, and Tanacetum); (Jaiswal et al., 2011).

Chicoricacid properties have been reported to include anti- cancer, anti-obesity, antiviral, and anti-diabetic (Queffelec et al., 2008; Tousch et al., 2008; Tsai et al., 2012; Xiao et al., 2013). For example, chicoric acid and its analogs have been claimed to possess anti-Human Immunodeficiency Virus (HIV) activity due to its involvement in HIV integrase inhibition, which could perhaps hinder HIV strain replication (Queffelec et al.,2008). Bel-Rhlid et al. (2012) reported a means of chicoric acid hydrolysis, facilitated by the probiotic bacterium (*Lactobacillus johnsonii*) after ingestion, but prior to absorption and metabolism. Due to the concerns surrounding all in vitro antioxidant measurements, chicoric acid antioxidant studies will not be summarized here, but the issues and research needed in this area have been well summarized elsewhere.

Caffeic Acid in Brain Research

We have reported here few current researches about caffeic acid and its derivatives effect on brain and neurodegenerative disorder. According to Tsai et al., 2015 reported that in neurodegenerative disorder, inflammation is occurred due to activation of Microglial cell. Effects of CAPE on microglia activation by using *in vitro* and *in vivo* models it found that CAPE significantly inhibited the expressions of inducible nitric oxide synthase (NOS), cyclooxygenase (COX)-2 and the production of nitric oxide (NO). Administration of CAPE resulted in increased expressions of hemeoxygenase (HO)-1and erythropoietin (EPO) in microglia. The phosphorylated adenosine monophosphate-activated protein kinase (AMPK)-α was further found to regulate the anti-inflammatory effects of caffeic acid. *In vivo* results from immunohistochemistry along with rotarod test also revealed the anti-neuroinflammatory effects of CAPE in microglia activation. The current study has evidenced several possible molecular determinants, AMPKα, EPO, and HO-1, in mediating anti-neuroinflammatory responses in microglial cells.

Hakan et al., 2015, has investigated the effects of Caffeic Acid Phenethyl Ester (CAPE) on proinflammatory cytokines, IL-1 and TNF, and explore its healing effect after acute spinal cord injury. CAPE suppressed TNF-and IL-1levels in the serum. In histopathological evaluation, it was detected that CAPE decreased hemorrhage and necrosis. CAPE suppresses the levels of proinflammatory cytokines, TNF-and IL-1, after acute spinal cord injury in the early phase and contributes to the healing process.

Santos et al., 2014 showed that CAPE has protective effect on neurotoxicity (in vitro and in vivo). CAPE protects PC12 cells from the cellular death induced by the dopaminergic neurotoxin MPP+ by increasing the network of neurites. CAPE induced the formation, elongation and ramification of neurites in PC12 cells non-stimulated with NGF (nerve growth factor) and inhibited the shortage of neurites induced by the dopaminergic neurotoxin. These effects were associated with increased expression of neuron-typical proteins responsible for axonal growth (GAP-43) and synaptogenesis (synaptophysin and synapsin I). It is noteworthy that, unlike neurotrophins, CAPE would be able to cross the blood brain barrier and exert its neurotrophic effects in the brain. This study corroborates the therapeutic potential of CAPE in neurodegenerative diseases while proposes the involvement of neuroplasticity in the mechanism of neuroprotection.

Silvia et al., 2013: CAPE is a derivative of a caffeic acid and its protective effect against the dopaminergic neuronal loss induced by 6-hydroxydopamine (6-OHDA) in rats. Inhibitory effect of CAPE on mitochondrial permeability transition (MPT), a mediator of neuronal death that triggers cytochrome c release an caspase-3 activation. Scavenging of reactive oxygen species (ROS) and metal chelation was demonstrated in the brain-affected areas of the rats treated with 6-OHDA and CAPE. it does not affect brain mitochondrial function and has ability to cross the blood–brain barrier, therefore in Parkinson's treatment CAPE can be a future molecule.

Yiş et al., 2013: they have investigated whether caffeic acid phenethyl ester exerts neuroprotective effects on the developing rat brain after status epilepticus. They showed that caffeic acid phenethyl ester significantly preserved the number of neurons in the CA1, CA3, and dentate gyrus regions of the hippocampus and the prefrontal cortex. It also diminished apoptosis in the hippocampus and the prefrontal cortex. In conclusion, this experimental study suggests that caffeic acid phenethyl ester administration may be neuroprotective in status epilepticus in the developing rat brain.

Zhang et al., 2007, reported that caffeic acid has protective effects against traumatic brain injury (induces neuron damage) therefore; they studied the effect of caffeic acid on the lesion in the early and late phases of cryoinjury in mice. As result caffeic acid in a particular dose reduced the lesion area and attenuated the neuron loss around the lesion core 1 to 28 days, but attenuated the neuron loss in the lesion core only 1 day after cryoinjury. Moreover, caffeic acid attenuated astrocyte proliferation, glial scar wall formation and glial fibrillary acidic protein (GFAP) protein expression in the late phase of cryoinjury. Caffeic acid also inhibited the reduction of superoxide dismutase activity and the increase in malondialdehyde content in the brain 1 day after cryoinjury. These results indicate that caffeic acid exerts a protective effect in traumatic brain injury, especially on glial scar formation in the late phase, which at least is associated with its anti-oxidant ability.

Kerman M et al., 2012; has reported that the therapeutic efficacy of CAPE with an experimental traumatic brain injury (TBI) model in rats. To date, no biochemical and histopathological changes of neurodegeneration in the frontal cortex after TBI in rats by CAPE treatment have been reported. The TBI significantly increased tissue malondialdehyde (MDA) levels, and significantly decreased tissue superoxide dismutase (SOD) and glutathione peroxidise (GPx) activities, but not tissue catalase (CAT) activity, when compared with controls. The administration of a single dose of CAPE (101 mol/kg) 15 min after the trauma has shown protective effect via decreasing significantly the elevated MDA levels and also significantly increasing the reduced antioxidant enzyme (SOD and GPx) activities, except CAT activity. The morphology of neurons in the CAPE treatment group was well protected. Treatment of CAPE markedly reduced the immunoreactivity of degenerating neurons. In conclusion, the CAPE treatment might be beneficial in preventing trauma-induced oxidative brain tissue damage, thus showing potential for clinical implications.

Oboh G 2013: This study sought to investigate and compare the interaction of caffeic acid and chlorogenic acid on acetylcholinesterase (AChE) and butyrylcholinesterase (BChE), and some pro-oxidants (FeSO4, sodium nitroprusside and quinolinic acid) induced oxidative stress in rat brain in vitro. It shows that caffeic acid and chlorogenic acid inhibited AChE and BChE activities in dose-dependent manner; however, caffeic acid had a higher inhibitory effect on AChE and BChE activities than chlorogenic acid. Combination of the phenolic acids inhibited AChE and BChE activities antagonistically. Furthermore, pro-oxidants such as, FeSO4, sodium nitroprusside and quinolinic acid caused increase in the malondialdehyde (MDA) contents of the brain which was significantly decreased dose-dependently by the phenolic acids. Inhibition of AChE and BChE activities slows down acetylcholine and butyrylcholine breakdown in the brain. Therefore, one possible mechanism through which the phenolic acids exert their neuroprotective properties is by inhibiting AChE and BChE activities as well as preventing oxidative stress-induced neurodegeneration. However, esterification of caffeic acid with quinic acid producing chlorogenic acid affects these neuroprotective properties.

Anwar J, 2012: The aim of this study was to investigate in vitro and in vivo whether caffeic acid alters the AChE activity and behavioral parameters in rats. It shows that caffeic acid increased the AChE activity in the cerebral cortex, cerebellum, hypothalamus, whole blood, and lymphocytes at different concentrations. In muscle, this compound caused an inhibition in the AChE activity at different concentrations. They also reported that caffeic acid decreased the AChE activity in the cerebral cortex and striatum and increased the activity of this enzyme in the cerebellum, hippocampus, hypothalamus, pons, lymphocytes, and muscles. The amount of 100 mg/kg of caffeic acid improved the step-down latencies in the inhibitory avoidance. Caffeic acid improved memory and interfered with the cholinergic signaling. As a natural and promising compound caffeic acid should be considered potentially therapeutic in disorders that involve the cholinergic system.

Khan M 2007, Ischemic stroke is a neurovascular disease treatable by thrombolytic therapy, but the therapy has to be initiated within 3 h of the incident. This therapeutic limitation stems from the secondary injury which results mainly from oxidative stress and inflammation. A potent antioxidant/anti-inflammatory agent, caffeic acid phenethyl ester (CAPE) has potential to mitigate stroke's secondary injury, and thereby widening the therapeutic window. They showed that CAPE protected the brain in a dose-dependent manner (1–10 mg/kg body weight) and showed a wide therapeutic window (about 18 h) in a rat model of transient focal cerebral ischemia and reperfusion. The treatment also increased nitric oxide and glutathione levels, decreased lipid peroxidation and nitrotyrosine levels, and enhanced cerebral blood flow. CAPE down-regulated inflammation by blocking nuclear factor kappa B activity. The affected mediators included adhesion molecules (intercellular adhesion molecule-1 and E-selectin), cytokines (tumor necrosis factor-a and interleukin-1b) and inducible nitric oxide synthase. Anti-inflammatory action of CAPE was further documented through reduction of ED1 (marker of activated macrophage/microglia) expression. The treatment inhibited apoptotic cell death by down-regulating caspase 3 and upregulating anti-apoptotic protein Bcl-xL. Conclusively, CAPE is a promising drug candidate for ischemic stroke treatment due to its inhibition of oxidative stress and inflammation, and its clinically relevant wide therapeutic window.

Caffeic Acid in Disease Control

Prostate cancer is the fifth most common cancer overall in the world. The majority of patients having prostate cancer are over 65 years old and the 5 years survival rate. In 1941, Dr. Charles Huggins discovered that deprivation of androgen caused regression of hormone-responsive metastatic prostate cancer (Lin et al., 2013). Androgen ablation therapy has become the primary treatment for metastatic prostate cancer. CAPE is a well-known NF-κB inhibitor. CAPE treatment at concentrations of 50 μM to 80 μM inhibits the activation of NF-κB via preventing the translocation of p65 unit of NF-κB and preventing the binding between NF-κB and DNA (Natarajan et al., 1996).

CAPE, via suppression of RANKL-induced NF-κB and NFAT activation, has dual effects on osteoclasts; that is, inhibiting osteoclastogenesis and inducing apoptosis. Given that many pathological bone diseases are associated with increased osteoclast formation and activation, our studies imply that this remarkable natural compound might be useful for the prevention or treatment of osteolytic bone diseases. (Ang et al., 2009).

A hepatocyte is a cell of the main parenchymal tissue of the liver. Hepatocytes make up 70-85% of the liver's mass. The hepatoprotective capacity of CAPE was investigated in numerous studies. CAPE treatment elevated tissue catalase (CAT) activity and improved ultrastructural changes associated with aging in the rat liver tissue (Esrefoglu et al., 2012). CAPE exhibits protection against lipid peroxidation, necrosis, aberrant cell proliferation and p65 activation. Akcam et al. (Akcam,) recently indicated that CAPE could protect against aflatoxin B1-induced hepatotoxicity via modulating free radical production. The hepatoprotective effect of CAPE was found to be associated with its antioxidative potential through reducing malondialdehyde (MDA) levels and elevating the reduced activities of superoxide dismutase (SOD) and glutathione peroxidase (GPx) enzymes in liver tissue.

Every year 20% of the people die cause of cardiovascular diseases (Mansour et al., 2012). In 2011, Motawi et al. (Motawi et al., 2011) reported that CAPE pretreatment protected against LPS induced cardiac stress in rats through upregulating heme oxygenase (HO) and downregulating TNF-a in cardiac tissues. Esrefoglu et al. demonstrated that both CAPE and GPx) (Tolba et al., 2013). Furthermore, Park and Min (Tolba MF,) showed that CAPE prevented diabetes-induced alterations in insulin like growth factor (IGF)-I and IGF-II expression in rat heart. It also improved diabetes induced oxidative-damage in cardiac tissues (Tolba et al., 2013). Taken together, these findings suggest its potential merit against diabetes-induced cardiac tissue damage. Moreover, CAPE exerted curative effect when administered postinfarction. This was evidenced via marked reduction of the dimensions of infarct area and cardiomyocyte injury parameters (LDH, CK-MB and troponin I). The suggested mechanism was attenuation of cardiomyocyte apoptosis via CAPE-inhibition of p38 MAPK activation and caspase-3 activity together with reduction of the proinflammatory cytokines (IL-1b and TNF-a) in cardiac tissues (Tolba et al., 2013). The free radical scavenging activity of CAPE also plays a role in preventing the apoptotic and cardio-protection (Tolba et al., 2013).

Lung cancer, is also known as carcinoma of the lung or pulmonary carcinoma. The vast majority (80–90%) of cases of lung cancer are due to long-term exposure to tobacco smoke. Lots of chemopreventive agents have been reported to interfere with the intracellular signal that is related to carcinogenesis, the proliferation of cancer cells, apoptosis, and cell migration. The effects of the representative chemopreventive agents on the transforming growth factor-beta (TGF-β)-induced invasive phenotype using A549 lung adenocarcinoma cells as a model system was investigated by using CAPE as therapeutic agent. CAPE effectively suppressed TGF- β-enhanced cell motility and TGF-β-induced Akt (protein kinase β) activation as well as a specific inhibitor of phosphatidyl inositol 3-kinase (PI3K)/Akt pathway, LY294002 (SHIGEOKA 2004). By this finding, the Authors suggest that CAPE can be applied not only as chemopreventive agent but also as a anti-metastatic therapeutic agent. The useful therapeutic dose of CAPE for A549 lung cancer cell line was found to be 6 μg/ml and the tolerance dose of normal lung fibroblast for CAPE appeared to be higher than the therapeutic dose (Chen et al., 2004). It is claimed that CAPE inhibits the growth of tumor cells using oxidative stress pathways connected to p53-independent pathways and observed to inhibit oxidative processes by decreasing the generation of intracellular hydrogen peroxide (H2O2) in A549 cells. The decrease in H2O2 production and GSH level of A549 cells was seen in a very rapid and profound manner17. Moreover, the depleting intracellular stores of GSH (reduced glutathione) by CAPE can render cells more susceptible to oxidative stress-induced apoptosis (Cotgreave et al., 1998).

Physical and Mental Fitness Influence Human Behavior

According to the World Health Organization (WHO) report most of the illness is behaviourally related. Current day 80% people are suffering from physical and mental stress as a result harassed, worried, anxious, and afraid and seek help and reassurance are common problem for modern society (Lele 2014). It is important to note that mental stress can induce silent ischemia more frequently than physical exercise in the same patient. With continuous monitoring of LVEF by ambulatory vest, and proved that arithmetic problems, speaking assignments are cause of mental stress. The prognostic value of mental stress testing in coronary artery disease (CAD) has been established. Mental stress can induce adverse cardiac events instable angina patients. Different Brain research studies has reported that rate of blood flow during the metal trace is reduced 47% whereas rate of flow significantly increased (15%) during laughter time.

Ornish et al., (1977) has explored their study the effect of diet and lifestyle modification on the reversal of coronary atherosclerosis using Positron emission tomography (PET) imaging. It showed that modification of cardiovascular risk factors that contribute to endothelial dysfunction improve patient outcome disproportionately to the regression in the anatomic atherosclerotic lesions. Patients with "mild" coronary artery lesions can have severe endothelial dysfunction. Long-term follow-up showed more cardiac events in such patients. Myocardial ischemia can occur during exercise in patients without significant epicardial vessel stenosis, due to endothelial dysfunction. Recent insights into vascular biology help us to understand how the benefit occurs in relation to endothelial dysfunction.

The brain and nervous system require specific nutrients to function properly, and the evidence is overwhelming that nutrient deficiencies can lead to aggression and violent behavior. Modern research in neuroscience has revealed many new insights into our complex mental processes and their regulation by an intricate network of electrical and chemical circuits neurotransmitters (both excitatory and inhibitory) connect the brain, the neuroendocrine, and the immune system (psycho-neuroendocrinology and psychoneuroimmunology). Cannabinoid CB2 receptors are expressed on cells of the peripheral immune system and are coupled to inhibition of adenylyl cyclase, suggesting a potential antiinflammatory role. Mental depression also depresses the immune system. Functional brain imaging with PET and MRI has revealed characteristic changes in blood flow, metabolism, and neurochemical activity in anxiety, depression, obsessive-compulsive neurosis, schizophrenia, and manic-depressive psychosis. A unique Fludeoxyglucose (FDG)-PET study of glucose metabolism of the brain (Lele 2014) has shown that the ancient Indian practice of meditation causes a generalized hypometabolism in the brain — the reverse happens in stress. The GABA, encephalin–endorphin– dynophrin, and endocannabinoid system becomes activated during stress. Cannabiniod CB1 receptors are found in high abundance in the brain, especially basal ganglia, cerebellum, hippocampus, and cerebral cortex.

Considerably lower expression is found in peripheral tissues including lungs, testes, uterus, and vascular tissue. Following agonist binding CB1, receptors cause inhibition of adenylcyclase, inhibition of the N and Q type voltage-operated calcium channels and stimulation of inwardly rectifying and A-type K^+ channels. Signaling by the endocannabinoid system represents a mechanism by which neurons can communicate backward across synapses to modulate their inputs. The endocannabinoids reduce pain and anxiety, cause sedation or slowing down, increase appetite, and promote food intake.

Functional brain studies of pain perception have shown that when a peripheral pain stimulus is applied, the contralateral thalamus gets activated and relays the pain sensation to the cingulate cortex, which also becomes activated. Studies during hypnosis have shown that while the thalamus continues to be activated, the cirgulate cortex is not; hence there is no awareness of pain — possibly due to activation of the cannabinoid system.

The lipid anandamide (arachidonyl–ethanolamide) exemplifies cannabinoids in the brain, analogous to the endogenous opiate receptor agonists. Their physiological role is increasingly appreciated, as Natural Healers of the body. Deficiencies cause mental symptoms ranging from poor concentration, depression and severe agitation to hallucinations. 39 Deficiencies are caused by pernicious anemia, an autoimmune condition; they are also found in vegetarians and vegans, those with low animal protein intake, and individuals with leaky gut (Pacholok et al., 2011).

Conclusion and Future Promises

After numerous experiments had been going on its claims that caffeic acid and its derivatives possesses a wide range of nutraceutical values. It is a promising protecting molecule, including antiinflammatory, neuroprotective, hepatoprotective and cardioprotective effects. Yet now most experimental studies have been done on animal model rather than human. In the future, it can be possible to explore lots of unrevealed area on human model. There caffeic acid and its derivative would be promising and challenging antioxidant compounds for the current generation.

Acknowledgment

The author's scholarship supported by Erasmus Mundus (IDEAS project). Author would like to give his special thank Dr. Subrata Ghosh Dept of Neurology, Genetics & Aging Research Unit, Harvard Medical School, Massachusetts, USA for scientific discussion with him.

References

Akcam M, Artan R, Yilmaz A, Ozdem S, Gelen T, et al. Caffeic acid phenethyl ester modulates aflatoxin B1-induced hepatotoxicity in rats. *Cell Biochem. Funct.*, doi: 10.1002/cbf.2957.

Anwar J, Spanevello RM, Thomé G. et al., (2012) "Effects of caffeic acid on behavioral parameters and on the activity of acetylcholinesterase in different tissues from adult rats," *Pharmacology Biochemistry and Behavior.* 103, (2), 386–394.

Ang ES, Pavlos NJ, Chai LY, et al. (2009) Caffeic acid phenethyl ester, an active component of honeybee propolis attenuates osteoclastogenesis and bone resorption via the suppression of RANKL-induced NF-kappaB and NFAT activity. *J Cell Physiol.* 221(3), 642–649.

Bassoli BK, Cassolla P, Borba-Murad GR et al., (2008) "Chlorogenic acid reduces the plasma glucose peak in the oral glucose tolerance test: effects on hepatic glucose release and glycaemia," *Cell Biochemistry and Function,* 26, 3, 320–328.

Bassil D., Makris DP, Kefalas P (2005) "Oxidation of caffeic acid in the presence of L-cysteine: isolation of 2-S-cysteinylcaffeic acid and evaluation of its antioxidant properties," *Food Research International,* 38, 4, 395–402.

Bel-Rhlid R, Page-Zoerkler N, Fumeaux R, Ho-Dac T, Chuat J, Sauvageat JL (2012) Hydrolysis of chicoric and caftaric acids with esterases and Lactobacillus johnsonii in vitro and in a gastrointestinal model. *J. Agric. Food Chem.* 60, 9236–9241.doi:10.1021/jf301317h

Belitz, H.D.; Grosh, W.; Schieberle, P. Coffee, Tea, Cocoa. Food Chemistry, New York; *Springer-Verlag:* Heidelberg, Berlin, Germany, 2009.

Bors W, Michel C, Schikora S (1995) Interaction of flavonoids with ascorbate and determination of their univalent redox potentials – A pulse-radiolysis study. *Free Radical Biology and Medicine,* 19, 45–52.

Budavari S (1989) The Merck Index, llth ed., Rahway, NJ, *Merck and Co.,* p. 248.

Cai M, Shin BY, Kim DH et al., (2011) "Neuroprotective effects of a traditional herbal prescription on transient cerebral global ischemia in gerbils," *Journal of Ethnopharmacology,* 138, 3, 723–730.

Cantarelli MA, Pellerano RG, Marchevsky EJ, Cami˜na JM, (2008) "Quality of honey from Argentina: study of chemical composition and trace elements," *Journal of the Argentine Chemical Society*, 96, 2, 33–41.

Chen MF, Wu CT, Chen YJ, Keng PC, Chen WC. (2004) Cell killing and radiosensitization by caffeic acid phenethyl ester (CAPE) in lung cancer cells. *J Radiation Res;* 45, 253-260

Chepulis LM, Starkey NJ, Waas JR, Molan PC (2009) "The effects of long-term honey, sucrose or sugar-free diets on memory and anxiety in rats," *Physiology and Behavior,* 97, (3), 359–368.

Clifford MN (1999) Chlorogenic acids and other cinnamates – nature, occurrence and dietary burden. *Journal of Science and Food Agriculture,* 79, 362–372.

Cotgreave IA, Gerdes RG (1998) Recent trends in glutathione biochemistry–glutathione-protein interactions: a molecular link between oxidative stress and cell proliferation? *Biochem Biophys Res Commun.* 242, 1-9

Dhungyal B, Koirala P, Sharma C, Jha DK (2014) Caffeic Acid - A Potent Phyto compound against Diabetes Mellitus A Review, *SMU Medical Journal,* 1, 152-160

Esrefoglu M, Iraz M, Ates B, Gul M (2012) Melatonin and CAPE are able to prevent the liver from oxidative damage in rats: an ultrastructural and biochemical study. *Ultrastruct. Pathol.* 36, 171–178.

Frank-Cannon TC, Alto LT, McAlpine FE, Tansey MG (2009) "Does neuroinflammation fan the flame in neurodegenerative diseases?" Molecular Neurodegeneration, 4, (1), article 47.

Gonthier MP, Remesy C, Scalbert A et al., (2006) "Microbial metabolism of caffeic acid and its esters chlorogenic and caftaric acids by human faecal microbiota in vitro," *Biomedicine and Pharmacotherapy*. 60, 9, 536–540.

Grodzinska-Zachwieja Z, Kahl W, Klimczak M (1973) Spectrophotometric investigation of changes of caffeic, chlorogenic and isochlorogenic acids under the influence of sorne physicochemical factors. *Pol. Pharmacol. Pharm.* 25, 299 – 305.

Hakan Ak, İsmail MD, Gülşen MD, Karaaslan MD, et al., (2015) The effects of caffeic acid phenethyl ester on inflammatory cytokines after acute spinal cord injury, *Ulus Travma Acil Cerrahi Derg,* 21, 2

Heimler D, Isolani L, Vignolini P, Romani A (2008) Polyphenolcontent and antiradical activity of *Cichoriumintybus* L. from biodynamic and conventional farming. *Food Chemistry*. 114, 765–770.doi:10.1016 /j.foodchem.2008. 10.010

Jaiswal R, Kiprotich J, Kuhnert N (2011) Determination of the hydroxyl -cinnamate profile of 12 members of the Asteraceae family. *Phytochemistry* 72, 781–790. doi:10.1016 /j.phytochem.2011.02.027

Janssen Chim Ica. Catalog (1991) *Handbook of Fine Chemicals, Beerse*, p. 449,

Khan M, Elango C, Ansari MA, Singh I, Singh AK, (2007) Caffeic acid phenethyl ester reduces neurovascular inflammation and protects rat brain following transient focal cerebral ischemia, *Journal of Neurochemistry,* 102, 365–377.

Kerman M, Kanter M, Kenan Cos¸kun K, Erboga M, Gurel A, (2012) Neuroprotective effects of Caffeic acid phenethyl ester on experimental traumatic brain injury in rats, *J Mol Hist* 43, 49–57.

Koriem KMM, Soliman RE (2014) Chlorogenic and Caftaric Acids in Liver Toxicity and Oxidative Stress Induced by Methamphetamine. *Journal of Toxicology,* 1-10 http://dx.doi.org/10.1155/2014/583494

Kwon SH, Lee HK, Kim JA et al., (2010) "Neuroprotective effects of chlorogenic acid on scopolamine-induced amnesia via anti-acetylcholinesterase and anti-oxidative activities in mice," *European Journal of Pharmacology,* 649, 210–217.

Laranjinha J, Lester P (2001) Redox cycles of caffeic acid with [alpha]- tocopherol and ascorbate. *Methods in Enzymology*, 335, 282–295.

Lee J, Scagel CF (2013) Chicoric acid: chemistry, distribution, and production, *Frontires Chemistry*, 1, 1- 17

Lele RD (2014) Prevention is better and cheaper than cure, *Journal of Mahatma Gandhi Institute of Medical Sciences.* 19, 1

Li Y, Shi W, Li Y. et al., (2008) "Neuroprotective effects of chlorogenic acid against apoptosis of PC12 cells induced by methylmercury," *Environmental Toxicology and Pharmacology,* 26, (1), 13–21.

Liang XC, Meng W, Zhong YL et al., (2013) "Effects of chlorogenic acid on mouse insulin resistance development induced by high fat emulsion," *Chinese Pharmacological Bulletin.* 29, 5, 654–658, (Chinese).

Lide DR, (1991) CRC Handbook of Chemistry and Physics, 72nd ed., *Boca Raton, FL, CRC Press,* p. 3-181.

Lin HP, Lin CY, Liu CC, Su LC, et al., (2013) Caffeic Acid Phenethyl Ester as a Potential Treatment for Advanced Prostate Cancer Targeting Akt Signaling, *Int. J. Mol. Sci., 14*, 5264-5283; doi:10.3390/ijms14035264

Luna MC, Tudela JA, Martinez-Sanchex A, Allende A, Marin A, Gill MI (2012) Long-term deficit and excess of irrigation influences quality and browning related enzymes and phenolic metabolism of fresh cut ice burg lettuce (Lactuca sativa L.). *Postharv. Biol. Tech.* 73, 37–45.doi: 10.1016/j.postharvbio.2012.05.011

Manach C, Scalbert A, Morand C, Remesy C, Jimenez L (2004) Polyphenols: food sources and bioavailability. *Am J of Cli Nutri.* 79, 727 - 747.

Mansour HH, Tawfik SS (2012) Early treatment of radiationinduced heart damage in rats by caffeic acid phenethyl ester. *Eur. J. Pharmacol.* 692, 46–51.

Manyi-Loh CE, Clarke AM, Ndip RN (2011) "An over view of honey: therapeutic properties and contribution in nutrition and human health," *African Journal of Microbiology Research,* 5, (8), 844–852.

McCarty MF, (2005) "A chlorogenic acid-induced increase in GLP-1 production may mediate the impact of heavy coffee consumption on diabetes risk," *Medical Hypotheses,* 64, 4, 848–853.

Meng S, Cao J, Feng Q, Peng J, and Hu Y, (2013) Roles of Chlorogenic Acid on Regulating Glucose and Lipids Metabolism: A Review, Evidence-Based Complementary and Alternative Medicine, 11, http://dx.doi.org /10.1155/2013/801457

Mishra RN (2011) "Rasayan—the ayurvedic perspective," *Research Journal of Pharmaceutical, Biological and Chemical Sciences.* 2,(4), 269–282.

Motawi TK, Darwish HA, Abd El Tawab AM (2011) Effects of caffeic acid phenethyl ester on endotoxin-induced cardiac stress in rats: a possible mechanism of protection. *J. Biochem. Mol. Toxicol.* 25, 84–94.

Natarajan K, Singh S, Burke TR, Grunberger, D, Aggarwal BB (1996) Caffeic acid phenethyl ester is a potent and specific inhibitor of activation of nuclear transcription factor NF-kappa B. *Proc. Natl. Acad. Sci. USA,* 93, 9090–9095

Nishimura H, Satoh A (2006) Antimicrobial and nematicidal substances from the root of chicory (Cichoriuminty bus). *Allelo chemicals Biol. Contr. Plant Path. Dis.* 2, 177–180.doi:10.1007/1-4020-4447-X_9

Oboh G, Agunloye OM, Akinyemi AJ, Ademiluyi AO, Adefegha SA, (2013) Comparative Study on the Inhibitory Effect of Caffeic and Chlorogenic Acids on Key Enzymes Linked to Alzheimer's Disease and Some Pro-oxidant Induced Oxidative Stress in Rats' Brain-In Vitro, *Neurochem Res.* 38, 413–419.

Ornish D, Scherwitz LW, Doody RS, Kesten D, McLanahan SM, Brown SE, et al. (1983) Effects of stress management training and dietary changes in treating ischaemic heart disease. *JAMA;* 249:54-9.

Othman ZN, Shafin R, Zakaria NH, Hussain N, Mohammad ZW (2011) "Improvement in immediate memory after 16 weeks of tualang honey (Agro Mas) supplement in healthy postmenopausal women," *Menopause,* 18, (11), 1219– 1224.

Pacholok SM, Stuart JJ. (2011) Could it be B12? An Epidemic of Misdiagnoses, 2nd edition. *Fresno: Quill Driver Books*

Queffelec C, Bailly F, Mbemba G, Mouscadet J, Hayes S, Debyser Z, et al., (2008) Synthesis and antiviral properties of some polyphenols related to Salvia genus. *Bioorg. Med. Chem. Lett.* 18, 4736–4740.doi:10.1016 /j.bmcl.2008.06.063

Rahman MM, Gan SH,2 and Khalil I (2014) Neurological Effects of Honey: Current and Future Prospects, Evidence-Based Complementary and Alternative Medicine, Article ID 958721, 1-13, http://dx.doi.org/10.1155 /2014/958721

Rartley RD, Jones E (1975) Effect of ultraviolet light on substituted cinnamic acids and theestimation of their cis and trans isomers by gas chromatography. *Chromatogr.* 107, 213 - 218.

Rodriguez de Sotillo DV, Hadley M, (2002) "Chlorogenic acid modifies plasma and liver concentrations of: cholesterol, triacylglycerol, and minerals in (fa/fa) Zucker rats," *Journal of Nutritional Biochemistry*, 13, 12, 717–726.

Santos MD, Almeida MC, Lopes NP,. de Souza GEP (2006) "Evaluation of the anti-inflammatory, analgesic and antipyretic activities of the natural polyphenol chlorogenic acid," *Biological and Pharmaceutical Bulletin,* 29, (11), 236–2240.

Santos NAG, Martins NM, Silva RB, Ferreira RS, Sisti FM, Santos AC (2014) Caffeic acid phenethyl ester (CAPE) protects PC12 cells from MPP+ toxicity by inducing the expression of neuron-typical proteins, *NeuroToxicology.* 45, 131–138.

Scarpati ML, Oriente G (1958) Chicoric acid (dicaffeyltartic acid): its isolation from chicory (Chicoriuminty bus) and synthesis. *Tetrahedron* 4, 43–48.doi: 10.1016/0040-4020(58)88005-9

Shen W, Qi R, Zhang J et al., (2012) "Chlorogenic acid inhibits LPS-induced microglial activation and improves survival of dopaminergic neurons," *Brain Research Bulletin,* 88, 5, 487–494.

Shimoda H, Seki E, Aitani M, (2006) "Inhibitory effect of green coffee bean extract on fat accumulation and body weight gain in mice," *BMC Complementary and Alternative Medicine,* 6, article 9.

Silvia RB, Santos NAG, Martins NM, Ferreira DAS, Barbosa F, et al., (2013) Caffeic acid phynethyl ester protects against the dopamnergic neuronal loss induced by 6-hydroxy dopamine in rats, *Neuroscience* 233 (2013) 86–94

Svobodova, A., Psotova, J. and Walterova, D. (2003) Natural phenolics in the prevention of UVinduced skin damage: A Review. *Biomed Pap Med Fac Univ Palacky Olomouc Czech Repub.* 147, 137 - 45.

Tsai CF, Kuo YH, Yeh WL, Yu-Ju et al., (2015) Regulatory Effects of Caffeic Acid Phenethyl Ester on Neuroinflammation in Microglial Cells, *Int. J. Mol. Sci.* 16, 5572-5589.

Tsai Y, Chiu C, Chen J, Chan K, Lin S (2012) Cytotoxic effects of Echinacea purpurea flower extracts and cichoric acid on human colon cancer cells through induction of apoptosis. *J. Ethnopharm.* 143, 914–919.doi: 10.1016/j.jep.2012.08.032

Tolba MF, Azab SS, Khalifa AE, Abdel-Rahman SZ, Abdel-Naim AB, Caffeic Acid Phenethyl Ester, a Promising Component of Propolis with a Plethora of Biological Activities: A Review on its Anti-inflammatory, Neuroprotective, Hepatoprotective, and Cardioprotective Effects, International Union of Biochemistry and Molecular Biology, 65,(8),2013, 699–709

Tousch D, Lajoix A, Hosy E, Azay-Milhau J, Ferrare K, et al., (2008) Chicoric acid, a new compound able to enhance insulin release and glucose uptake. *Biochem. Biophys. Res. Comm.* 377, 131–135.doi: 10.1016/j.bbrc. 2008. 09.088

Xiao H, Wang J, Yuan L, Xiao C, Wang Y, Liu X (2013) Chicoric acid induces apoptosis in 3T3-L1 preadipocytes through ROS-mediated PI3K/Akt and MAPK signalling pathways. *J. Agric. Food Chem.* 61, 1509–1520.doi: 10.1021/jf3050268

Yang JQ, Zhou QX, Liu BZ, He BC (2008) "Protection of mouse brain from aluminum-induced damage by caffeic acid," *CNS Neuroscience &Therapeutics,* 14, (1), 10–16.

Yashin A, Yashin Y, Wang JY and Boris N Antioxidant and Antiradical Activity of Coffee, Antioxidants 2013, 2, 230-245; doi:10.3390 /antiox2040230

Yiş U, Topçu Y, Özbal S, Tuğyan K, Bayram E, Karakaya P, Yilmaz O, Kurul SH, (2013) Caffeic acid phenethyl ester prevents apoptotic cell death in the developing rat brain after pentylenetetrazole-induced status epilepticus, *Epilepsy & Behavior*. 29, 275–280

Yukawa GS, Mune M, Otani H et al., (2004) "Effects of coffee consumption on oxidative susceptibility of low-density lipoproteins and serum lipid levels in humans," *Biochemistry* (Moscow), 69, 1, 70–74.

Zhang L, Zhang W P, Chen KD, Qian X D, Fang S H, Wei E Q (2007), Caffeic acid attenuates neuronal damage, astrogliosis and glial scar formation in mouse brain with cryoinjury. *Life Sciences* 80, 530–537

Zhao Y, Wang J, Ballevre O, Luo H, Zhang W (2012) "Antihypertensive effects and mechanisms of chlorogenic acids," *Hypertension Research*, 35, 4, 370–374.

In: Caffeic Acid
Editor: Leanna Vaughn
ISBN: 978-1-63483-121-5

Chapter 2

ROLE OF CAFFEIC ACID IN WINEMAKING

Jose Luis Aleixandre-Tudo[1,2] and Jose Luis Aleixandre[1]
[1] Universidad Politecnica de Valencia, Spain
[2] Stellenbosch University, South Africa

ABSTRACT

Caffeic acid is a hydroxycinnamic acid that is involved in a large number of chemical interactions with other grape and wine components. Caffeoylglucoside anthocyanins have shown increased stability compared with monoglucoside anthocyanins because their acylated structures allows for intramolecular copigmentation. During wine aging, the phenyl pyranonathocyanin Pinotin A is formed from the direct reaction of caffeic acid with malvidin3-glucoside. This compound shows specific red color characteristics (i.e., hypsochromic shift from 520 nm) and a higher resistance to pH changes and SO_2 addition (i.e., increased stability over time) than malvidin-3-glucoside. In addition, intermolecular copigmentation interactions between anthocyanins and caffeic acid through weak hydrophobic forces exhibit both a hyperchromic (increased maximum absorbance) and a bathochromic shift towards bluish hues. These interactions increase the anthocyanin extraction during the early stages of winemaking and are thought to be the first step in the stabilization of wine color. On the other hand, caffeic acid also plays an important role in white wine browning reactions. Furthermore, caffeic acid elicits moderate puckering astringency and bitterness; however, its impact on overall wine taste remains unclear. Finally, caffeic acid could also be used as a potential inhibitor of aroma loss during aging with a consequent reduction in the SO_2 addition during winemaking. The participation of caffeic acid in a large number of mechanisms strongly verifies the important role of this molecule in wine chemistry.

1. INTRODUCTION

Wine and grape phenolic compounds are divided into two main families: flavonoids and non-flavonoids. Anthocyanins, flavanols and flavonols belong to the flavonoid family. Hydroxycinnamic and hydroxybenzoic acids, together with stilbenes (resveratrol isomers), belong to the non-flavonoid group. Hydroxycinnamates represents the major phenols in grape juice and hydroxycinnamic acids exist as esters of tartaric acids in the vacuoles of skins and pulp cells of the grape berries. Hydroxycinnamates due to the hydroxyl groups and the potential for delocalization of the ring p-electrons are antioxidant hydrophilic phenols. These molecules influence wine color, astringency, bitterness, oxidation level and turbidity.

p-Coutaric, caftaric and fertaric acids are found in grapes and wines with caftaric acid being the most predominant, followed by *p*-coutaric and fertaric acids. In wine, large differences in the proportions and amounts of each ester have been reported (Ong & Nagel, 1978; de Villiers et al., 2005). In aqueous acidic solutions, such as wine, the esters of tartaric acids are rapidly hydrolyzed to simple hydroxycinnamic acids, such as coumaric, caffeic and ferulic (Waterhouse, 2002). Therefore, caffeic acid (3,4-dihydroxycinnamic acid) is formed in wine after acid hydrolysis from its hydroxycinnamate ester precursor, caftaric acid. The caffeic acid content does not exceed 10 mg/L in the most common red wines and grape varieties (Rodriguez-Delgado et al., 2002; Landrault et al., 2001; Soleas et al., 1997, Schwarz et al., 2004, Rossouw & Marais, 2003). Only the South African autochthonous Pinotage (a cross between Pinot Noir and Cinsault) was able to express caffeic acid concentrations as high as 77 mg/L (Schwarz et al., 2003). This organic acid has a significant metabolic function in plant tissue with a variety of physiological and pharmacological properties (Robbins, 2003).

The health benefits of caffeic acid have also been extensively reported in the literature (Moghaddam et al., 2007; Nagaoka et al., 2002; Ye et al., 2010; Saija et al., 2000). Biological activities including antioxidant, antibacterial, anti-inflammatory and anti-tumor effects have been widely reported (Wang et al., 2012; Oh et al., 2013; Khan et al., 2012; Weng et al., 2012). In particular, its ability to prevent cholesterol oxidation by increasing the resistance of low-density lipoproteins has been researched by many authors (Castelluccio et al., 1995; Marinova & Yanishlieva, 2003; Sroka & Cisowski, 2003). In addition, regarding its physiological activities, caffeic acid acts as a plant defense mechanism against infection predation and plays an important role in growth and reproduction (Robbins, 2003; Wu et al., 2000).

During the winemaking process, skin, pulp and seed phenolic compounds in grape berries are extracted. Numerous self-interactions and interactions between the extracted phenolic compounds and other grape- and wine-derived components take place during winemaking and aging. These interactions have a significant effect on wine organoleptic properties. Anthocyanins, due to their water solubility, are extracted during the early stages of fermentation. These compounds are responsible for the red wine color and provide diverse color properties. Their conformation and/or interaction with other compounds also influence other sensory attributes (e.g., astringency) as well as the color stability of the wine. In addition, color is one of the most important sensory parameters for wine consumer acceptance, and wine chromatic characteristics can be enhanced by color stabilization.

Anthocyanin stabilization during aging has been a prolific area of research. As mentioned, they undergo a large number of interactions, including degradation reactions. Wine color stability is affected by many factors, as anthocyanins are influenced by pH, temperature, light and the presence of enzymes (Timberlake & Bridle, 1976; Brouillard, 1982; Mazza & Brouillard, 1990; Morais et al., 2002; Sarni-Manchado et al., 1996, Wrolstad, 2000, Castañeda-Ovando et al., 2009). In addition, the presence of other compounds in the wine matrix, such as flavanols, flavonols, phenolic acids and metals, can also influence its reactivity and stability. Additionally, the concentration and structure of the anthocyanin pool and the ratio between the anthocyanins and tannins have also been proposed as important factors in the final wine anthocyanin levels and conformations (Singleton et al., 1992; Fulcrand et al., 2004; Kilmister et al., 2014).

Hydroxycinnamic acids *p*-coumaric and caffeic are also found in grapes as acyl substituents of the anthocyanins. Increased color stability of acylated anthocyanins has been reported when compared with non-acylated ones. Cinnamic acids represent the main acyl group in acylated anthocyanins, and the stabilization has been attributed to the formation of intramolecular copigmentation complexes. The intramolecular copigmentation between the cinnamoyl residues and the anthocyanins is thermodynamically favored over the intermolecular copigmentation (between anthocyanins and external copigments). The intramolecular copigmentation ability of caffeic acid with anthocyanins was demonstrated for the first time by Dimitri-Markovic et al. (2000). The authors also highlighted that the feasibility of the process is driven by several factors, such as the pH, copigment/pigment ratio or temperature. Later research also confirmed that the antioxidant capacity of the anthocyanin:caffeic acid complex was greater than the antioxidant capacity of the anthocyanin alone (Dimitri-Markovic et al., 2003a, b).

In addition, phenolic acids and specifically caffeic acid have been identified as good cofactors or copigments that actively participate in intermolecular copigmentation interactions. During vinification, an intense anthocyanin extraction take place, but some of the initial red color is lost during aging. Copigmentation has been proposed as a phenomenon that contributes to color stability. Intermolecular copigmentation is thought to play an important role in anthocyanins extraction and color stability during the early stages of winemaking, serving perhaps as a first step in the formation of more stable pigments during wine aging (Aleixandre-Tudo et al., 2013; Schwarz et al., 2005; Brouillard & Dangles, 1994; Liao et al., 1992; Darias-Martin et al., 2002). An extended study was conducted in recent years regarding caffeic acid as a copigment in vitro and in vivo (Eiro & Heinonen, 2002; Gris et al., 2007; Gomez-Miguez et al., 2006, Baranac et al., 1997; Mirabel et al., 1999; Darias-Martin et al., 2001, 2002, Aleixandre-Tudo et al., 2013, Alvarez et al., 2009; Malien-Aubert et al., 2001). Moreover, published research demonstrated a higher degree of anthocyanin extraction during winemaking together with improved red wine color stabilization through aging when caffeic acid was added before the fermentation process (Darias-Martin et al., 2001, 2002).

Another example of a reaction between anthocyanins and phenolic acids relates to the formation of pyranoanthocyanins. Pyranoanthocyanins can be formed through a reaction between anthocyanins and compounds containing a polarizable double bond. These compounds were initially reported to form from the anthocyanins interaction with secondary yeast metabolites (acetaldehyde or pyruvic acid), but it has more recently been proven that hydroxycinnamic acids (e.g., caffeic acid) or their 4-vinylphenols decarboxylation products can react with anthocyanins (Escribano-Bailon & Santos-Buelga, 2012).

On the other hand, hydroxycinnamic acids are the major phenolic compounds in white wine and may contribute to its organoleptic properties. They have been reported to be involved in enzymatic and non-enzymatic browning reactions during white winemaking. After crushing, hydroxycinnamates are rapidly oxidized to quinones by the polyphenol oxidase enzyme. Quinones may further participate in non-enzymatic reactions with other phenolic compounds, such as flavanols, thus increasing the brown color in white wine. As previously mentioned, caftaric acid, the precursor of caffeic acid, is the major hydroxycinnamate found in white grapes. Therefore, higher levels of caffeic acids are expected in white wines and may be involved in browning reactions (Franquet-Griell et al., 2012; Danilewicz et al., 2003, 2007; Clark & Scollary, 2002; du Toit et al., 2006; Li et al., 2005).

In addition, hydroxycinnamic acids have been associated with changes in some wine sensory attributes. Caffeic acid was observed to elucidate astringency and bitterness (Hufnagel & Hofman, 2008), and its astringent sub-qualities have been described and compared with other phenolic compounds (Ferrer-Gallego et al., 2014). A synergistic effect has also been reported when mixtures of hydroxycinnamic acids (e.g., caffeic + coumaric acids) were evaluated by sensory analysis. These findings might question the assertion reported in previous papers where, at the concentrations found in wine (far below the perception thresholds), they appear to have no astringent or bitter taste (Verette et al., 1984).

Finally, caffeic acid has also been identified as a protective agent against aroma loss during storage. Different concentrations of caffeic acid, either alone or in combination with other aroma preservatives, have been added to model solutions as well as red and white wines, and the aroma profiles were investigated using gas chromatography. The results showed the potential preventative effect of caffeic acid against aroma loss during aging. The implications that these findings have on wine health are remarkable because a possible decrease in the use of SO_2 can also be considered. Caffeic acid has thus been proposed as a possible agent that can be used to reduce the SO_2 levels in wine (Roussis et al., 2005, 2007, 2013; Lambropoulos & Roussis, 2007; Roussis & Sergianitis, 2008).

As previously mentioned, caffeic acid is highly involved in wine chemistry and takes part in a large number of interactions that strongly influence wine organoleptic properties. In this research chapter, the effects that caffeic acid has on wine color (properties and stability), astringency, bitterness and aroma are reported and discussed. Based on the available published information and with the final goal of enlightening the wine research community, a summary of the most relevant research findings are summarized here.

2. Interactions of Caffeic Acid with Anthocyanins and Its Effect on Wine Color Properties and Stability

Copigmentation interactions among anthocyanins and copigments were proposed for the first time in studies performed by Robinson & Robinson (1931). Despite this report, its mechanism is still not perfectly understood. Copigmentation refers to the interaction between anthocyanins and other copigments or cofactors (generally colorless phenolic compounds) (Boulton, 2001; Escribano-Bailon & Santos Buelga, 2012). A complex molecule with increased stability is formed, and the equilibrium of the anthocyanins is shifted towards the colored forms. Copigmentation protects anthocyanins from the nucleophilic attack of water molecules and, therefore, from a wine color loss (Quina et al., 2009, Martin-Bueno et al., 2012). A large number of molecules are proposed as possible copigments of grapes and wine anthocyanins. Flavonoids, alkaloids, amino acids, organic acids, polysaccharides, metals and anthocyanins are known to be involved in copigmentation interactions.

Intramolecular copigmentation interactions through the acylation of the anthocyanin sugar moieties by caffeic acid are well established today. Initially, a study performed by Wulf

& Nagel (1978) demonstrated the presence of malvidin3-(6-caffeoyl)-glucoside in wine. Later on, Baldi et al. (1995) reported the formation of the 3-caffeoylglucosides of peonidin in grapes. Vidal et al. finally reported the presence of the caffeoylglucosides of cyanidin, petunidin and delphinidin in 2004. Acylated anthocyanins have shown higher stabilities than the non-acylated anthocyanins, thus strongly influencing wine color (Cevallos-Casals & Cisneros-Zevallos, 2004; Giusti & Wrolstad, 2003; Malien-Aubert et al., 2001; Bakowska-Barczak, 2005). On the other hand, intermolecular loose copigmentation reactions involving caffeic acid as a copigment have also been described. In this case, these weak hydrophobic interactions are believed to play an important role in the initial red wine color as well as on the anthocyanin extraction during the early stages of fermentation.

In addition, caffeic acid can also react with the anthocyanins through a cycloaddition reaction to form pyranoanthocyanins. The pathway leading to the formation of Pinotin A, a compound that originates from the direct reaction of caffeic acid with malvidin-3-glucoside, is well established today. Theses pigments also have a strong influence on the color of red wine and are thought to be crucial for wine color stability.

2.1. Anthocyanin-Caffeic Acid Intramolecular Copigmentation

Intramolecular copigmentation may take place when the copigment is part of the anthocyanin molecule. Organic acids, aromatic acyl groups or flavonoids interact with the anthocyanin glycosyl moiety, stabilizing the anthocyanin chromophore. Caffeic acid is involved in acylation reactions, which lead to the formation of the caffeoylglucoside anthocyanins. Its effects on wine color and stability are discussed in the next section.

2.1.1. Anthocyanin Acylation by Caffeic Acid

Aromatic acids may be involved in acylated interactions with the monoglucoside anthocyanin sugar residues. Aliphatic, hydroxybenzoic and hydroxycinnamic acids, such as caffeic acid, may participate in the acylation phenomenon. The acyl substituents are commonly bound at the C3 sugar, esterified to the 6-OH or less frequently to the 4-OH group of the sugars.

Due to the presence of a hydroxycinnamic acid acyl substituent, the appearance of an absorption band from 310-360 nm is detected in the UV-Vis spectra. This fact might be an indication of anthocyanin acylation. Moreover, in a study performed by Giusty & Wrolstad (2003), the presence of cinnamic acid acylation resulted in a bathochromic shift of the pigment in the visible λ_{max} towards higher wavelengths (bluish color) coupled with increased absorptivity (hyperchromic effect). Although these results were reported for the acylation of the anthocyanin pelargonidin, using cinnamic acid as the acylating agent, similar results might be expected for the acylation of grapevine anthocyanins with caffeic acid. These authors also reported that aromatic acids have better stabilization effects than aliphatic ones.

The molecular structure plays a crucial role in the stability of wine pigments. Anthocyanins found in wine are thus more stable as their methylation increases (Brouillard, 1982). Taking into account the bathochromic and hyperchromic effect of pelargonidin acylation by cinnamic acid, it can be suggested that an increased stability in the wine color due to the acylation of malvidin-3-glucoside with caffeic acid might occur, although this needs to be confirmed with appropriate experiments.

In addition, the increased stabilization of acylated anthocyanins is due to the stacking of the acyl groups with the pyrylium ring of the flavylium cation, avoiding or hindering the nucleophilic attack of water and the later formation of a colorless pseudobase or chalcone form of the anthocyanin. This stabilization is stronger when complexes are formed from the acyl substitution in the B ring, rather than in the A ring (Yoshida et al., 2002). Color fading is also better protected by acylated anthocyanins when the pH increases (Giusty & Wrolstad, 2003, Cevallos-Casals & Cisneros-Zevallos, 2004). The presence of more than one acyl group, the position of the acyl group on the anthocyanin and the position of the sugar moieties might also influence the stability of the copigmented structures (Malien-Aubert et al., 2001; Eiro & Heinonen, 2002). Acylated anthocyanins have also shown higher stability against light degradation, high temperatures and SO_2 addition (Gris et al., 2005). Acetylated anthocyanins have thus been proposed as a source of food colorant based on their improved stability (Bakowska-Barczak, 2005).

2.2. Anthocyanins-Caffeic Acid Intermolecular Copigmentation Reactions

Intramolecular copigmentation reactions have been extensively studied and the elucidation and characterization of most of them are currently well established (Escribano-Bailon & Santos Buelga, 2012). However, intermolecular copigmentation of anthocyanins occurring in grapes and wines has not been elucidated yet (Boulton, 2001). Intermolecular copigmentation is defined as loose non-covalent π-π stacking association (through hydrogen bonds or van der Waals interactions) between the anthocyanin-colored forms (flavylium cation and quinoidal form) and the copigments (normally colorless organic compounds) with the planar polarizable nuclei of the anthocyanins. This results in a structure that is protected against attack by water at position 2 of the pyrylium nucleus, which gives rise to the hemiketal and chalcone forms. This protection is due to the surrounding sugar molecules, which are joined by external hydrogen bridges (Asen et al., 1972; Mazza & Brouillard, 1990, Awika, 2008; Gauche et al., 2010). In addition, a change in the chromatic properties of the solution is observed. Visually, the color gets more intense (hyperchromic effect) and a change in hue color takes place (bathochromic effect). The wavelength of the maximum absorbance in the visible region of the spectrum is higher than that of the non-interacted free anthocyanin. In addition, the bathochromic shift is thought to be due to the interaction of the copigment with the anthocyanin in its first excited state, which is stronger than in the ground-state anthocyanin (Alluis et al., 2000, Malien-Aubert et al., 2001).

Several factors have been postulated to influence the extent of the copigmentation phenomenon. Temperature, pH, alcohol content, the presence of metals, the natures of the anthocyanin chromophore and the copigment, the copigment/pigment molar ratio and the pigment absolute concentration influence the magnitude and extent of the copigmentation (Dangles et al., 1993; Asen et al., 1972; Baranac et al., 1996; Santos-Buelga et al., 1995; Gomez-Miguez et al., 2006).

Caffeic acid has been investigated as copigment or cofactor participating in intermolecular copigmentation interactions. In this way, several studies have been conducted where caffeic acid was added to different anthocyanin solutions. Caffeic acid has been postulated (within the hydroxycinnamic acids) to be a strong copigment (Eiro & Heinonen, 2002) showing good stability of the anthocyanins (Gris et al., 2007). In addition, hydroxycinnamic acids have been proposed as better copigments than flavan-3-ols and some proanthocyanidins but worse than flavonols (Gomez-Miguez et al., 2006, Baranac et al., 1997; Mirabel et al., 1999). Caffeic acid has also shown a strong copigmentation effect when it was added prior to the fermentation process, influencing red wine color stabilization during aging (Darias-Martin et al., 2001, 2002, Aleixandre-Tudo et al., 2013, Alvarez et al., 2009).

2.2.1. Addition of Caffeic Acid in Model Wine Solutions

The influence of the addition of different phenolic copigments on the color of malvidin-3-glucoside (the major anthocyanin found in grapes and wines) in model wine solutions at a copigment/pigment molar ratio 1:1 was investigated by Gomez-Miguez et al. (2006). Compounds pertaining to the main phenolic groups were tested. Catechin, epicatechin and procyanidin B2 (flavanols), caffeic and coumaric acids (hydroxycinnamic acids) and myricitrin and quercitrin (flavonols) were considered in the study. Caffeic acids showed intermediate copigmentation efficacy, whereas flavan-3-ols were determined to be less effective copigments and flavonols were the best.

Concerning hydroxycinnamic acids, a progressive hyperchromic shift was observed suggesting that copigmentation is not necessarily an immediate effect and might take place more gradually for less effective copigments. In addition, a slight shift towards bluish hues was first produced (suggesting intermolecular copigmentation), but later on, the shift was moved towards the orange/yellow region of the spectrum suggesting the formation of new pigments with pyranoanthocyanin structures. This fact was proved with the appearance of a new pigment from two weeks of storage in the solution containing caffeic acid. This new peak was identified as the 4-vinylcatechol pyranoanthocyanin adduct of malvidin-3-glucoside. In the formation of these compounds, a bond between the C4 of the anthocyanins and the C2 position of the caffeic acid would be established giving rise to an intermediate carbonium compound that can be attached to form the new pyran ring. The final steps would consist in decarboxylation and oxidation of the pyran moieties (Schwarz et al., 2003b).

It was also observed that the decrease in the malvidin-3-glucoside was similar in the solution containing caffeic acid to the decrease in the solution containing only anthocyanin. Despite this, the color is better maintained in the caffeic acid/malvidin-3-glucoside solution. The protection of the anthocyanin against hydration was proposed as a preventive degradation mechanism (Kunsagi-Mate et al., 2009). In addition, the smaller hydration constant of the pyranoanthocyanins allows them to appear in a greater proportion in colored forms in weakly acidic conditions (Vivar-Quintana et al., 2002, Sarni-Machado et al., 1996), thus highly influencing the wine color properties.

Other studies investigated the intermolecular copigmentation interactions of five different anthocyanins and five different phenolic acids (Eiro & Heinonen, 2002). The copigments, ferulic, caffeic, rosmarinic, chlorogenic and gallic acids were added in a model wine solution containing cyanidin-3-glucoside, pelargonidin-3-glucoside, malvidin-3-glucoside, a cyanidin acylated anthocyanin with *p*-coumaric acid and a trisaccharide cyanidin derivatized anthocyanin. Different molar ratios were also tested. Within phenolic acids, caffeic acid was included in the best copigments group together with ferulic acid. The simple cinnamic acid chemical structure of both compounds might explain their ability to act as copigments. In addition, the copigmentation effect was stronger with the malvidin-3-glucoside anthocyanin. It was also suggested that the presence of an acylation extension or the increased number of sugar moieties are enough to prevent intermolecular copigmentation favoring intramolecular interactions (Eiro & Heinonen, 2002). The authors also reported a decline in the bathochromic shift of the caffeic acid solutions after 6 months of storage. This fact might be related to the stability of the intermolecular copigmentation interactions (Boulton, 2001) and the formation of new reaction products (Schwarz et al., 2004).

The copigmentation phenomenon is affected by the molar ratio between the anthocyanins and the copigments (Malien-Aubert et al., 2001). At the molar ratio anthocyanin/copigment 1:1, copigmentation takes place (Gomez-Miguez et al., 2006), although a higher concentration of copigments induces a larger absorbance increase at 520 nm in malvidin-3-glucoside solutions with phenolic acids (Mazza & Brouillard, 1990; Gris et al., 2007). However, when the copigment concentration exceeds certain levels, no further copigmentation interactions should be expected (Hosino et al., 1980; Eiro & Heinonen, 2002).

A further step in the model wine additions is the use of grape extracts instead of pure isolated anthocyanins (Malien-Aubert et al., 2001, Gris et al., 2007; Gauche et al., 2010). The color stability of commercial anthocyanin-based extracts from different sources was analyzed and the phenolic composition of the extracts was also taken into account. Intramolecular copigmentation was favored in those extracts rich in acylated anthocyanins, with increased protection of the red chromophore when anthocyanins were diacylated. In colorants lacking acylated anthocyanins intermolecular copigmentation reactions play a crucial role in pigment stabilization. Moreover, the copigment/pigment molar ratio showed its influence on the color with increased stability at increased copigment concentrations (Malien-Aubert et al., 2001). The relevance of all these findings is of great importance and reveals the important role played by caffeic acid in wine color.

In another study, anthocyanins extracted from Cabernet Sauvignon grapes were copigmented with caffeic acid at a 1:1 molar ratio (Gris et al., 2007). The results showed that, in general, better color stability of the caffeic acid/anthocyanin solutions under the different storage conditions. Anthocyanin degradation was faster at higher temperatures (Gris et al., 2007; Morais et al., 2002) and caffeic acid added solutions were able to increase pigment stability when compared with the solutions without caffeic acid. The presence of light also had an important effect on the color; solutions stored in the dark showed higher pigment stabilities (Gris et al., 2007).

The effects of organic acid addition to crude Cabernet Sauvignon grape skin extracts at different pH values were evaluated by Gauche et al. (2010). For every acid used (caffeic, ferulic, gallic and tannic) the maximum hyperchromic and bathochromic shifts where observed at pH 3.3. When anthocyanins only exist in the flavylium cation form (pH1-2), the hyperchromic effect is only due to the interactions between the colored anthocyanin and the copigment. At higher pH values, the role of the copigments also consists of preventing the formation of a non-colored pseudobase and avoiding anthocyanin color loss via hydration (Gris et al., 2007; Wilska-Jeszka & Korzuchowska, 1996; Boulton, 2001).

In addition, the alcohol content has also been proposed as a key factor in the copigmentation phenomenon (Hermosin-Gutierrez, 2003). Previous research reported that the extent of the copigmentation is hampered by increased alcohol concentrations (Dangles & Broullard, 1992; Ribereau-Gayon et al., 1983; Brouillard et al., 1991). However, an increase in the redness of wine after the addition of alcohol to red wine was reported by Berg & Akiyoshi in 1975 (Mirabel et al., 1999). This effect was explained by the alcohol dehydration effect over the chalcone and hemiketal forms giving rise to the flavylium cation form. Early research showed that copigmentation interactions are better protected above 8% alcohol

(Kunsagi-Mate et al., 2007, 2009). This was explained by the rotation of the water-ethanol complexes that surround the copigmented pigment. At alcohol levels below 8%, the more polar water molecules point towards the copigmented stacks, while above 8% alcohol, the complexes rotate by 180°, resulting in the non-polar carbon side of the ethanol pointing towards the stacking, which increases the stabilization of the copigmented molecule (Kunsagi-Mate et al., 2007, 2008). Later, these authors studied the formation kinetics of malvidin-3-glucoside with caffeic acid and catechin in different molecular environments. The interactions between anthocyanins and caffeic acid were reported to be more likely to occur than those with catechin due to the increased energy barrier of the association in the latter case, which is supported by the lower frequency factor observed (Kunsagi-Mate et al., 2009). In the first case, the copigmentation complexes were formed in water and later altered after dilution with ethanol. In the second case, the complexes were already formed in a binary solvent (water/ethanol). Faster formation of the complexes was observed in the first case, which simulates traditional fermentation, when interactions probably start in water media before alcohol formation. Thereafter, the alcohol content might act as a stabilizing agent for the new copigmentation complexes.

In summary, the addition of caffeic acid to model wine solutions containing anthocyanins highlighted the occurrence of a large number of interactions that will highly influence and further define wine color properties and stability. Moreover, the extent of the interactions is affected by many factors. The copigment/pigment molar ratio, temperature, the presence or absence of light, pH and alcohol content are identified as important parameters to take into account if these stabilization reactions are to be encouraged. On the other hand, the observed effects need to be confirmed by "in vivo" caffeic acid additions to wines.

2.2.2. Addition of Caffeic Acid during Winemaking

Caffeic acid was added before fermentation at different dosages and colorimetric and phenolic measures demonstrated that the effects observed in model wine solutions were also present in wine. Darias-Martin et al. (2001) added 120 mg/L of caffeic acid or catechin to two red wines. It was observed that caffeic acid addition enhanced wine color up to 60% (whereas catechin increased only 10%). The authors also observed an increased in the extraction, i.e., more pigment was dissolved during fermentation, and higher color provided by the copigmented anthocyanins. In further studies, the same authors investigated the effect of different dosages of caffeic acid during storage. The results showed color enhancement after copigment addition and during storage suggesting that copigmentation could be a first step in the formation of more stable pigments (Darias-Martin et al., 2002, Mirabel et al., 1999). Interestingly, the maximum absorbance was hypsochromically shifted towards lower wavelengths for Listan Negro wine whereas Negramoll wine showed the opposite behavior with a bathochromic shift. It can be suggested that color stabilization and anthocyanin protection against oxidation are strongly linked to caffeic acid addition.

Furthermore, caffeic acid was added to Cabernet Sauvignon and Pinot Noir grapes after crushing. The copigmentation effect was more pronounced in the Pinot Noir (also showing a bathochromic shift) than in the Cabernet Sauvignon wines (Bloomfield et al., 2003). The lack of acylated anthocyanins in the Pinot Noir grapes could have favored the interaction between the copigments and the anthocyanin with subsequent changes in the color properties. Forty days after storage, the copigmented caffeic acid of Cabernet Sauvignon wines showed a hypochromic effect with a decrease in the maximum absorption. However, at 60 and 90 days after storage, a hyperchromic effect was observed. In addition, no hypso- or bathochromic shifts were identified throughout the experiment. The importance of the phenolic composition in the success of the copigmentation and the fact that the copigmentation phenomenon does not immediately occur were proved in this study (Bloomfield et al., 2003).

Caffeic acid addition prior to fermentation was also investigated in Tempranillo and Cabernet Sauvignon wines (Schwarz et al., 2003b). Analyses were performed after fermentation and at 9 months of storage. Moreover, the formation of pyranoanthocyanins was also investigated at 21 months of storage. Hypochromic effects were detected for the two varieties studied when the addition of caffeic acid was evaluated. These results are in agreement with those reported by Bloomfield et al. (2003) in Cabernet Sauvignon wines but contradict those previously described by Darias-Martin et al. (2001, 2002) for the pre-fermentative addition of caffeic acid to wines made with Listan Negro and Negramoll. No increases in the extent of the copigmentation were observed when caffeic acid was added to wines, whereas an increase in the polymerization degree was detected. These results were in agreement with the color analysis because wines with caffeic acid had less saturated colors and weak purple hues.

During aging, a decrease in monomeric anthocyanins was observed and was attributed to the formation of polymeric pigments. This also gave rise to a complete disappearance of copigmentation interactions at approximately 40 weeks after the end of the fermentation. The anthocyanin/copigment molar ratio varies during aging, which strongly influences the copigmentation interactions (Boulton, 2001). The wine color turned lighter and was less saturated, confirming the fading of the hyperchromic and bathochromic effects. This author also mentioned that the monomeric anthocyanin profile might play a critical role in the extent of the copigmentation phenomenon. The higher *p*-coumaroylated anthocyanin presence could diminish the effect of the added copigments in Tempranillo wines. The anthocyanin monoglucosides and polymeric pigments strongly contributed to the overall wine color, but the contribution of pyranoanthocyanins to it is not well established (Schwarz et al., 2003b).

During winemaking, the phenolic compounds from the grape skins and seeds are dissolved in the wine. The extent of the extraction depends on many factors. Within those factors, the grape cultivar, climatic and soil conditions, ripening level, winegrowing and winemaking techniques will influence the presence of phenolic compounds in future wines. Several winemaking techniques have been reported to influence grape phenolic extractability (Sacchi et al., 2005). Pre-fermentative cold maceration has been found to increase phenolic levels in wines. Anthocyanins (water-soluble compounds) are extracted during the early maceration step in the absence of ethanol. The low temperature needed in the maceration step can be achieved by several methods. The use of low temperatures is one of the most extended methods. In addition, dry ice can also be used as a cooling agent. The goal when applying this technique relies on delayed fermentation, which favors the extraction of compounds from the solid parts of the berry.

Moreover, the addition of small controllable amounts of oxygen to wine at different stages of the fermentation has been reported to influence the content and evolution of the phenolic fraction. Numerous reactions have been associated with oxygen addition through acetaldehyde linkages (Es Safi et al., 1999; Fulcrand et al., 1996; Saucier et al., 1997; Fulcrand et al., 2006; Es Safi et al., 2002, Es Safi et al., 2003; Liu & Pilone, 2000).

Copigments, including caffeic acid, were added pre-fermentation to Tempranillo wines elaborated with different cold maceration techniques (Alvarez et al., 2009). Caffeic acid addition seemed to increase the color intensity and total anthocyanin content, which is probably related to the also observed increase in the copigmented anthocyanin fraction. These effects were more pronounced in dry ice cold macerated wines. The freezing effect and the breakage of the cell walls could have caused an increased phenolic solubilization followed by an increase in the interactions between the phenolic compounds. Interestingly, the combined caffeic acid addition and cryo-maceration decreased wine astringency, as evaluated by the method reported by Llaudy et al. (2004) and by sensory analysis. The inclusion of some astringent phenolics into copigmentation stacks and the possible extraction of polysaccharides, followed by an increased combination with tannins could have caused a decrease in the wine astringency perception (Alvarez et al., 2009).

Varietal and seasonal differences influence grape phenolic composition and, therefore, the extent of the copigmentation phenomenon (Aleixandre-Tudo, 2012). In further studies, the effect of caffeic acid addition in combination with different winemaking techniques was investigated again in Tempranillo wines (Aleixandre-Tudo et al., 2013). Pre-fermentative cold maceration techniques and the effect of micro-oxygenation through acetaldehyde addition were tested for up to 12 months of storage and over three consecutive vintages. Pre-fermentative cooling techniques in combination with caffeic acid did not have a significant effect on wine color. Only an increase in anthocyanin extraction was observed at the end of the fermentation process, but the color properties were not improved after wine aging. On the other hand and despite the positive effect on phenolics and color observed in acetaldehyde added wines, the combined effect of this technique with caffeic acid addition was not significant. In addition, caffeic acid was also added after fermentation. Intermolecular copigmentation interactions were non-significant and the intramolecular copigmentation in combination with pyranoanthocyanin formation was proposed as the mechanism responsible for the observed results.

2.3. Pyranoanthocyanins Formation with Caffeic Acid

Anthocyanins can be converted to pyranoanthocyanins during winemaking and aging (Escribano-Bailon & Santos-Buelga, 2012; Freitas & Mateus et al., 2011). For that reaction to take place, the anthocyanin flavylium cation needs to interact with double bond activated compounds by electron-donating substituents. These compounds were first identified in wines in 1996 by Cameira dos Santos et al. Also in 1996, Fulcrand et al. elucidated their structures, which consisted of a pyran ring at C-4 and a hydroxyl group at position 5. Pyranoanthocyanins can be formed from enol precursors and result from the reaction of anthocyanins with pyruvic acid (Vitisin A) or acetaldehyde (Vitisin B). The addition of the double bond of the enol form to the flavylium C4, followed by rehydration and rearomatization has been proposed as the mechanism yielding the formation of the new pigments (flavanyl-pyranoanthocyanins). Other types of pyranoanthocyanins can be formed from the addition of the 8-vinylflavanol, obtained after the acid-catalyzed cleavage, to the anthocyanin (vinylpyranoanthocyanins). In addition, portisins, compounds isolated from port wine, are derived from carboxypyranoanthocyanins. In this case, the flavanol and the pyranoanthocyanin are bridged by a vinyl substituent (vinylpyranoanthocyanins). Finally, phenyl pyranoanthocyanins result from the interaction between the anthocyanins and the 4-vinylphenol, 4-vinylcatechol, 4-vinylguaiacol and 4-vinylsyringol adducts. The vinylphenol-related adducts were thought to be formed by enzymatic decarboxylation of the corresponding cinnamic acids (*p*-coumaric, caffeic, ferulic and sinapic) (Freitas & Mateus et al., 2011). However, later research showed that yeast strains are able to decarboxylate *p*-coumaric and ferulic acid but none of them were capable of transforming either caffeic or sinapic acid (Lao et al., 1997; Schwarz et al., 2003b). Moreover, it was reported that certain wine constituents (especially flavan-3-ols and oligomeric proanthocyanidins) strongly inhibited the decarboxylation of cinnamic acids (Chatonnet et al., 1993). Schwarz et al. (2003b) observed that the interaction between vinylphenols and anthocyanins proceed rather slowly and requires years of storage to complete. Based on this observation, the authors speculated about the 4-vinylphenol reactivity. These molecules would be consumed rapidly, and thus, their gradual formation over prolonged periods of time would be unlikely to occur during wine aging. It was concluded that the carboxy-lyase enzyme might be inactivated by some wine components, and its activity would not be significant in the anthocyanin-vinylphenol adduct formation. The direct reaction between free phenolic acids and anthocyanins was finally elucidated and is currently accepted as the main mechanism for the formation of phenyl pyranoanthocyanins (Schwarz et al., 2003a).

In comparison to single anthocyanins, all classes of pyranoanthocyanins (except portisins, which shows a maximum absorption bathochromically shifted towards bluish colors) possess hypsochromically shifted absorption maxima, which result in more orange-red colors. The formation of the pyran ring protects the newly formed molecule against the nucleophilic attack and the subsequent colorless carbinol base formation. It has also been extensively observed that the pyran ring increases the stability of the pyranoanthocyanins color at varying pH values and protects them from SO_2 bleaching (Schwarz et al., 2003b; Sarni-Machado et al., 1996, Fulcrand, 1998; Morata et al., 2007).

In the early stages of winemaking, the formation of phenyl pyranoanthocyanins is mainly due to the interactions between anthocyanins and vinylphenol or vinylguaiacol adducts, which are the products of the enzymatic decarboxylation of free acids (*p*-coumaric and ferulic). It has been stated that, after an increase in the formation of these compounds observed during the first months of winemaking, a stable period of two years takes place (Monagas et al., 2005). After 2-3 years of storage, due to the direct reactions between hydroxycinnamic acids and anthocyanins, a second increase is observed (Schwarz et al., 2004). The formation of these compounds during wine aging depends on the availability of both anthocyanins and the low molecular weight phenolic reagents. On the other hand, in the specific case of the carboxypyranoanthocyanins, these compounds reach their maximum concentrations after approximately 5 months of storage (Schwarz et al., 2004).

Regarding caffeic acid, the pigmented compound Pinotin A has been recently isolated and characterized (Schwarz et al., 2003a) from Pinotage wines. Pinotin A is formed from the direct reaction between free caffeic acid and malvidin-3-glucoside without the need for any previous decarboxylation by wine yeast. The effect of this compound on color and wine stability is expected to be similar as the ones mentioned above for the pyranoanthocyanins and specifically for the phenyl pyranoanthocyanins.

In addition, the factors that influence the formation of Pinotin A have been investigated in Pinotage wines by Schwarz et al. (2004). The formation of these compounds is influenced by the two partners, malvidin-3-glucoside and caffeic acid. Initially, for up to 2.5 years of storage, the concentration of the anthocyanin decreases. This can be explained by degradation reactions and/or conversion into polymeric pigments with other wine components. Pinotin A formation is therefore hampered by other side reactions. Taking into account the lower concentrations of caffeic acid when compared with other reactants, it can be understood why the formation of Pinotin A is not favored at this point. This scenario changes during further aging (from 2.5 to 5-6 years). Side interactions became less frequent due to decreases in the reactants and self-associations. At the same time, the levels of caffeic acid remain stable, and thus, the caffeic acid/anthocyanin molar ratio increases with a consequent increase in the exponential formation of Pinotin A (Schwarz et al., 2004). It was also observed that a minimum concentration of malvidin-3-glucoside (5-10 mg/L) is required to maintain Pinotin A formation at a higher rate than polymeric pigment formation (Rentzsch et al., 2007). Finally, prediction analysis highlighted that the presence of caffeic acid (cofactor) is more important than the presence of the anthocyanins with regards to this new pigment formation (Schwarz et al., 2004).

3. Protective Effect of Caffeic Acid in Wine Aroma Composition

Oxidation or other chemical reactions are responsible for aroma loss during wine aging. Ester levels in wine are influenced by hydrolysis and esterification reactions. Terpenes are also affected by oxidation and other reactions. Alcohols have been reported to be very stable compounds during storage but may also undergo oxidation reactions to aldehydes. Finally, fatty acids are affected by hydrolysis and auto-oxidation reactions. Several studies dealing with the addition of caffeic acid as inhibitor of the decline of volatile compounds during storage have been recently reported (Roussis et al., 2005, 2007, 2013; Lambropoulos & Roussis, 2007; Roussis & Sergianitis, 2008).

Caffeic acid was added at 100 ppm to a model wine solution, Xinomavro red wines and Muscat white wines. The decrease in the volatile esters and terpenols after two days of oxidative storage was prevented by the caffeic acid addition (Roussis et al., 2005). Later studies tested the combined effect of caffeic acid (CFA) with the strong anti-browning agent N-acetyl-cysteine (NAC) in Debina white wines and in model wine solutions (Sergianitis & Roussis, 2008). The decrease in the volatile acetate esters, ethyl esters and terpenes was prevented by 10 (NAC) + 30 (CFA) mg/L after 20 months of wine aging. On the contrary, this mixture did not have any effect on higher alcohols and fatty acids levels. Moreover, a mixture of the two compounds at low concentrations (1.25 (NAC) + 3.75 (CFA) mg/L) inhibited the decrease of several volatile esters and terpenes after two months. Low concentrations of CFA and NAC also inhibited the decrease in the levels of isoamyl acetate, ethyl hexanoate and linalool.

The effects of caffeic and gallic acids as inhibitors of aroma loss were also observed in white wine and wine model solutions at 60 mg/L (Lambropoulos & Roussis, 2007). The authors reported that caffeic acid was active at 7.5 mg/L while gallic acid was active at 15 mg/L and that the sulfur dioxide (SO_2) sometimes increased the inhibitory effect of the two acids. Later, glutathione was tested in combination with caffeic and gallic acids at different SO_2 concentrations (Roussis & Sergianitis, 2008). The sulfhydryl groups contained in the glutathione molecule have been reported to be good inhibitors of enzymatic and non-enzymatic browning in foods. Glutathione + caffeic acid was always more effective than mixtures with gallic acid, preserving isoamyl acetate, ethyl hexanoate and linalool. Both mixtures were also more effective than SO_2 additions at 20 or 40 mg/L. Moreover, in some cases, additive effects of the three compounds were observed. The author concluded that the role of SO_2 in the stabilization of wine aroma may be limited, and reductions in its levels could be considered.

In addition, the effect of mixtures of glutathione, caffeic acid and gallic acid in a Cabernet Sauvignon-Merlot red wine blend with low SO_2 levels was investigated (Roussis et al., 2013). After 36 months of storage, wines showed higher levels of most volatile compounds (i.e., the decrease was limited) when compared with a low SO_2 level in wine. Moreover, treated wines showed almost the same levels of volatile compounds than wines with high SO_2 levels (35 mg/L). Sensory analysis confirmed the ability of the mixture to protect sensory quality because low sulfur wines without the mixture were rated worse by the judges. It was suggested that some of the SO_2 could be replaced by a mixture of other antioxidants. Similar results were observed when a mixture of glutathione and caffeic acid was tested in white Debina wines with low SO_2 levels, confirming that a reduction of SO_2 is possible with the addition of caffeic acid alone or in combination with other antioxidants (Roussis et al., 2007). The exact mechanism that prevents aroma loss during aging was not fully elucidated yet, but the authors attribute its effect to the well-reported antioxidant activity of these molecules.

4. Role of Caffeic Adic in Wine Bitternes and Astringency

It was initially believed that hydroxycinnamates including caffeic acid had an important influence on wine taste, associated with bitterness and astringency (Singleton & Noble, 1976; Nagel et al., 1979; Herrik & Nagel, 1985). A sensory assessment of caffeic acid at 120 mg/L in a non-bitter and non-astringent Chardonnay wine and also in aqueous ethanol solutions was later conducted (Verette et al., 1988). No significant differences in Chardonnay wine taste were observed, although caffeic acid was detected in aqueous ethanol solutions after duo-tri tests. A synergistic effect was not found when caffeic acid and *p*-coumaric acid were tested together. Finally, caffeic acid was also neither described as bitter or astringent. The authors suggested that attention should move towards flavonoid compounds in order to control bitterness and astringency in white wines, although non-phenolic components are also mentioned as possible contributors to white wine taste (Verette et al., 1988).

In contrast, a study was recently conducted where different fractions of red wine phenolic compounds were identified and characterized based on their taste properties (Hufnagel & Hofmann, 2008). Compounds were separated based on their perceived mouth feel sensations and classified into three groups. The first group was composed of compounds elucidating velvety astringency (mostly flavonols), the second group was defined as imparting puckering astringency (hydroxybenzoic and hydroxycinnamic acids together with a polymeric fraction with molecules higher than 5 kDa), and finally, the third group included bitter and astringent compounds (flavan-3-ols, procyanidin dimers B1, B2 and B3, procyanidin trimer C1 and ethyl esters of phenolic acids). Therefore, caffeic acid was defined as a puckering astringent compound. Its astringency taste threshold in water was determined to be 13 mg/L. The half-tongue test (Scharbert et al., 2004) was used for the evaluations of the solutions at increased concentrations of the specific compounds. Moreover, caffeic acid ethyl esters were included in the third group (bitter and astringent compounds), with an astringent taste threshold of 58 mg/L and a bitter taste threshold of 229 mg/L. These interesting results clearly showed the important role played by low molecular weight phenolics, which strongly contribute to the bitter and astringent sensations because they have been found to have stronger impacts on taste than other compounds, such as procyanidin monomers, dimers and trimers.

Finally, the sensory evaluation of the bitterness and astringency sub-qualities of caffeic acid were also recently evaluated (Ferrer-Gallego et al., 2014). Aqueous solutions adjusted to pH 3.60 and concentrations of 2 g/L were used in the experiments. Phenolic acids were described as producing over-moderate astringency with slight bitterness. The caffeic acid sub-qualities were defined between weak and moderate for velvety, drying, harsh and unripe. Moreover, caffeic acid showed higher scores for the persistence and intensity of astringency together with the dynamic sub-quality than *p*-coumaric acid (a compound also in the hydroxycinnamic acids family). In addition, a synergistic effect on the bitterness and astringency was also observed when a mixture of caffeic and *p*-coumaric acid was tested. The authors explained this effect by increased chemical variability (which might facilitate protein binding with different families of salivary proteins) in combination with a possible activation of different receptors (which will also elicit different mouth feel characteristics).

Astringency perception is a very complex phenomenon and the mechanism responsible for astringency and bitterness elucidation is currently not perfectly understood. It was thought that water-soluble polymers of certain size were required to elicit astringency after complexation with proteins (Bate-Smith and Swain, 1962; Harbertson et al., 2002; Sarnekis et al., 2006). However, later results clearly show the importance of low molecular phenolic compounds, such as caffeic acid, on wine mouth feel sensations. Elucidating their roles in the wine organoleptic characteristics is a great challenge for future research.

5. Participation of Caffeic Acid in Wine Oxydative Browning

Oxidative browning of grape must and wine can be due to enzymatic or non-enzymatic reactions (Li et al., 2008). In the early stages of winemaking, the enzymatic oxidation of phenols occurs in the presence of atmospheric oxygen and polyphenol oxidase enzymes. Enzymatic browning in grape must, which occurs at quicker rates than non-enzymatic browning (du Toit et al., 2006), is strongly correlated with its hydroxycinnamic acids content. The enzymatic oxidation of caffeic acid and its former ester caftaric acid has been extensive studied (Cheynier et al., 1991). The kinetic and spectroscopic information explains that an electrophilic ortho-quinone is formed from the oxidation of hydroxycinnamic acids, followed by non-enzymatic coupled oxidative reactions with flavanols. In addition, ortho-quinones derived from flavanols oxidation undergo further reactions giving rise to polymeric phenolic structures (Guyot et al., 1996) with increased detrimental brown coloration of wine.

On the other hand, non-enzymatic browning phenomenon, which is thought to constitute the major reason for wine browning (Li et al., 2008), uses *o*-diphenols, such as caffeic acid and its esters, among other phenolic compounds (including flavan-3-ols, anthocyanins and gallic acid) as the most susceptible compounds taking part in oxidation browning reactions.

This process is very slow due to the acidic conditions of wine and is highly dependent on the presence of iron or copper metals (Danilewicz et al., 2003, Palacios-Macias et al., 2001; Clark & Scollary, 2002, Waterhouse and Laurie, 2006). One of the most important pathways in browning reactions involves the oxidation of phenols (copper or iron catalyzed) and subsequent polymerization of the oxidized products. During the process, unstable semi-quinone free radicals are produced from *o*-diphenols in combination with the reduction of oxygen to hydrogen peroxide (H_2O_2). Further pigment reactions give rise to colored products with higher molecular masses, or reduction reactions regenerate the original phenols. In addition, phenols, such as caffeic acid, can combine with quinones and produce dimers or polymers due to their high electronic affinities. These low redox potential compounds are much easier to oxidize, thus accelerating the polymerization reactions of phenolic materials (Boulton et al., 2001; Li et al., 2005, Zhai et al., 2001). Moreover, during the oxidation process, more H_2O_2 is produced and more ethanol is oxidized to acetaldehyde in the presence of transition metals. On the other hand, hydroxyl radicals are formed by the association of H_2O_2 with ferrous iron (Fe^{2+}) (also known as Fenton reaction) giving rise to more reactive oxygen species. The important role played by SO_2, i.e., limiting the oxidation of ethanol through interactions with H_2O_2 and other saturated hydroxyl compounds, is widely accepted today (Boulton et al., 2001; Danilewicz, 2007).

Moreover, acetaldehyde and glyoxylic acid have also been reported as agents that lead to browning reactions (Timberlake & Bridle, 1976; Wildenradt & Singleton, 1974; Es Safi et al., 1999; Fulcrand et al., 1996; Saucier et al., 1997; Oszmianski et al., 1996; Fulcrand et al., 1997). The acetaldehyde produced during the oxidation of ethanol reacts with phenolic molecules through a nucleophilic addition of the flavanol to the carbocation derived from acetaldehyde protonation and will be ethyl-linked to a second flavanol molecule leading to a flavanol dimer. The acetaldehyde ethyl linkages are not stable and lead to cleavage and repolymerization (Fulcrand et al., 2006). An increase in the yellow color is observed as the condensation degree increases (Lopez-Toledano et al., 2004). Moreover, glyoxylic acid formed from tartaric acid is also involved in browning reactions through flavanol interactions. In this specific case, a carboxy-methine bridge is now responsible for the yellow dimer formation by dehydration and subsequent oxidation process (Fulcrand et al., 1997, Es Safi et al., 2000, 1999).

Wine phenolic compounds have great antioxidant activities but they are also susceptible to oxidation during winemaking, thus leading to browning phenomena. This oxidative process adversely influences the wine sensory properties. Oxidation entails color loss, flavor and aroma degradation and increased astringency perception. The changes in the phenolic composition, especially the role of hydroxycinnamic acids and flavanols in browning reactions, were investigated in white wines (Fernandez-Zurbano et al., 1995, 1998). Decreases in the concentrations of the hydroxycinnamic acids were observed, but they were reported to be unrelated to the extent of wine browning. On the other hand, the flavanol contents seemed to explain the changes in color due to browning reactions. Further studies investigating the influence of redox-active polyphenols (such as caffeic acid among others) showed strong evidence that epicatechin was the principal browning agent. This highlights the fact that wine browning seems to be driven mostly by flavanols rather than by phenolic acids (Sioumis et al., 2006).

Furthermore, it has been suggested that the oxidation coloration of white wines must be examined considering an increase in the absorbance at 420 nm (yellow) in combination with the absorbance at 500 nm (red) (Skoroumonis et al., 2005). Since the ability of caffeic acid to increase the red coloration after oxidation of a model medium containing iron and a flavanol was reported by George et al. (2006), it has been suggested that the oxidation of caffeic acid may participate in further browning polymerization reactions with flavanols (Lutter et al., 2007). The enzymatic and non-enzymatic oxidation processes of caffeic acid have been extensively studied (Cheynier & Moutounet, 1992; Antolovich et al., 2004; Fulcrand et al., 1994; Taziki et al., 2001). Dihydroxybenzaldehyde is thought to be a potential precursor for nucleophilic flavanol polymerization reactions in the acidic conditions of wine causing color changes in the solution (Lopez-Toledano et al., 2004). In addition, the combination of caffeic acid and catechin in a white wine-like medium at pH 3.60 showed the appearance of a catechin dimer linked through a dihydroxybenzenemethine linkage, which resulted in increased absorbance at 440 nm (brown color) and the appearance of some other yellow/red pigments (Lutter et al., 2007; Clark et al., 2008). The formation of these compounds was more favored in the presence of iron (Li et al., 2008).

CONCLUSION

Caffeic acid has been found to directly or indirectly participate in a large number of interactions with other grape and wine constituents. The important role that this molecule plays in the organoleptic properties of wine has been extensively proven. Caffeic acid appears to participate in a larger number of intra- and intermolecular copigmentation interactions that affect red wine color. Although contradictory results on the effects of caffeic acid addition in wine have been found, its effects on the increase in the wine color intensity and stability is widely recognized. Anthocyanin acylation and pyranoanthocyanin formation as well as weak copigmentation interactions are thought to be primarily responsible for the mentioned stability. The extent of the effects of caffeic acid on wine seems to depend on the grape variety, which is probably related to the particular phenolic composition expressed by each cultivar. Further research is needed in this area in order to fully understand the conditions under which caffeic acid addition induces the desired effects on wines. The influence that some winegrowing and winemaking techniques have on the extent of caffeic acid behavior is currently not perfectly understood, and new studies on these matters will give wine scientists and producers valuable information.

In addition, research has shown the importance of caffeic acid when it participates in enzymatic and non-enzymatic browning reactions. Compounds with different absorbance features have been identified, and a better approach toward understanding the extent of the browning phenomenon has been established. Moreover, although the process of understanding the effects that caffeic acid have on wine taste has already started, challenging research is still required to better understand the role of this compound in the complex astringency and bitterness mechanisms. Finally, very promising results have been reported regarding the inclusion of caffeic acid, alone or in combination with other compounds, as a preventative agent of aroma loss during aging. The implications of these findings are of great importance for the wine industry because a decrease in the levels of SO_2 during winemaking and aging can be considered.

REFERENCES

Aleixandre-Tudo, J.L. Efecto de la adición de copigmentos sobre la composición fenólica de vinos tintos de Tempranillo elaborados con diferentes técnicas de vinificación. PhD. Universidad Politecnica de Valencia. Marzo 2012.

Aleixandre-Tudó, J.L., Alvarez, I., Lizama, V., Garcıa, M.J., Aleixandre, J.L. & Du Toit, W.J. (2013). Impact of caffeic acid addition on phenolic composition of Tempranillo wines from different winemaking techniques. *Journal of Agricultural and Food Chemistry.* 61(49), 11900–11912.

Álvarez, I., Aleixandre, J.L., García, M.J., Lizama, V. & Aleixandre-Tudó, J.L. (2009). Effect of the prefermentative addition of copigments on the polyphenolic composition of Tempranillo wines after malolactic fermentation. *European Food Research and Technology.* 228(4), 501–510.

Alluis, B., Pérol, N., El-hajji, H. & Dangles, O. (2000). Water-soluble flavonol (3-hydroxy-2-phenyl-4H-1-benzopyran-4-one) derivatives: chemical synthesis, coloring, and antioxidant properties. *Helvetica Chimica Acta.* 83, 428–443.

Antolovich, M., Bedgood, D.R., Bishop, A.G., Jardine, D., Prenzler, P.D. & Robards, K. (2004). LC–MS investigation of oxidation products of phenolic antioxidants. *Journal of Agriculture and Food Chemistry.* 52, 962–971.

Asen, S., Stewart, R.N. & Norris, K.H. (1972). Copigmentation of anthocyanins in plant tissues and its effect on color. *Phytochemistry.* 11, 1139 – 11145.

Awika, J.M. (2008). Behavior of 3-deoxyanthocyanidins in the presence of phenolic copigments. *Food Research International.* 41, 532–538.

Bakowska-Barczak, A. (2005). Acylated anthocyanins as stable, natural food colorants – a review. *Polish Journal of Food Nutrition and Science.* 14/55(2), 107–116.

Baldi, A., Romani, A., Mulinacci, N., Vincieri, F.F. & Casetta, B. (1995). HPLC/MS application to anthocyanis of Vitis vinifera L. *Journal of Agricultural and Food Chemistry.* 43, 2104-2109.

Baranac, J.M., Petranovic, N.A. & Dimitric-Markovic J.M. (1996). Spectrophotometric study of anthocyan copigmentation reactions. *Journal of Agricultural and Food Chemistry.* 44:1333-1336 (1996).

Baranac, J.M., Petranovic, N.A. & Dimitric-Markovic J.M. (1997a). Spectrophotometric study of anthocyan copigmentation reactions. *Journal of Agricultural and Food Chemistry.* 45, 1694-1697.

Baranac, J.M., Petranovic, N.A. & Dimitric-Markovic J.M. (1997b). Spectrophotometric study of anthocyan copigmentation reactions. *Journal of Agricultural and Food Chemistry.* 45, 1698-1700.

Baranac, J.M., Petranovic, N.A. & Dimitric-Markovic J.M. (1997c). Spectrophotometric study of anthocyan copigmentation reactions. *Journal of Agricultural and Food Chemistry.* 45, 1701-1703.

Bate-Smith, E.C. & Swain, T. (1962). In: Mason, H.S., Florkin, A.M. (Eds.), Comparative Biochemistry. *Springer,* Berlin, p. 764.

Berg, H.W. & Akiyoshi, M.A. (1975). On the Nature of Reactions Responsible for Color Behavior in Red Wine: A Hypothesis. *American Journal of Enology and Viticulture.* 26, 134– 143.

Bloomfield, D.G., Heatherbell, D.A. & Pour Nikfardjam, M.S. (2003). Effect of pcoumaric acid on the colour in red wine. *Mitteilungen Klosterneuburg,* 53, 195–198.

Boulton, R. (2001). The copigmentation of Anthocyanins and Its Role in the Color of Red Wine: A Critical Review. *American Journal of Enology and Viticulture.* 52, 67–87.

Brouillard, R. & Dangles, O. (1994). Anthocyanin molecular-interactions: The first step in the formation of new pigments during wine aging? *Food Chemistry.* 51, 365-371.

Brouillard, R. (1982). Chemical structure of anthocyanins, In: Anthocyanins as food colors. (Markakis, P. (Ed.)), Academic Press, New York.

Brouillard, R., Wigand, M.C., Dangles, O. & Cheminat, A. (1991). pH and solvent effect on the copigmentation reaction of malvin with polyphenols, purine and pyrimidine derivatives. *Journal of the Chemical Society Perkin Trans.* 2, 1235–1241.

Cameira dos Santos, P.J., Brillouet, J.M., Cheynier, V. & Moutounet, M. (1996). Detection and partial characterisation of new anthocyanin-derived pigments in wine. *Journal of the Science of Food and Agriculture,* 70, 204-208.

Castañeda-Ovando, A., Pacheco-Hernández, M.L., Páez-Hernández, M.E., Rodríguez, J.A. & Galán-Vidal, C.A. (2009). Chemical studies of anthocyanins: A review. *Food Chemistry,* 113, 859–871.

Cevallos-Casals, B.A. & Cisneros-Zevallos, L. (2004). Stability on anthocyanin-based aqueous extracts of Andean purple corn and red- fleshed sweet potato compared to synthetic and natural colorants. *Food Chemistry.* 86, 69–77.

Clark, A.C. & Scollary, G.R. (2002). Copper(II)-mediated oxidation of (+)-catechin in a model white wine system. *Australian Journal of Grape and Wine Research.* 8(3), 186–195.

Clark, A.C., Pedretti, F., Prenzler, P.D. & Scollary, G.R. (2008). Impact of ascorbic acid on the oxidative colouration and associated reactions of a model wine solution containing (+)-catechin, caffeic acid and iron. *Australian Journal of Grape and Wine Research.* 14, 238–249.

Chatonnet, P., Dubourdieu, D., Boidron J.N. & Lavigne. V. (1993). Synthesis of Volatile Phenols by Saccharomyces cerevisiae in Wines. *Journal of the Science of Food and Agriculture.* 62, 191-202.

Cheynier, V. & Moutounet, M. (1992). Oxidative reactions of caffeic acid in model solutions containing polyphenol oxidase. *Journal of Agriculture and Food Chemistry.* 40, 2038–2044.

Cheynier, V. & Ricardo da Silva, J.M. (1991). Oxidation of Grape Procyanidins in Model Solutions Containing trans-Caffeoyltartaric Acid and Polyphenol Oxidase. *Journal of Agricultural and Food Chemistry.* 39, 1047-1049.

Dangles, O. & Brouillard, R. (1992). A spectroscopic method based on the anthocyanin copigmentation interaction and applied to the quantitative study of molecular complexes. *Journal of Chemical Society Perkin Trans.* 2, 247-257.

Dangles, O., Saito, N. & Brouillard, R. (1993). Kinetic and thermodynamic control of flavylium hydration in the pelargonidin--cinnamic acid complexation. Origin of the extraordinary flower colour diversity of Pharbitis nil. *Journal of the American Chemical Society.* 115, 3125-32.

Danilewicz, J.C. (2003). Review of reaction mechanisms of oxygen and proposed intermediate reduction products in wine: Central role of iron and copper. *American Journal of Enology and Viticulture.* 54(2), 73–85.

Danilewicz, J.C. (2007). Interaction of sulfur dioxide, polyphenols, and oxygen in a wine-model system: Central role of iron and copper. *American Journal of Enology and Viticulture.* 58(1), 53–60.

Darias-Martín, J., Carrillo, M., Díaz, E. & Boulton, R.B. (2001). Enhancement of wine colour by prefermentation addiction of copigments. *Food Chemistry.* 73, 217–220.

Darias-Martín, J., Martín, B., Carrillo, M., Lamuela, R., Diaz C. & Boulton, R. B. (2002). The effect of caffeic acid on the color of red wine. *Journal of Agricultural and Food Chemistry.* 50 (7), 2062-2067.

de Villiers, A., Majek, P., Lynen, F., Crouch, A., Lauer, H. & Sandra, P. (2005). Classification of South African red and white wines according to grape variety based on the non-coloured phenolic content. *European Food Research and Technology.* 221, 520–528.

Dimitric-Markovic, J.M., Ignjatovic, L.M., Markovic, D.A. & Baranac, J.M. (2003a). Antioxidant capabilities of some organic acids and their co-pigments with malvin – Part I. *Journal of Electroanalytical Chemistry.* 553, 169–175.

Dimitric-Markovic, J.M., Ignjatovic, L.M., Markovic, D.A. & Baranac, J.M. (2003b). Antioxidant capabilities of some organic acids and their co-pigments with malvin – Part II. *Journal of Electroanalytical Chemistry.* 553, 177–182.

Dimitric-Markovic, J.M.D., Petranovic, N.A. & Baranac, J.M. (2000). Spectrophotometric study of the copigmentation of malvin with caffeic and ferulic acids. *Journal of Agricultural and Food Chemistry,* 48, 5530–5536.

Du Toit, W.J., Marais, J., Pretorius, I.S. & du Toit, M. (2006). Oxygen in must and wine: A review. *South African Journal of Enology and Viticulture,* 27, 76–94.

Eiro, M.J. & Heinonen, M. (2002). Anthocyanin color behavior and stability during storage: Effect of intermolecular copigmentation. *Journal of Agricultural and Food Chemistry.* 50, 7461–7466.

Escribano-Bailon, M.T. & Santos-Buelga, C. (2012). Anthocyanin Copigmentation - Evaluation, Mechanisms and Implications for the Colour of Red Wines. *Current Organic Chemistry.* 16(6), 715-723.

Es-Safi, N.E., Cheynier, V. & Moutounet, M. (2002). Role of aldehydic derivatives in the condensation of phenolic compounds with emphasis on the sensorial properties of fruit-derived foods. *Journal of Agricultural and Food Chemistry.* 50(20), 5571–5585.

Es-Safi, N.E., Cheynier, V. & Moutounet, M. (2003). Effect of copper on oxidation of (+)-catechin in a model solution system. *International Journal of Food Science and Technology.* 38(2), 153–163.

Es-Safi, N.E., Fulcrand, H., Cheynier, V. & Moutounet, M. (1999). Competition between (+)-catechin and (-)-epicatechin in acetaldehyde-induced polymerization of flavonols. *Journal of Agricultural and Food Chemistry.* 47, 2088–2095.

Es-Safi, N.E., Guerneve, C., Cheynier, V. & Moutounet, M. (2000). New phenolic compounds formed by evolution of (+)-catechin and glyoxylic acid in hydroalcoholic solution and their implication in color changes of grape-derived foods. *Journal of Agricultural and Food Chemistry.* 48, 4233–4240.

Fernandez-Zurbano, P., Ferreira, V., Escudero, A. & Cacho, J. (1998). Role of hydroxycinnamic acids and flavanols in the oxidation and browning of white wines. *Journal of Agricultural and Food Chemistry.* 46(12), 4937–4944.

Fernandez-Zurbano, P., Ferreira, V., Pena, C., Escudero, A., Serrano, F. & Cacho, J. (1995). Prediction of oxidative browning in white wines asa function of their chemical composition. *Journal of Agricultural and Food Chemistry*. 43(11), 2813–2817.

Ferrer-Gallego, R., Hernández-Hierro, J.M., Rivas-Gonzalo, J.C. & Escribano-Bailón, MT. (2014). Sensory evaluation of bitterness and astringency sub-qualities of wine phenolic compounds: synergistic effect and modulation by aromas. *Food Research International*. 62, 1100–1107.

Franquet-Griell, H., Checa, A., Núnez, O., Saurina, J., Hernández-Cassou, S. & Puignou, L. (2012). Determination of polyphenols in Spanish wines by capillary zone electrophoresis: application to wine characterization by using chemometrics, *Journal of Agricultural and Food Chemistry*. 60, 8340–8349.

Freitas, V. & Mateus, N. (2011). Formation of pyranoanthocyanins in red wines: a new and diverse class of anthocyanin derivatives. *Analytical Bioanalytical Chemistry*. 401, 1463–1473.

Fulcrand, H., Atanasova, V., Salas, E. & Cheynier, V. (2004). The fate of anthocyanins in wines: Are there determining factors? In A. L. Waterhouse & J. A. Kennedy (Eds.), Red Wine Color: Revealing the Mysteries (ACS Symposium Series 886, pp. 68–88). Washington: American Chemical Society.

Fulcrand, H., Benabdeljalil, C., Rigaud, J., Cheynier, V. & Moutounet, M. (1998). A new class of wine pigments generated by reaction between pyruvic acid and grape anthocyanins. *Phytochemistry*. 47, 1401-1407.

Fulcrand, H., Cameira dos Santos, P.J., Sarni-Manchado, P., Cheynier, V. & Fabre-Bonvin, J. (1996). Structure of new anthocyanin-derived wine pigments. *Journal of Chemical Society Perkin Trans 1*. 735–739.

Fulcrand, H., Cheminat, A., Brouillard, R. & Cheynier, V. (1994). Characterisation of compounds obtained by chemical oxidation of caffeic acid in acidic conditions. *Phytochemistry*. 35, 499–505.

Fulcrand, H., Cheynier, V., Oszmianski, J. & Moutounet, M. (1997). An oxidized tartaric acid residue as a new bridge potentially competing with acetaldehyde in flavan-3-ol condensation. *Phytochemistry*. 46(2), 223–227.

Fulcrand, H., Dueñas, M., Salas, E., & Cheynier, V. (2006). Phenolic reactions during winemaking and aging. *American Journal of Enology and Viticulture*. 57, 289–297.

Gauche, C., Malagoli, D.S. & Luiz, M.T.Z. (2010). Effect of pH on the copigmentation of anthocyaninfrom Cabernet Sauvigon grape extract with organic acid. *Scientia Agricola* (Pricaba, Braz), Vol. 67 No. 1, pp. 41-6.

George, N., Clark, A.C., Prenzler, P.D. & Scollary, G. R. (2006). Factors influencing the production and stability of xanthylium cation pigments in a model white wine system. *Australian Journal of Grape and Wine Research*. 12, 57–68.

Giusti, M.M. & Wrolstad, R.E. (2003). Acylated anthocyanins from edible sources and their applications in food systems. *Biochemical Engineering Journal*. 14, 217–25.

Gomez-Mıguez, M., Gonzalez-Manzano, S., Escribano-Bailon, M.T., Heredia, F.J. & Santos-Buelga, C. (2006). Influence of different phenolic copigments on the color of malvidin 3-glucoside. *Journal of Agricultural and Food Chemistry*. 54, 5422-5429.

Gomez-Miguez, M., Gonzalez-Miret, M. & Heredia, F. (2007). Evolution of color and anthocyanin composition of Syrah wines elaborated with pre-fermentative cold maceration. *Journal of Food Engineering*. 79, 271-278.

Gris, E.F., Ferreira, E.A., Falcao, L.D. & Bordignon-Luiz, M.T. (2007). Caffeic acid copigmentation of anthocyanins from Cabernet Sauvignon grape extract in model system. *Food Chemistry.* 1289–1296.

Guyot, S., Vercauteren, J. & Cheynier, V. (1996). Structural determination of colourless and yellow dimers resulting from (+)-catechin coupling catalysed by grape polyphenoloxidase. *Phytochemistry.* 42, 1279–1288.

Harbertson, J.F., Kennedy, J.A. & Adams, D.O. (2002). Tannin in skins and seeds of Cabernet Sauvignon, Syrah, and Pinot noir berries during ripening. *American Journal of Enology and Viticulture.* 53, 54-59.

Hermosin-Gutierrez, I. (2003). Influence of Ethanol Content on the Extent of Copigmentation in a Cencibel Young Red Wine. *Journal of Agricultural and Food Chemistry.* 51, 4079-4083.

Herrick, I.W. & Nagel, C.W. (1985). The caffeoyl tartrate content of white Riesling wines from IUPAC 1976 Nomenclature of cyclitols. *Biochemistry Journal.* 153, 23-31.

Hoshino, T., Matsumoto, U. & Goto, T. (1980). The stabilizing effect of the acyl group on the copigmentation of acylated anthocyanins with Cglucosylflavones. *Phytochemistry.* 19, 663–667.

Hufnagel, J.C. & Hofmann, T. (2008). Quantitative reconstruction of the non-volatile sensometabolome of a red wine. *Journal of Agricultural and Food Chemistry.* 56, 9190–9199.

Khan, A.Q., Khan, R., Qamar, W., Lateef, A., Ali, F., Tahir, M., Rehman, M.U. & Sultana, S. (2012). Caffeic acid attenuates 12-O-tetradecanoyl-phorbol-13-acetate (TPA)-induced NF-kB and COX-2 expression in mouse skin: abrogation of oxidative stress, inflammatory responses and proinflammatory cytokine production. *Food and Chemical Toxicology.* 50, 175–183.

Kilmister, R.L., Mazza, M., Baker, N.K., Faulkner, P. & Downey, M.C. (2014). A role for anthocyanin in determining wine tannin concentration in Shiraz. *Food Chemistry.* 152, 475–482

Kunsági-Máté, S., Kumar, A., Sharma, P., Kollár, L. & Nikfardjam, M.P. (2009). Effect of molecular environment on the formation kinetics of complexes of malvidin-3-O-glucoside with caffeic acid and catechin. *The Journal of Physical Chemistry B.* 113, 7468–7473.

Kunsági-Máté, S., Ortmann, E., Kollár, L. & Nikfardjam, M.P. (2007). Effect of the solvatation shell exchange on the formation of malvidin-3-O-glucoside–ellagic acid complexes. *Journal of Physical Chemistry B.* 111(40), 11750–11755.

Kunsági-Máté, S., Stampel, E., Kollár, K. & Nikfardjam, M. (2008). The effect of the oxidation state of iron ions on the competitive complexation of malvidin by caffeic or ellagic acid. *Food Research International.* 41, 693–696.

Lambropoulos, I. & Roussis, I.G., (2007). Inhibition of the decrease of volatile esters and terpenes during storage of a white wine and a model wine medium by caffeic acid and gallic acid. *Food Research International.* 40, 176-181.

Landrault, N., Poucheret, P., Ravel, P., Gasc, F., Cros, G. & Teissedre, P.L. (2001). Antioxidant capacities and phenolics levels of French wines from different varieties and vintages. *Journal of Agricultural and Food Chemistry.* 49, 3341–3348.

Li, H., Guo, A. & Wang, H. (2008). Mechanisms of oxidative browning of wine. *Food Chemistry.* 108, 1–13.

Li, H., Wang, H., Yuan, C. & Wang, S. (2005). Wine chemistry. *Beijing: Scientific Publishing Company.*

Liao, H., Cai, Y. & Haslam, E. (1992). Polyphenol interactions. 6. Anthocyanin co-pigmentation and colour changes in red wines. *Journal of the Science of Food and Agriculture.* 59, 299-305.

Liu, S.Q. & Pilone, G.J. (2000). An overview of formation and roles of acetaldehyde in winemaking with emphasis on microbiological implications. *International Journal of Food Science and Technology.* 35(1), 49–61.

Liu, S.Q. & Pilone, G.J. (2000). An overview of formation and roles of acetaldehyde in winemaking with emphasis on microbiological implications. *International Journal of Food Science and Technology.* 35(1), 49–61.

Lopez-Toledano, A., Villano-Valencia, D., Mayen, M., Merida, J. & Medina, M. (2004). Interaction of yeasts with the products resulting from the condensation reaction between (+)-catechin and acetaldehyde. *Journal of Agricultural and Food Chemistry.* 52(8), 2376–2381.

Lutter, M., Clark, A.C., Prenzler, P.D. & Scollary, G.R. (2007). Oxidation of caffeic acid in a wine-like medium: Production of dihydroxybenzaldehyde and its subsequent reactions with (+)-catechin. *Food Chemistry.* 105, 968–975.

Llaudy, M.C., Canals, R., Canals, J.M., Rozés, N., Arola L. & Zamora F. (2004). New method for evaluating astringency in red wine. *Journal of Agricultural and Food Chemistry.* 52, 742-746.

Malien-Aubert, C., Dangles, O. & Amiot, M.J. (2001). Color stability of commercial anthocyanin-based extracts in relation to the phenolic composition. Protective effects by intra- and intermolecular copigmentation. *Journal of Agricultural and Food Chemistry.* 49, 170–176.

Martín-Bueno, J., Sáez-Plaza, P., Ramos-Escudero, F., Jiménez, A.M., Fett, R. & Asuero, A.G. (2012). Analysis and Antioxidant Capacity of Anthocyanin Pigments. Part II: Chemical Structure, Color, and Intake of Anthocyanins. *Critical Reviews in Analytical Chemistry.* 42:2, 126-151.

Mazza, G. & Brouillard, R. (1990). The mechanism of co-pigmentation of anthocyanins in aqueous solutions. *Phytochemistry.* 29, 1097–1102.

Mazza, G.Y. & Brouillard, R. (1987). Recent developments in the stabilization of anthocyanins in food products. *Food Chemistry.* 25, 207-225.

Mirabel, M., Saucier, C., Guerra, C. & Glories, Y. (1999). Copigmentation in model wine solutions: Occurrence and relation to wine aging. *American Journal of Enology and Viticulture.* 50, 211–218.

Mirabel, M., Saucier, C., Guerra, C. & Glories, Y. (1999). Copigmentation in Model Wine Solutions: Occurrence and Relation to Wine Aging. *American Journal of Enology and Viticulture.* 50(2), 211-218.

Moghaddam, A.B., Ganjali, M.J., Dinarvand, R., Norouzi, P., Saboury, A.A. & Moosavi-Movahedi, A.A. (2007). Electrochemical behavior of caffeic acid at single-walled carbon nanotube:graphite-based electrode. *Biophysical Chemistry* 128, 30–37.

Monagas, M., Gomez-Cordoves, C. & Bartolome, B. (2005). Evolution of polyphenols in red wines from Vitis vinifera L. during aging in the bottle. I. Anthocyanins and pyroanthocyanins. *European Food Research Technology.* 220, 607 614.

Morais, H., Ramos, C., Forgacs, E., Cserhati, T. & Oliviera, J. (2002). Influence of storage conditions on the stability of monomeric anthocyanins studied by reversed-phase high-performance liquid chromatography. *Journal of Chromatography B.* 770(1-2), 297–301.

Morata, A., González, C. & Suárez-Lepe, J.A. (2007) Formation of vinylphenolic pyranoanthocyanins by selected yeasts fermenting red grape musts supplemented with hydroxycinnamic acids. *International Journal of Food Microbiology.* 116, 144–152.

Nagaoka, T., Banskota, A.H., Tezuka, Y., Saiki, I. & Kadota, S. (2002). Selective Antiproliferative Activity of Caffeic Acid Phenethyl Ester Analogues on Highly Liver-Metastatic Murine Colon 26-L5 Carcinoma Cell Line. *Bioorganic & Medicinal Chemistry* 10, 3351–3359.

Nagel, C.W., Baranowski, J.D., Wulf, L.W. & Powers, J.R. (1979). The hydroxycinnamic acid-tartari acid ester content of musts and grape varieties grown in the Pacific Northwest. *American Journal of Enology and Viticulture.* 30, 198-201.

Oh, J., Jo, H., Cho, A.R., Kim, S. & Han, J. (2013). Antioxidant and antimicrobial activities of various leafy herbal teas. *Food Control* 31, 403–409.

Ong, B.Y. & Nagel, C.W. (1978). Hydroxycinnamic acid-tartaric acid ester content in mature grapes and during the maturation of white Riesling grapes. *American Journal of Enology and Viticulture.* 29(4), 277-281.

Oszmianski, J., Cheynier, V. & Moutounet, M. (1996). Iron-catalyzed oxidation of (+)-catechin in model systems. *Journal of Agricultural and Food Chemistry.* 44(7), 1712–1715.

Palacios-Macias, V.M., Caro-Pina, I. & Perez Rodriguez, L. (2001). Factors influencing the oxidation phenomena of sherry wine. *American Journal of Enology and Viticulture.* 52(2), 151–155.

Quina, F.H., Moreira, P.F., Vautier-Giongo, C., Rettori, D., Rodrigues, R.F., Freitas, A.A., Silva, P.F. & Macanita, A.L. (2009). Photochemistry of Anthocyanins and Their Biological Role in Plant Tissues. *Pure Applied Chemisty.* 81(9), 1687–1694.

Rentzsch, M. Schwarz, M., Winterhalter, P. & Hermosin-Gutierrez. I. (2007). Pyranoanthocyanins: an overview on structures, occurrence and pathways of formation. *Journal of Agricultural and Food.Chemistry.* 55, 4883–4888.

Ribéreau-Gayon, P., Pontallier, P. & Glories, Y. (1983). Some interpretations of colour changes in young red wines during their conservation. *Journal of the Science of Food and Agriculture.* 34, 505-516.

Robbins, R.J. (2003). Phenolic acids in foods: An overview of analytical methodology. Journal of Agricultural and Food Chemistry. 51, 2866-2887.

Robinson, G.M. & Robinson, R. (1931). A survey of anthocyanins. *Biochemical Journal.* 25, 1687.

Rodriguez-Delgado, M.A., Gonzalez-Hernandez, G., Conde-Gonzalez, J.E. & Perez-Trujillo, J.P. (2002). Principal component analysis of the polyphenol content in young red wines. *Food Chemistry.* 78, 523–532.

Rossouw, M. & Marais, J. (2003). Pinotage polyphenol profile. *Wynboer.* 172, 113-115.

Roussis, I.G. & Sergianitis, S. (2008). Protection of some aroma volatiles in a model wine medium by sulphur dioxide and mixtures of glutathione with caffeic acid or gallic acid. *Flavour and Fragrance Journal.* 23, 35-39.

Roussis, I.G., Lambropoulos, I. & Papadopoulou, D., (2005). Inhibition of the decline of volatile esters and terpenols during oxidative storage of Muscat white and Xinomavro-red wine by caffeic acid and N-acetyl-cysteine. *Food Chemistry*. 93, 485-492.

Roussis, I.G., Lambropoulos, I. & Tzimas, P., (2007). Protection of volatiles in a wine with low sulfur dioxide by caffeic acid or glutathione. *American Journal of Enology and Viticulture*. 58, 274-278.

Roussis, I.G., Patrianakou, M. & Drossiadis, A. (2013). Protection of Aroma Volatiles in a Red Wine with Low Sulphur Dioxide by a Mixture of Glutathione, Caffeic Acid and Gallic Acid. *South African Journal of Enology and Viticulture*. 34, 262-265.

Sacchi, K.L., Bisson, L.F. & Adams, D.O. (2005). A review of the effect of winemaking techniques on phenolic extraction in red wines. *American Journal of Enology and Viticulture*. 56, 197–206.

Saija, A., Tomaino, A., Trombetta, D., De Pasquale, A., Uccella, N., Barbuzzi, T., Paolino, D. & Bonina, F. (2000). In vitro and in vivo evaluation of caffeic and ferulic acids as topical photoprotective agents. *International Journal of Pharmaceutics*. 199, 39–47.

Santos-Buelga, C., Bravo-Haro, S. & Rivas-Gonzalo, J. C. (1995). Interactions between catechin and malvidin 3-monoglucoside in model soluctions. *Zeitschrift für Lebensmittel-Untersuchung und Forschung*. 201, 269-274.

Sarneckis, C. J., Dambergs, R. G., Jones, P., Mercurio, M., Herderich, M. J. & Smith, P. A. (2006). Quantification of condensed tannins by precipitation with methyl cellulose: development and validation of an optimized tool for grape and wine analysis. *Australian Journal of Grape and Wine Research*. 12, 39–49.

Sarni-Manchado, P., Fulcrand, H., Souquet, J.M., Cheynier, V. & Moutounet, M. (1996). Stability and color of unreported wine anthocyanin-derived pigments. *Journal of Food Science*. 61, 938-941.

Saucier, C., Little, D. & Glories, Y. (1997). First evidence of acetaldehyde–flavanol condensation products in red wine. *American Journal of Enology and Viticulture*. 48, 370–373.

Scharbert, S., Holzmann, N. & Hofmann, T. (2004). Identification of the astringent taste compounds in black tea infusions by combining instrumental analysis and human bioresponse. *Journal of Agricultural and Food Chemistry*. 52, 3498–3508.

Schwarz, M., Hofmann, G. & Winterhalter. P. (2004). Investigation on anthocyanins in wines from Vitis vinifera cv. Pinotage: Factors influencing the formation of pinotin A and its correlation with wine age. *Journal of Agricultural and Food Chemistry*. 52, 498-504.

Schwarz, M., Jerz, G. & Winterhalter, P. (2003a). Isolation and structure of pinotin A, a new anthocyanin derivative from Pinotage wine. *Vitis*. 42, 105-106.

Schwarz, M., Picazo-Bacete, J.J., Winterhalter, P. & Hermosin-Gutierrez, I. (2005). Effect of copigments and grape cultivar on the colour of red wines fermented after the addition of copigments. *Journal of Agricultural and Food Chemistry*. 53(21), 8372–8381.

Schwarz, M., Wabnitz, T.C. & Winterhalter, P. (2003b). Pathway leading to the formation of anthocyanin-vinylphenol adducts and related pigments in red wines. *Journal of Agricultural and Food Chemistry*. 51, 3682-3687.

Sergianitis, S. & Roussis, I.G. (2008). Protection of volatile esters and terpenes during storage of a white wine and a model wine medium by a mixture of N-acetyl-cysteine and caffeic acid. *European Food Research and Technology*. 227, 643-7.

Singleton, V.L. & Noble, A.C. (1976). Wine flavor and phenolic substances. In: Phenolic Sdfur and Nitrogen Compounds in Food Flavors, eds Charalambous G & Katz I. ACS Symposium Series 26, American Chemical Society, Chicago, 111, pp 47-70.

Singleton, V.L. & Trousdale, E.K. (1992). Anthocyanin–tannin interactions explaining differences in polymeric phenols between white and red wines. *American Journal of Enology and Viticulture*. 43(1), 63–70.

Sioumis, N., Kallithraka, S., Makris, D.P. & Kefalas, P. (2006). Kinetics of browning onset in white wines: Influence of principal redox-active polyphenols and impact on the reducing capacity. *Food Chemistry*. 94(1), 98–104.

Skouroumounis, G.K., Kwiatkowski, M.J., Francis, I.L., Oakey, H., Capone, D. L., Peng, Z., Duncan, B., Sefton, M.A. & Waters, E.J. (2005). Influence of ascorbic acid on the composition, colour and flavour properties of a Riesling and a wooded Chardonnay wine over five years' storage. *Australian Journal of Grape and Wine Research*. 11, 355–368.

Soleas, G.J., Dam, J., Carey, M. & Goldberg, D.M. (1997). Toward the Fingerprinting of Wines: Cultivar-Related Patterns o Polyphenolic Constituents in Ontario Wines. *Journal of Agricultural and Food Chemistry*. 45, 3871−3880.

Tazaki, H., Taguchi, D., Hayashida, T. & Nabeta, K. (2001). Stable isotope-labelling studies on the oxidative coupling of caffeic acid via o-quinone. *Bioscience, Biotechnology and Biochemistry*. 65, 2613–2621.

Timberlake, C.F. & Bridle, P. (1976). Interactions between anthocyanins, phenolic compounds, and acetaldehyde and their significance in red wines. *American Journal of Enology and Viticulture*. 27(3), 97–105.

Verette, E., Noble, A.C. & Somers, T.C, (1988). Hydroxycinnamates of Vitis vinifera: Sensory assessment in relation to bitterness in white wines. *Journal of the Science of Food and Agriculture*. 45, 267-272.

Vidal, S., Hayasaka, Y., Meudec, E., Cheynier, V. & Skouroumounis, G. (2004). Fractionation of Grape Anthocyanin Classes Using Multilayer Coil Countercurrent Chromatography with Step Gradient Elution. *Journal of Agricultural and Food Chemistry*. 52, 713−719.

Vivar-Quintana, A.M., Santos-Buelga, C. & Rivas-Gonzalo, J.C. (2002). Anthocyanin-derived pigments and colour of red wines. *Analytica Chimica Acta*. 458, 147-155.

Wang, F. & Yang, J. (2012). A comparative study of caffeic acid and a novel caffeic acid conjugate SMND-309 on antioxidant properties in vitro. *Journal of Food Science and Technology*. 46, 239–244.

Waterhouse, A.L. & Laurie, V.F. (2006). Oxidation of Wine Phenolics: A Critical Evaluation and Hypotheses. *American Journal of Enology and Viticulture*. 57(3), 306-313.

Waterhouse, A.L. (2002). Wine phenolics. Annals of the New York Academy of Science. 957, 21-26

Weng, C. & Yen, G. (2012). Chemopreventive effects of dietary phytochemicals against cancer invasion and metastasis: phenolic acids, monophenol, polyphenol, and their derivatives. *Cancer Treatments Reviews*. 38, 76–87.

Wildenradt, H.L. & Singleton, V.L. (1974). The production of aldehydes as a result of oxidation of polyphenolic compounds and its relation to wine aging. *American Journal of Enology and Viticulture*. 25(2), 119–126.

Wilska-Jeszka, J. & Korzuchowska, A. (1996). Anthocyanins and chlorogenic acid copigmentation. Influence on the colour of strawberry and chokeberry juices. *Zeitschrift für Lebensmittel-Untersuchung und Forschung. A.* 203, 38–42.

Wrolstad, R.E. (2000). Anthocyanins. In F. J. Francis, & G. J. Lauro (Eds.), Natural Food Colorants (pp. 237–252). New York: Marcel Dekker. Chp.11.

Wu, H. Haig, T. Prately, J. Lemerle, D. & An, M. (2000). Allelochemicals in wheat (*Triticum aestiVum* L.): Variation of phenolic acids in root tissues. *Journal of Agricultural and Food Chemistry. 48*, 5321-5325.

Wulf, L.W. & Nagel, C.W. (1978). High-pressure liquid chromatographic separation of anthocyanins in Vitis vinifera. *American Journal of Enology and Viticulture.* 29, 42-49.

Ye, J.C., Hsiao, M.W., Hsieh, C.H., Wu, W.C., Hung, Y.C. & Chang, W.C. (2010) Analysis of caffeic acid extraction from ocimum gratissimum linn. by high performance liquid chromatography and its effects on a cervical cancer cell line. *Taiwan Journal of Obstetrics and Gynecology* 49(3), 267-271.

Yoshida K., Okuno R., Kameda K. & Kondo T. (2002). Prevention of UV-light induced E,Z-isomerization of caffeoyl residues in the diacylated anthocyanin, gentiodelphin, by intramolecular stacking. *Tetrahedron Letters.* 43, 6181–6184.

Zhai, H., Du, J., Guan, X., Qiao, X. & Pan, Z. (2001). Cultivating and processing technologies for wine grapes. *Beijing: China Agricultural Press.*

Reviewed by:
Raul Ferrer Gallego
Departament of Chemistry and Biochemistry
Universidade do Porto
Andrea Versari
Dipartimento di Scienze e Tecnologie Agro-Alimentari
Università di Bologna

In: Caffeic Acid
Editor: Leanna Vaughn

ISBN: 978-1-63483-121-5

Chapter 3

THE COMPREHENSIVE UNDERSTANDING OF ANTI-DIABETIC AND ANTI-OBESITY EFFECTS OF COFFEE INTAKE VIA NUTRIGENOMICS

Wanping Aw[1,2], Huijuan Jia[1]*, Shoko Takahashi[1], Shinji Fukuda[2] and Hisanori Kato[1,3]

[1]Corporate Sponsored Research Program "Food for Life," Organization for Interdisciplinary Research Projects, The University of Tokyo, Japan

[2]Institute for Advanced Biosciences, Keio University, Japan

[3]Department of Applied Biological Chemistry, Graduate School of Agricultural and Life Sciences, The University of Tokyo, Japan

ABSTRACT

In recent years, there has been rapid progress in high-throughput technologies and platforms to simultaneously assay global gene expression, mRNA, proteins, metabolites and gut microbiota leading to the promotion of "nutrigenomics" in the field of food and nutrition science. Nutrigenomics, which comprises of genomics, transcriptomics, proteomics, metabolomics, metagenomics and other omics technologies, is a discipline in which all available information about the genome and other biological molecules is effectively utilized to unveil every detail of the interactions between diets and the human body. Coffee, one of the most popular beverages in the world, contains major anti-oxidative phenolic compounds: caffeic acid (CA) and chlorogenic acids (CGA) which is an ester of CA and (-) quinic acid. Epidemiologic findings have associated coffee drinking to numerous health benefits ranging from protecting against metabolic diseases, and non-metabolic diseases like depression and skin cancer. In this chapter, we describe in detail the elucidation of the molecular mechanisms of coffee and its constituents: CA

* Corresponding author: Huijuan Jia, Project Lecturer, Corporate Sponsored Research Program "Food for Life," Organization for Interdisciplinary Research Projects, The University of Tokyo, 1-1-1, Yayoi, Bunkyo-ku Tokyo 113-8657, Japan. Tel.: +81-3-5841-5116; Fax: +81-3-5841-5116, Email: ginajhj@gmail.com

and CGA; dietary supplementation on metabolic diseases like obesity and Type 2 Diabetes (T2D) by utilizing the nutrigenomic approach. We believe that it would be able to provide evaluation of the effects of coffee and its constituents on host physiology, which is expected to be useful and informative, shaping the future of coffee and the other nutraceutical's research.

Introduction

Coffee, a popular beverage worldwide, is greatly enjoyed by many people due to its aromatic flavors and delicious taste. Coffee is a mixture of physiologically active compounds whose concentrations can be influenced by preparation methodologies (e.g., roasting and brewing) or agricultural factors [1]. Phenolic polymers, chlorogenic acid (CGA), minerals, water, caffeine, organic acids, sugars, lipids and aromatic compounds are contained in the water-soluble constituents of coffee. In addition, triacylglycerols, tocopherols, and esters of diterpene alcohols and fatty acids are present in the lipid fraction of coffee. Even though caffeine is regarded as the major active component of coffee [2], water soluble components such as CGA [3] and caffeic acid (CA); and lipid soluble components like cafestol [4] and kahweol [5] are reported to possess potent antioxidant activity, which can be maintained even after coffee beans are roasted [6].

Numerous research investigations, epidemiological studies and meta-analyses have revealed the inverse correlation between coffee intake and incidence of various diseases including but not limited to diabetes mellitus [7], obesity [8, 9], cardiovascular health [10], non-alcoholic fatty liver disease [11, 12], Parkinson's disease [13, 14], and Alzheimer's disease [15, 16]. By the year 2025, the number of people estimated to be afflicted with Type 2 Diabetes (T2D) will increase by 65% to reach approximately 380 million individuals globally, with the greatest burden shouldered by the lower-and middle income countries of the Asia-Pacific region. T2D results in considerable morbidity and mortality in those affected and is often associated with hefty economic, health, and societal costs [17, 18]. In addition, when compared with unaffected individuals, those with T2D are at greatly elevated risk of other chronic illnesses, including cardiovascular disease, in which cases T2D patients have twice the risks of experiencing cardiac arrest or stroke [19, 20]. Therefore, it is of considerable importance to identify the possible modifiable risk factors for the primary prevention of T2D. Various dietary and lifestyle factors identified out of which obesity has been labelled as one of the most fundamental causes of risk to T2D [21]. Obesity is a systemic inflammatory response and many inflammatory mediators exhibit expression patterns that interfere with insulin action [22-25]. Glucose and macronutrient intake results in oxidative stress and inflammatory changes. As such, this chronic over nutrition leading to obesity might be a pro-inflammatory state with oxidative stress [26, 27]. In addition, elevated levels of inflammatory markers like TNF-α and IL-6 which is associated with obesity and T2D, might interfere with insulin action by suppressing insulin signal transduction, resulting in insulin resistance which in turn promotes inflammation [28]. Inhibition of pathways that regulate inflammatory responses either pharmacologically or genetically has been found to protect experimental animals and human subjects against diet-induced insulin resistance [29, 30].

THE APPLICATION OF NUTRIGENOMICS IN FOOD AND NUTRITION RESEARCH

Recently, dramatic improvements in high-throughput technologies and platforms in global mRNA, proteins, metabolites and intestinal microbiota assessment has allowed researchers in food and nutrition science to witness a rapid expansion of the field of nutrigenomics. Nutrigenomics which encompasses applying transcriptomics, proteomics, metabolomics, metagenomics and other omics technologies; is a discipline in which all available information about the genome and other biological molecules is effectively utilized to comprehensively unveil detailed interactions between dietary habits, nutrition status and host physiology [31, 32]. In this chapter, we will describe the current status in applying nutrigenomics to elucidate the molecular mechanisms of coffee intake on obesity and T2D *in vivo* (Figure 1).

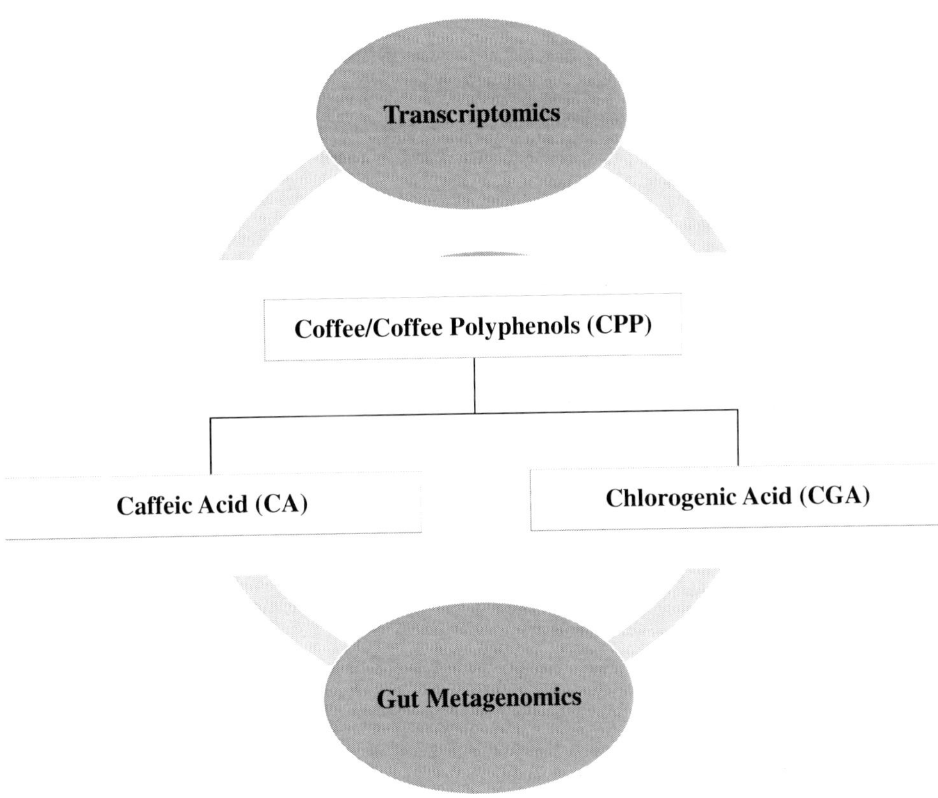

Figure 1. Overview of the Chapter: The applications of nutrigenomics to elucidate the anti-diabetic and anti-obesity effects of coffee intake and coffee polyphenols.

TRANSCRIPTOMIC EVALUATION OF COFFEE INTAKE

Out of the above mentioned omics methodologies, transcriptomics is most widely employed in food research due to the many merits of the DNA microarray technology. DNA microarray allows researchers to comprehensively understand gene expression using established protocols, and data are highly reliable and reproducible [33-36]. Apart from DNA microarray, real-time quantitative PCR is also commonly employed to study the influence of dietary intervention on the expression of selected genes [37-42].

Effects of Caffeic Acid on Obesity

In a report to investigate the influence of phenolic CA in high-fat diet (HFD) induced obese mice and its underlying mechanisms on adipose and hepatic lipid lipogenesis [38], C57BL/6J mice were administered a normal diet or a HFD (20% fat, w/w) with or without CA (0.02% and 0.08%, w/w) for 6 weeks. Body weight, visceral fat mass, plasma aspartate aminotransferase (GOT) and alanine aminotransferase (GPT) levels, fatty acid synthase activity, and concentrations of free fatty acids were evidently lowered by CA intervention as when compared to the HFD group. In addition, CA also lowered both plasma and hepatic triglyceride and cholesterol levels. Moreover, cholesterol biosynthesis was significantly inhibited by CA as evidenced by 3-hydroxy-3-methylglutaryl CoA reductase in the liver. Lipogenesis via sterol regulatory element-binding protein (SREBP)-1 c and its target enzyme fatty acid synthase were also suppressed. Increased phosphorylation of AMP-activated protein kinase and decreased acetyl carboxylase, a downstream target of AMPK, which are related to fatty acid β-oxidation in the liver were also observed. These results indicate that CA exhibits a significant potential as an anti-obesity agent by suppression of lipogenic enzymes and hepatic lipid accumulation [38]. Ohara *et al.* [41] developed an anti-obesity agent by examining the combined effects of glucosyl hesperidin (G-hesperidin) and caffeine in HFD-induced obese KK mice fed a low-fat diet with or without G-hesperidin, caffeine, or their combination for 14 days. Decrease in body weight and significantly lower adipose tissue weight were observed in the mice fed combined diets but not in the G-hesperidin-fed or caffeine-fed mice. Results from hepatic DNA microarray evaluation suggested that the feeding of G-hesperidin + caffeine was associated with lower lipogenesis. The anti-lipogenic effects of G-hesperidin + caffeine were then re-examined in fasted–refed KK mice. Hepatic triglyceride levels were significantly lower in the mice fed combined diets during the refeeding period but not in the mice fed each alone. Moreover, hepatic gene expression levels related to lipogenesis, such as SREBP-1c or fatty acid synthase, were significantly lower in the mice fed G-hesperidin + caffeine compared with that in the control mice. This suggested that the combination of G-hesperidin + caffeine intervention is effective in obesity control partly via the inhibition of hepatic lipogenesis [41]. In another study [42], CA induced a significant reduction of the blood glucose and glycosylated hemoglobin levels than controls in C57BL/KsJ-db/db mice. In addition, plasma insulin, C-peptide, and leptin levels in mice fed CA were significantly higher than those of the control group, whereas the plasma glucagon concentration was lower. The observed increase in plasma insulin by CA could be attributed to an antidegenerative effect on the islets. CA also markedly increased glucokinase activity, glucokinase gene expression and glycogen content. At the same time, CA intervention also simultaneously lowered glucose-6-phosphatase and phosphoenolpyruvate carboxykinase activities and their respective gene expression levels which was also accompanied by a reduction in the glucose transporter 2 expression in the liver. In addition, CA significantly increased superoxide dismutase, catalase, and glutathione peroxidase activities and their gene expression levels. Hydrogen peroxide and thiobarbituric acid reactive substances concentrations in the erythrocyte and liver of db/db mice were also decreased. This indicates that CA plays a significant role as an antidiabetic agent by suppressing a progression of T2D as suggested by an improvement in hepatic glucose output and enhanced adipocyte glucose uptake, insulin secretion, and antioxidant capacity [29].

Effects of Chlorogenic Acid on Obesity

Ma *et al.* [39] assessed the preventive and therapeutic effects of CGA on obesity and obesity-related liver steatosis and insulin resistance. Two sets of experiments were conducted. Firstly, 6-week old C57BL/6 mice were fed a regular chow or HFD (60% kJ/fat) for 15 weeks with twice intra-peritoneal (IP) injection of CGA (100 mg/kg) or DMSO (carrier solution) weekly. In the other experiment, obese mice weighing an average of 50 g were treated by CGA (100 mg/kg, IP, twice weekly) or DMSO for 6 weeks. CGA treatment attenuated HFD-induced hepatic steatosis and insulin resistance. Quantitative PCR analysis revealed that CGA intervention lowered gene expression levels of peroxisome proliferator-activated receptor (PPAR)γ, cluster of differentiation 36, fatty acid binding protein 4, and alpha-1,3-mannosyl-glycoprotein 2-beta-N-acetylglucosaminyltransferase in the liver. CGA treatment also improved hepatic inflammation and white adipose tissue accompanied by lower expression levels of including F4/80, cluster of differentiation 68, cluster of differentiation molecule 11b, cluster of differentiation molecule 11b, and tumor necrosis factor alpha, monocyte chemotactic protein-1 and C-C chemokine receptor type 2 encoding inflammatory proteins. The results of this study suggest that drinking coffee is beneficial in maintaining metabolic homeostasis when on a HFD [39].

Effects of Coffee Polyphenols on Obesity

The effects of coffee polyphenols (CPP) on diet-induced body fat accumulation were investigated [43]. C57BL/6J mice were fed either a control diet, a HFD (30%), or a HFD supplemented with 0.5 to 1.0% CPP for a period lasting 2–15 weeks. In groups that were supplemented with CPP, significantly reduced body weight gain, abdominal and liver fat accumulation, and infiltration of macrophages into adipose tissues were observed. Energy expenditure evaluated by indirect calorimetry was significantly increased in CPP-fed mice. Hepatic gene expression levels of SREBP-1c, acetyl-CoA carboxylase-1 and -2, stearoyl-CoA desaturase-1, and pyruvate dehydrogenase kinase-4 were significantly lower in CPP-fed mice than in HFD induced obese mice. Moreover, CPP suppressed the expression of these genes in Hepa 1–6 cells, followed by an increase in microRNA-122. Structure-activity relationship studies of nine quinic acid derivatives isolated from CPP was performed in Hepa 1–6 cells and results suggested that mono- or di-caffeoyl quinic acids (CQA) are the active substances in contributing to the beneficial effects of CPP. Furthermore, CPP and 5-CQA decreased the nuclear active form of SREBP-1 protein, acetyl-CoA carboxylase activity, and cellular malonyl-CoA levels. These findings from this study indicate that CPP enhances energy metabolism and reduces lipogenesis by downregulating SREBP-1c and related molecules, thereby leading to lowered body fat accumulation [43].

Effects of Coffee Intake on Obesity

In order to investigate the risk-reducing effects of coffee in metabolic syndrome, male C57BL/6J mice were raised for 8 weeks on either a normal diet (N group), a HFD (14% beef fat and 14% lard) group, or a HFD with 1.1% decaffeinated (HF+DC group) or 1.1% caffeine-containing instant coffee (HF+CC group) [44]. Hepatic and adipose tissue were analyzed using DNA microarray. Despite similar food intake between groups, the body weights of mice in the HF+DC and HF+CC groups were mostly intermediate between the N and HF groups. Mesenteric fat weight was significantly lower in the HF+DC group than in the HF group and tended to become lower in the HF+CC group than in the HF group. Serum GOT and GPT levels were significantly lower in the HF+DC and HF+CC groups than in the HF group. Inflammatory cytokine interleukin-1β gene expression in liver was up-regulated in the HF group and significantly down-regulated in the HF+DC and HF+CC groups. Gene expression level of monocyte chemoattractant protein-1 in white adipose tissue was also significantly suppressed by DC intervention [44]. Our group elucidated the molecular mechanisms involved in the ameliorative effects of caffeinated coffee (CC), decaffeinated coffee (DC) and unroasted caffeinated green coffee (GC) on skeletal muscle gene expression profiles and their relationships in eight-week-old male C57BL6 mice which were raised for 9 weeks *ad libitum* on a normal diet, a HFD (60% kcal fat), or HFD containing 2% freeze-dried CC, or DC, or GC [8]. Total RNA and protein were extracted from skeletal muscle and subjected to microarray and western blotting analyses, respectively. Results from comprehensive gene analysis and validation via quantitative real-time PCR revealed that coffee intake down-regulated the anti-inflammatory genes activating transcription factor 3, FBJ osteosarcoma oncogene, heat shock protein 1A, heat shock protein 1B and synuclein, gamma and the inflammation-associated insulin signaling genes stearoyl-coenzyme A desaturase 1 and secreted phosphoprotein 1. In addition, coffee intervention mitigated the insulin resistance by decreasing plasma glucose levels during an insulin tolerance test and by increasing tyrosine phosphorylation of insulin receptor substrate-1 (IRS-1), p85/IRS-1 complex and pAkt/PKB (protein kinase B) as observed by western blotting. These results indicate the positive effects of coffee intake on impaired insulin signaling, inflammation and obesity, thereby providing a new perspective on the prevention of obesity and T2D [8]. In another study by the same group, the effects of CC, DC and GC on the livers of C57BL/6J mice fed a HFD (60% kcal fat), and to elucidate the detailed physiological responses to coffee intake via transcriptomic evaluation using DNA microarrays [45]. The three types of coffee suppressed the weight gain and fat accumulation induced by a HFD. Hepatic expression levels of PPARγand its related molecules were downregulated. Additionally, protein levels of hepatic PPAR were revealed to be lower in all of the coffee-fed groups compared to the high-fat group by Western blotting. Since overexpression of PPARγ in the liver has been implicated in adipogenic hepatic steatosis, it was hypothesised that the changes in expression levels of PPARγ and its target molecules, which play pivotal roles in adipogenic hepatic steatosis, could be involved in the anti-obesity effect of coffee [45]. In addition, Yamauchi *et al.*, studied the anti-diabetic effects of coffee and caffeine intervention in spontaneously diabetic KK-A(y) mice [46]. KK-A(y) mice were given regular drinking water as controls or 2-fold diluted coffee for 5 weeks and it was observed that coffee intake ameliorated the development of hyperglycemia and improved insulin sensitivity. Gene expression levels of monocyte chemoattractant protein-1, interleukin-6, and tumor necrosis factor alpha in white adipose tissue; adipose tissue monocyte chemoattractant protein-1 concentration, and serum interleukin-6 levels in the coffee group were lower as compared to controls. Both coffee and caffeine ingestion improved the fatty liver. Caffeine ingestion as drinking water also caused an amelioration of hyperglycemia. These results suggest that coffee intake suppresses hyperglycemia by improving insulin sensitivity, partly due to lowered inflammatory cytokine expression and improved fatty liver. Moreover, caffeine may be one of the effective antidiabetic compounds in coffee [46].

Proteomic Evaluation of Coffee Intake

Typically, proteome analyses comprises of separation, quantification, and identification of proteins [47]. In traditional proteome analytical methodologies, proteins in biological samples are separated using two-dimensional gel electrophoresis, in which separated proteins are visualized and quantified after silver or fluorescent staining. A more sophisticated strategy based on the same principle is the differential imaging gel electrophoresis method, in which proteins of different samples are pre-labeled with different fluorescent dyes [48]. Recently, a novel, tandem mass-spectrometry (MS) based approach for the relative quantification of proteins, relying on isobaric tag for relative and absolute quantitation (iTRAQ) has been reported. Due to the isobaric mass design of the iTRAQ reagents, differentially labeled proteins do not differ in mass; accordingly, their corresponding proteolytic peptides appear as single peaks in MS scans [48]. The evaluation of coffee dietary interventions in obesity and T2D prevention via proteomic analysis has yet to be reported. As such, we would introduce a paper which investigated the molecular mediators contributing to coffee's beneficial effects in nonalcoholic fatty liver disease (NAFLD) using proteomic technologies. Salamone *et al.* [49] employed proteomic analysis and elucidated the stress proteins mediating the antioxidant effects of coffee. They established the contribution of specific coffee-induced endoplasmic reticulum and mitochondrial chaperones ensuring correct protein folding and degradation in the liver. Decaffeinated espresso coffee or vehicle were administered to rats fed an HFD for 12 weeks and then examined the effects of coffee intervention on NAFLD by using two-dimensional polyacrylamide gel electrophoresis proteomic analysis integrated with mass spectrometry. In animals which were fed only high-fat and water, panacinar steatosis, lobular inflammation, and mild fibrosis were observed. On the other hand, rats that were fed HFD (58% kcal fat) and coffee-containing drinking water exhibited only mild steatosis. Coffee intake upregulated expression levels of the endoplasmic reticulum chaperones glucose-related protein 78 and protein disulfide-isomerase A3; and mitochondrial chaperones heat stress protein 70 and DJ-1 in the liver. Proteomic evaluation also revealed that coffee consumption induces the expression of master regulators of redox status (i.e., peroxiredoxin 1, glutathione S-transferase α2, and D-dopachrome tautomerase). Coffee intake was also associated with lowered expression of electron transfer flavoprotein subunit α, a component of the mitochondrial respiratory chain, involved in *de novo* lipogenesis [49].

GUT MICROBIOME EVALUATION OF COFFEE INTAKE

The gut microbiota consists of all the microorganisms inhabiting the gastrointestinal tract, which is an intricate ecosystem between the gut microbiota and host eukaryotic cells [50]. The gut microbiota which consists of four main bacterial phyla: *Firmicutes*, *Bacteroidetes*, *Actinobacteri*a and *Proteobacteria* perform roles that define the host health [51] can vary in composition along the digestive tract and adapt through life accordingly to the lifestyle and dietary status of the host [52]. Currently, the gut microbiome can be comprehensively evaluated using next-generation whole DNA sequencing and 16S rRNA gene sequencing [53-55]. Although studies have reported numerous effects of coffee on human health, there are not many studies that have examined its specific effects on gut microbiota. Nakayama *et al.* [56] aimed to clarify the effects of coffee and galacto-oligosaccharide (GOS) consumption on gut microbiota and host responses. Mice were given coffee and GOS and the intestinal bacterial counts were measured with quantitative RT-PCR. In the proximal colon, *Escherichia coli* and *Clostridium* spp. counts were significantly decreased although *Bifidobacterium* spp. counts increased remarkably in the same area. In another experiment, results showed that *E. coli* growth was inhibited only by a coffee agar, and not GOS-containing agar. Both coffee and GOS intervention remarkably increased aquaporin 8 expression in the proximal colon demonstrating that coffee intake can modulate gut microbiota and increase aquaporin 8 expression, both of which are important in maintaining intestinal balance [56]. In another study examining the effects of coffee intake in healthy adults [57], sixteen healthy adult volunteers consumed 3 cups of coffee daily during 3 weeks. Fecal samples were collected before and after the consumption of coffee, and the impact of the ingestion of the product on the intestinal bacteria as well as the quantification of specific bacterial groups was assessed using PCR. Although no significant influence on fecal profiles of the dominant microbiota were observed after coffee intake, the numbers of *Bifidobacterium* spp. increased after the 3-week test period. In addition, in certain test subjects, there was a moderate increase in the metabolic activity of *Bifidobacterium* spp which is commonly associated with its role in body weight management and anti-obesity effects [57].

METABOLOMICS INTEGRATED MULTI-OMICS EVALUATION OF COFFEE INTAKE

In this section, we would discuss several studies which integrated other omics technologies with metabolomics evaluation. Our group conducted a multi-omics study to study the changes in hepatic proteome and metabolome in C57BL/6J mice fed a HFD (60% kcal fat) containing three types of coffee (caffeinated, decaffeinated and green unroasted coffee). In proteome evaluation, an up-regulation of isocitrate dehydrogenase (a key enzyme in the TCA cycle) and its related proteins, suggesting increased energy generation were observed. In the metabolome, there was an up-regulation of metabolites involved in the urea cycle, with which the transcriptome data were highly consistent [45], indicating increased energy expenditure. Since the TCA cycle and the urea cycle are directly connected by mutually providing each other's intermediates, they are likely be accelerated in a concerted manner. The up-regulation of these pathways might result in increased ATP turnover, which is related to the alterations of lipid metabolism which is important part in the suppressing on obesity, inflammation, and hepatosteatosis [58]. In another study to determine if chronic coffee intake could improve gut microbiota and metabolomic profile changes induced by a HFD [59], male Sprague–Dawley rats were randomized to chow (12% kcal fat) or high-fat (60% kcal fat) diet. Each group was then further divided into water or caffeinated coffee for 10 weeks. In high-fat-fed rats that fed on caffeinated coffee, lowered body weight, adiposity, liver triglycerides and energy intake were observed. However, despite favorable body composition, rats that fed on coffee displayed profound systemic insulin resistance, likely due to caffeine. Coffee intake lowered the *Firmicutes* (F)-to-*Bacteroidetes* (B) ratio and *Clostridium* cluster XI associated with high-fat feeding but also resulted in augmented levels of Enterobacteriaceae. Coffee intake induced significant changes in the serum metabolome, increasing levels of aromatic and circulating short-chain fatty acids whilst lowering levels of branched-chain amino acids [59].

CONCLUSION

In this chapter, we have shared many reported possible molecular mechanisms underlying the beneficial effects of coffee intake or coffee polyphenols in the prevention of obesity, obesity associated insulin resistance and NAFLD (Table 1). It is evident that nutrigenomics is an optimal approach to provide scientific evidence in evaluating the effects of coffee and its constituents on host physiology. We believe that nutrigenomics is a useful and informative evaluation approach and can shape the future of coffee and the other nutraceutical's research.

Table 1. Summary of omics evaluation of coffee intake

No.	Type of omics evaluation and reference no.	Name of compound	Summary of results
1.	Transcriptomics [38]	CA	Body weight, visceral fat mass, plasma GOT and GPT levels, fatty acid synthease activity, and concentrations of free fatty acids were significantly lowered by CA intervention. CA also lowered both plasma and hepatic triglyceride and cholesterol levels. CA inhibited cholesterol biosynthesis. Lipogenesis via SREBP-1c and its target enzyme fatty acid synthase were also suppressed. Increased phosphorylation of AMP-activated protein kinase and decreased acetyl carboxylase, a downstream target of AMPK, was also observed.
2.	Transcriptomics [41]	CA + G-hesperidin	Decrease in body weight and significantly lower adipose tissue weight were observed in the mice fed combined diets but not in the G-hesperidin-fed or caffeine-fed mice. Results from hepatic DNA microarray evaluation suggested that the feeding of G-hesperidin + caffeine was associated with lower lipogenesis. In fasted–refed KK mice, hepatic triglyceride levels were significantly lower in the mice fed combined diets during the refeeding period but not in the mice fed each alone. Hepatic gene expression levels related to lipogenesis, such as SREBP-1c or fatty acid synthase, were significantly lower in the mice fed G-hesperidin + caffeine compared with that in the control mice.

Table 1. (Continued)

No.	Type of omics evaluation and reference no.	Name of compound	Summary of results
3.	Transcriptomics [42]	CA	CA induced a significant reduction of the blood glucose and glycosylated hemoglobin levels in C57BL/KsJ-db/db mice. Plasma insulin, C-peptide, and leptin levels in mice fed CA were significantly higher than those of the control group, whereas the plasma glucagon concentrations was lower. CA also markedly increased glucokinase activity, glucokinase gene expression and glycogen content. CA intervention also simultaneously lowered glucose-6-phosphatase and phosphoenolpyruvate carboxykinase activities and their respective gene expression levels which was also accompanied by a reduction in the glucose transporter 2 expression in the liver. CA significantly increased superoxide dismutase, catalase, and glutathione peroxidase activities and their gene expression levels. Hydrogen peroxide and thiobarbituric acid reactive substances concentrations in the erythrocyte and liver of db/db mice were also decreased.
4.	Transcriptomics [39]	CGA	CGA treatment attenuated HFD-induced hepatic steatosis and insulin resistance. Quantitative PCR analysis revealed that CGA intervention lowered gene expression levels of PPARγ, cluster of differentiation 36, fatty acid binding protein 4, and alpha-1,3-mannosyl-glycoprotein 2-beta-N-acetylglucosaminyltransferase in the liver. CGA treatment also improved hepatic inflammation and white adipose tissue accompanied by lower expression levels of including F4/80, cluster of differentiation 68, cluster of differentiation molecule 11b, cluster of differentiation molecule 11b, and tumor necrosis factor alpha, monocyte chemotactic protein-1 and C-C chemokine receptor type 2 encoding inflammatory proteins.
5.	Transcriptomics [43]	CPP	CPP significantly reduced body weight gain, abdominal and liver fat accumulation, and infiltration of macrophages into adipose tissues were observed. Energy expenditure evaluated by indirect calorimetry was significantly increased in CPP-fed mice. Hepatic gene expression levels of SREBP-1c, acetyl-CoA carboxylase-1 and -2, stearoyl-CoA desaturase-1, and pyruvate dehydrogenase kinase-4 were significantly lower in CPP-fed mice than in HFD induced obese mice. CPP suppressed the expression of these genes in Hepa 1–6 cells, followed by an increase in microRNA-122. Structure-activity relationship studies of nine quinic acid derivatives isolated from CPP was performed in Hepa 1–6 cells and results suggested that mono- or di-caffeoyl quinic acids (CQA) are the active substances in contributing to the beneficial effects of CPP. CPP and 5-CQA decreased the nuclear active form of SREBP-1, acetyl-CoA carboxylase activity, and cellular malonyl-CoA levels.
6.	Transcriptomics [44]	CC and DC	Body weights of mice in the HF+DC and HF+CC groups were mostly intermediate between the N and HF groups. Mesenteric fat weight was significantly lower in the HF+DC group than in the HF group and tended to become lower in the HF+CC group than in the HF group. Serum GOT and GPT levels were significantly lower in the HF+DC and HF+CC groups than in the HF group. Inflammatory cytokine interleukin-1β gene expression in liver was up-regulated in the HF group and significantly down-regulated in the HF+DC and HF+CC groups. Gene expression level of monocyte chemoattractant protein-1 in white adipose tissue was also significantly suppressed by DC intervention.

No.	Type of omics evaluation and reference no.	Name of compound	Summary of results
7.	Transcriptomics [8]	CC, DC and GC	Coffee intake down-regulated the anti-inflammatory genes activating transcription factor 3, FBJ osteosarcoma oncogene, heat shock protein 1A, heat shock protein 1B and synuclein, gamma and the inflammation-associated insulin signaling genes stearoyl-coenzyme A desaturase 1 and secreted phosphoprotein 1. Coffee intervention mitigated the insulin resistance by decreasing plasma glucose levels during an insulin tolerance test and by increasing tyrosine phosphorylation of insulin receptor substrate-1 (IRS-1), p85/IRS-1 complex and pAkt/PKB (protein kinase B) as observed by western blotting.
8.	Transcriptomics [45]	CC, DC and GC	The three types of coffee suppressed the weight gain and fat accumulation induced by HFD. Hepatic expression levels of PPARγ and its related molecules were downregulated. Protein levels of hepatic PPAR were revealed to be lower in all of the coffee-fed groups compared to the high-fat group by Western blotting.
9.	Transcriptomics [46]	Coffee and caffeine	Gene expression levels of monocyte chemoattractant protein-1, interleukin-6, and tumor necrosis factor alpha in white adipose tissue; adipose tissue monocyte chemoattractant protein-1 concentration, and serum interleukin-6 levels in the coffee group were lower as compared to controls. Both coffee and caffeine ingestion improved the fatty liver. Caffeine ingestion as drinking water also caused an amelioration of hyperglycemia.
10.	Proteomics [49]	DC	Coffee intake upregulated expression levels of the endoplasmic reticulum chaperones glucose-related protein 78 and protein disulfide-isomerase A3; and mitochondrial chaperones heat stress protein 70 and DJ-1 in the liver. Proteomic evaluation also revealed that coffee consumption induces the expression of master regulators of redox status (i.e., peroxiredoxin 1, glutathione S-transferase α2, and D-dopachrome tautomerase). Coffee intake with also associated with lowered expression of electron transfer flavoprotein subunit α, a component of the mitochondrial respiratory chain, involved in de novo lipogenesis
11.	Gut metagenomics [56]	Coffee + GOS	*E. coli* growth was inhibited only by a coffee agar, and not GOS-containing agar. Both coffee and GOS intervention remarkably increased aquaporin 8 expression in the proximal colon demonstrating that coffee intake can modulate gut microbiota and increase aquaporin 8 expression, both of which are important in maintaining intestinal balance.
12.	Gut metagenomics [57]	Coffee	No significant influence on fecal profiles of the dominant microbiota were observed after coffee intake. The numbers of *Bifidobacterium spp*. increased after the 3-week test period. In addition, in certain test subjects, there was a moderate increase in the metabolic activity of *Bifidobacterium spp*
13.	Proteomics and metabolomics [58]	CC, DC and GC	In proteome evaluation, an up-regulation of isocitrate dehydrogenase (a key enzyme in the TCA cycle) and its related proteins, suggesting increased energy generation were observed. In the metabolome, there was an up-regulation of metabolites involved in the urea cycle, with which the transcriptome data were highly consistent [45], indicating increased energy expenditure.
14.	Gut metagenomics and metabolomics [59]	CC	In high-fat-fed rats that fed on caffeinated coffee, lowered body weight, adiposity, liver triglycerides and energy intake were observed. Coffee intake lowered the *Firmicutes* (F)-to-*Bacteroidetes* (B) ratio and *Clostridium Cluster XI* associated with high-fat feeding but also resulted in augmented levels of *Enterobacteria*. Coffee intake induced significant changes in the serum metabolome, increasing levels of aromatic and circulating short-chain fatty acids whilst lowering levels of branched-chain amino acids

REFERENCES

[1] Dórea, J.G. and T.H.M. da Costa, Is coffee a functional food? *British Journal of Nutrition*, 2007. 93(06): p. 773.

[2] Glade, M.J., Caffeine-Not just a stimulant. *Nutrition,* 2010. 26(10): p. 932-8.

[3] Sato, Y., et al., In vitro and in vivo antioxidant properties of chlorogenic acid and caffeic acid. *Int. J. Pharm*, 2011. 403(1-2): p. 136-8.

[4] Lee, K.J. and H.G. Jeong, Protective effects of kahweol and cafestol against hydrogen peroxide-induced oxidative stress and DNA damage. *Toxicol Lett*, 2007. 173(2): p. 80-7.

[5] Lam, L.K., V.L. Sparnins, and L.W. Wattenberg, Effects of derivatives of kahweol and cafestol on the activity of glutathione S-transferase in mice. *Journal of Medicinal Chemistry* 1987 30(8): p. 1399-1403.

[6] Anese, M. and M.C. Nicoli, Antioxidant properties of ready-to-drink coffee brews. Journal *of Agricultural and Food Chemistry*, 2003 51(4): p. 942-946.

[7] Salazar-Martinez, E., et al., Coffee consumption and risk for type 2 diabetes mellitus. *Annals of Internal Medicine*, 2004 140(1): p. 1-8.

[8] Jia, H., et al., Coffee intake mitigated inflammation and obesity-induced insulin resistance in skeletal muscle of high-fat diet-induced obese mice. *Genes Nutr*, 2014. 9(3): p. 389.

[9] Rustenbeck, I., et al., Effect of chronic coffee consumption on weight gain and glycaemia in a mouse model of obesity and type 2 diabetes. *Nutr Diabetes*, 2014. 4: p. e123.

[10] Bidel, S., et al., Coffee consumption and risk of total and cardiovascular mortality among patients with type 2 diabetes. *Diabetologia,* 2006. 49(11): p. 2618-26.

[11] Vitaglione, P., et al., Coffee reduces liver damage in a rat model of steatohepatitis: the underlying mechanisms and the role of polyphenols and melanoidins. *Hepatology,* 2010. 52(5): p. 1652-61.

[12] Catalano, D., et al., Protective role of coffee in non-alcoholic fatty liver disease (NAFLD). *Dig. Dis. Sci.*, 2010. 55(11): p. 3200-6.

[13] Cavin, C., et al., Induction of Nrf2-mediated cellular defenses and alteration of phase I activities as mechanisms of chemoprotective effects of coffee in the liver. *Food Chem. Toxicol.*, 2008. 46(4): p. 1239-48.

[14] Joghataie, M.T., et al., Protective effect of caffeine against neurodegeneration in a model of Parkinson's disease in rat: behavioral and histochemical evidence. *Parkinsonism Relat. Disord*, 2004. 10(8): p. 465-8.

[15] Arendash, G.W., et al., Caffeine protects Alzheimer's mice against cognitive impairment and reduces brain beta-amyloid production. *Neuroscience*, 2006. 142(4): p. 941-52.

[16] Tohda, C., T. Kuboyama, and K. Komatsu, Search for natural products related to regeneration of the neuronal network. *Neurosignals,* 2005. 14(1-2): p. 34-45.

[17] Sullivan, P.W., V. Ghushchyan, and R.H. Ben-Joseph, The effect of obesity and cardiometabolic risk factors on expenditures and productivity in the United States. *Obesity (Silver Spring)*, 2008. 16(9): p. 2155-62.

[18] Ringborg, A., et al., Resource use and costs of type 2 diabetes in Sweden - estimates from population-based register data. *Int. J. Clin. Pract.*, 2008. 62(5): p. 708-16.

[19] Huxley, R., F. Barzi, and M. Woodward, Excess risk of fatal coronary heart disease associated with diabetes in men and women: meta-analysis of 37 prospective cohort studies. *BMJ,* 2006. 332(7533): p. 73-8.

[20] Woodward, M., et al., The effects of diabetes on the risks of major cardiovascular diseases and death in the Asia-Pacific region. *Diabetes Care,* 2003. 26(2): p. 360-366.

[21] Hu, F.B., et al., Diet, lifestyle, and the risk of type 2 diabetes mellitus in women. *The New England Journal of Medicine*, 2001 345(11): p. 790-797.

[22] Pradhan, A.D., et al., C-reactive protein, interleukin 6, and risk of developing type 2 diabetes mellitus. *JAMA*, 2001 286(3): p. 327-324.

[23] Barzilay, J.I., et al., The relation of markers of inflammation to the development of glucose disorders in the elderly: the Cardiovascular Health Study. *Diabetes* 2001 50(10): p. 2384-2389.

[24] Han, T.S., et al., Prospective study of C-reactive protein in relation to the development of diabetes and metabolic syndrome in the Mexico City Diabetes Study. *Diabetes Care*, 2002 25(11): p. 2016-2021.

[25] Pradhan, A.D., et al., C-reactive protein is independently associated with fasting insulin in nondiabetic women. *Arterioscler. Thromb. Vasc. Biol.*, 2003. 23(4): p. 650-5.

[26] Esposito, K., Inflammatory Cytokine Concentrations Are Acutely Increased by Hyperglycemia in Humans: Role of Oxidative Stress. *Circulation,* 2002. 106(16): p. 2067-2072.

[27] Dandona, P., et al., The suppressive effect of dietary restriction and weight loss in the obese on the generation of reactive oxygen species by leukocytes, lipid peroxidation, and protein carbonylation. *The Journal of Clinical Endocrinology & Metabolism*, 2001 86(1): p. 355-362.

[28] Dandona, P., et al., Insulin inhibits intranuclear nuclear factor kappaB and stimulates IkappaB in mononuclear cells in obese subjects: evidence for an anti-inflammatory effect? *The Journal of Clinical Endocrinology & Metabolism*, 2001 86(7): p. 3257-3265.

[29] Schenk, S., M. Saberi, and J.M. Olefsky, Insulin sensitivity: modulation by nutrients and inflammation. *J. Clin. Invest.*, 2008. 118(9): p. 2992-3002.

[30] Wellen, K.E., Inflammation, stress, and diabetes. *Journal of Clinical Investigation*, 2005. 115(5): p. 1111-1119.

[31] Kato, H., Nutrigenomics: the cutting edge and Asian perspectives. *Asia Pacific Journal of Clinical Nutrition*, 2008. 17(Suppl 1): p. 12-15.

[32] Müller, M. and S. Kersten, Nutrigenomics: goals and strategies. *Nature Reviews Genetics*, 2003 Apr;4(4):315-22. 4(4): p. 315-322.

[33] Jia, H., et al., Transcriptional profiling in rats and an ex vivo analysis implicate novel beneficial function of egg shell membrane in liver fibrosis. *Journal of Functional Foods*, 2013. 5(4): p. 1611-1619.

[34] Jia, H., et al., DNA microarray analysis identified molecular pathways mediating the effects of supplementation of branched-chain amino acids on CCl4-induced cirrhosis in rats. *Mol. Nutr. Food Res.*, 2013. 57(2): p. 291-306.

[35] Imamura, W., et al., Adverse effects of excessive leucine intake depend on dietary protein intake: a transcriptomic analysis to identify useful biomarkers. *Journal of Nutritional Science and Vitaminology*, 2013. 59(1): p. 45-55.

[36] Chang, W.C., et al., Beneficial effects of soluble dietary Jerusalem artichoke (Helianthus tuberosus) in the prevention of the onset of type 2 diabetes and non-alcoholic fatty liver disease in high-fructose diet-fed rats. *Br. J. Nutr.*, 2014. 112(5): p. 709-17.

[37] Wu, T., et al., Enhanced effect of daytime restricted feeding on the circadian rhythm of streptozotocin-induced type 2 diabetic rats. *The American Journal of Physiology - Endocrinology and Metabolism*, 2012 302(9): p. E1027-1035.

[38] Liao, C.C., et al., Prevention of diet-induced hyperlipidemia and obesity by caffeic acid in C57BL/6 mice through regulation of hepatic lipogenesis gene expression. *J. Agric. Food Chem.*, 2013. 61(46): p. 11082-8.

[39] Ma, Y., M. Gao, and D. Liu, Chlorogenic Acid improves high fat diet-induced hepatic steatosis and insulin resistance in mice. *Pharm. Res.*, 2015. 32(4): p. 1200-9.

[40] Kim, M.J., et al., The Effect of Caffeine on 3T3-L1 Adipocyte Differentiation : A Nutrigenomical Approach. *Korean Journal of Nutrition*, 2005. 38(8): p. 649-655.

[41] Ohara, T., et al., A combination of glucosyl hesperidin and caffeine exhibits an anti-obesity effect by inhibition of hepatic lipogenesis in mice. *Phytother Res.*, 2015. 29(2): p. 310-6.

[42] Jung, U.J., et al., Antihyperglycemic and antioxidant properties of caffeic acid in db/db mice. *J. Pharmacol Exp. Ther.*, 2006. 318(2): p. 476-83.

[43] Murase, T., et al., Coffee polyphenols suppress diet-induced body fat accumulation by downregulating SREBP-1c and related molecules in C57BL/6J mice. *The American Journal of Physiology - Endocrinology and Metabolism*, 2011 300(1): p. E122-133.

[44] Fukushima, Y., et al., Effects of coffee on inflammatory cytokine gene expression in mice fed high-fat diets. *J. Agric. Food Chem.*, 2009. 57(23): p. 11100-5.

[45] Takahashi, S., et al., Coffee intake down-regulates the hepatic gene expression of peroxisome proliferator-activated receptor gamma in C57BL/6J mice fed a high-fat diet. *Journal of Functional Foods*, 2014. 6: p. 157-167.

[46] Yamauchi, R., et al., Coffee and caffeine ameliorate hyperglycemia, fatty liver, and inflammatory adipocytokine expression in spontaneously diabetic KK-Ay mice. *J. Agric. Food Chem.*, 2010. 58(9): p. 5597-603.

[47] Swatton, J.E., et al., Protein profiling of human postmortem brain using 2-dimensional fluorescence difference gel electrophoresis (2-D DIGE). *Mol Psychiatry*, 2004. 9(2): p. 128-43.

[48] Wiese, S., et al., Protein labeling by iTRAQ: a new tool for quantitative mass spectrometry in proteome research. *Proteomics*, 2007. 7(3): p. 340-50.

[49] Salomone, F., et al., Coffee enhances the expression of chaperones and antioxidant proteins in rats with nonalcoholic fatty liver disease. *Transl Res*, 2014. 163(6): p. 593-602.

[50] Fukuda, S. and H. Ohno, Gut microbiome and metabolic diseases. *Semin. Immunopathol.*, 2014. 36(1): p. 103-14.

[51] Qin, J., et al., A human gut microbial gene catalogue established by metagenomic sequencing. *Nature*, 2010. 464(7285): p. 59-65.

[52] Xu, J., et al., Evolution of symbiotic bacteria in the distal human intestine. *PLoS Biol.*, 2007 5(7): p. e156.
[53] Zhu, X.Y., et al., 16S rRNA-Based Analysis of Microbiota from the Cecum of Broiler Chickens. *Applied and Environmental Microbiology*, 2002. 68(1): p. 124-137.
[54] Gill, S.R., et al., Metagenomic analysis of the human distal gut microbiome. *Science*, 2006 312(5778): p. 1355-1359.
[55] Dethlefsen, L., et al., The pervasive effects of an antibiotic on the human gut microbiota, as revealed by deep 16S rRNA sequencing. *PLoS Biol.*, 2008. 6(11): p. e280.
[56] Nakayama, T. and K. Oishi, Influence of coffee (Coffea arabica) and galacto-oligosaccharide consumption on intestinal microbiota and the host responses. *FEMS Microbiol. Lett.*, 2013. 343(2): p. 161-8.
[57] Jaquet, M., et al., Impact of coffee consumption on the gut microbiota: a human volunteer study. *Int. J. Food Microbiol.*, 2009. 130(2): p. 117-21.
[58] Takahashi, S., et al., An integrated multi-omics study revealed metabolic alterations underlying the effects of coffee consumption. *PLoS One*, 2014. 9(3): p. e91134.
[59] Cowan, T.E., et al., Chronic coffee consumption in the diet-induced obese rat: impact on gut microbiota and serum metabolomics. *J. Nutr. Biochem.*, 2014. 25(4): p. 489-95.

In: Caffeic Acid
Editor: Leanna Vaughn

ISBN: 978-1-63483-121-5

Chapter 4

ANTIMICROBIAL PROPERTIES AND POTENTIAL USEFULNESS OF CAFFEIC ACID

M. J. Rodríguez-Vaquero*, C. V. Vallejo, O. A. Sosa, F. M. Saguir and P. A. Aredes-Fernandez

Facultad de Bioquímica, Química y Farmacia – Universidad Nacional de Tucumán, Ayacucho 471 – CONICET. Tucumán. Argentina

ABSTRACT

The resistance of microorganisms to drugs in pharmaceutical industry and the food contamination and spoilage by microorganisms in food industry are serious problems. Control of these microorganisms is of importance not only for industry but also for government health agencies. As a result of this problematic, all these years an intensive search for new compounds able to replace currently used drugs was performed. The utilization of phenolic compounds as antimicrobial agents or a possible pharmaceutical drug would provide additional benefits due to their health beneficial properties. Caffeic acid is one of the most abundant hydroxycinnamic acids in the human diet and chlorogenic acid (5-O-caffeoyl-quinic acid), an ester of caffeic acid with quinic acid, has been received considerable attentions for their potential biological effects. Chlorogenic acid has antioxidant property and probable roles in diseases such as cancer, parkinson and cardiovascular as well as antimicrobial activity. In this chapter, we present the last researches that indicate that chlorogenic acid of different plant sources or derived products could be a possible candidate to be utilized in different industries. The advantages in the usefulness of chlorogenic acid as antimicrobial agent are discussed in relation with its antimicrobial effectivity and its health benefits.

CHEMICAL STRUCTURE AND PRINCIPAL SOURCES

Caffeic acid, 3, 4-dihydroxycinnamic acid, is a widespread phenolic compound from the group of hydroxycinnamates, which is derived biosynthetically from phenylalanine in plants.

*Author to whom correspondence should be addressed. E-mail: mariajo@fbqf.unt.edu.ar

It occurs naturally in many agricultural products such as fruits, vegetables and coffee [1]. Hydroxycinnamic acids in general have a C6-C3 skeleton and differences in the substituents of the aromatic ring yield different phenolic acids, including ρ-coumaric, caffeic, ferulic and sinapic acids. Figure 1 shows the molecular structure of caffeic acid, which contain a catechol group with an α, β,-unsaturated carboxylic acid chain.

Phenolic compounds are strong antioxidant reagents necessary for vegetal cell functioning, and caffeic acid is one of the many phenolics considered to be an important part of the plant defense mechanism against infection predation [2, 3].

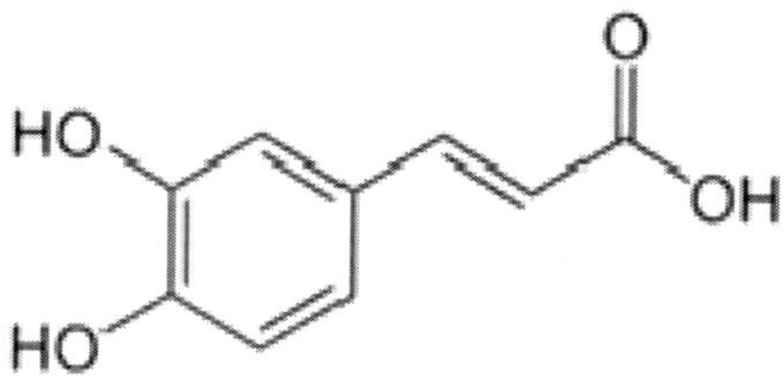

Figure 1. Molecular structure of caffeic acid.

Caffeic acid is found in relatively high concentrations in many plants as a key intermediate in the biosynthesis of lignins. Caffeic acid and its naturally occurring derivative caffeic acid phenethyl ester are found in high concentrations in propolis of the honeybee.

Hydroxycinnamic acids (e.g., ferulic and caffeic acids) are widely distributed in plants and their products such as cereals, coffee, fruits, vegetables etc. [4]. These are also present in fruits, tubers and the herbaceous parts of many vegetable species [5]. They occur naturally in combination with other compounds usually in the form of esters.

Polyphenols, including caffeic acid, are often found in herbs [6], teas [7, 8, 9], coffee, wine [10, 11, 12, 13], vegetables, fruits and fruit juice [14].

Coffee is one of the most consumed beverages in the world, and it has aroused more and more interest in the past few years because of its high content of hydroxycinnamic acids [15]. In olive mill wastes, caffeic acid is present in high proportions, which are a major potential source of phenolics considering that the annual production exceeds 7 million tonnes and the medium content of phenolics is approx 1.0–1.8% [16].

Caffeic acid also was found in the outer bran layer of wheat seeds [17, 18], in wheat caryopses [19] and in rice root exudates [20].

Biological Activities of Caffeic Acid

Since 1980, several studies have shown that lower risk of chronic diseases was correlated with a diet rich in fruit and vegetables [21, 22].

The hydroxycinnamic acids have attracted considerable attention due to their various biological and pharmacological activities. Many epidemiological studies have indicated that long-term coffee consumption is associated with a reduced risk of developing type 2 diabetes mellitus [15, 23]. Some authors demonstrated that phenolic compounds, included caffeic acid, possess different beneficial effects on human health, such as anti-bacterial, antioxidant, anti-thrombotic, anti-allergic, anti-inflammatory and anti-hypertensive activities [24, 25, 26, 27, 28, 29, 30].

Da Silva, Stradiotto& Oliveira [31] reported that caffeic acid derivatives are inhibitors of human immunodeficiency virus type 1 (HIV-1) integrase.

Epilepsy is characterized by unprovoked episodes of aberrant synchronous excitation of brain regions that disrupt normal functioning and cause successive seizures [32, 33]. The World Health Organization reported that about 50 million people are affected worldwide, and approximately 70–80% of patients with new-onset epilepsy enter remission when they are treated with antiepileptic drugs currently prescribed, so antiepileptic drugs fail to control seizures in 20–30% of patients, or present troubling side effects [34, 35, 36]. Rodrigues Coelho et al., [37] demonstrated that caffeic acid have neuroprotective activity, which can be important in the prevention and management of various neurological disorders, such as epilepsy, but new investigations about the mechanisms involved in the protective effect of these compounds must carried out.

Epidemiological studies indicate that a rich diet in fruit and vegetable reduces cancer risk in humans, suggesting that certain dietary constituents may therefore be effective in cancer prevention [38], so phenolic compounds have been investigated for their potential use as cancer chemopreventive agents [39]. Caffeic acid esters, such as caffeic acid phenethyl and benzyl esters, display selective antiproliferative activity against some types of cancer cells [40]. Etzenhouser et al., [41] proposed that the molecular basis of this action may be connected to the inhibition of the nuclear transcription factor NK kappa B.

Natural products are widely being used in the form of dietary supplements for their possible health promoting effects such as their potential in preventing cancer, atherosclerosis, heart disease, etc. [42, 43, 44, 45]. Natural products are rich in antioxidants with possible therapeutic benefits. In recent years, caffeic acid and its derivatives have attracted considerable attention due to its antioxidative activity [27, 29, 46, 47, 48, 49]. An antioxidant compound possesses the ability to scavenge primary and secondary radicals, to inhibit free-radical-induced membrane damage and potential to bind iron that can prevent radical formation [50, 51]. Hydroxycinnamic acid derivatives are a major category of phenolic acids that may provide important benefits as antioxidants. The molecular structure of caffeic acid is responsible for its efficient interaction with several types of oxidant radicals [52]. The o-dihydroxy structure is the natural radical site, producing the o-semiquinone after one-electron donation. The lateral double bond conjugated with the catechol group leads to an extensive electron delocalisation, increasing the stability of the o-semiquinone radical and, the antioxidant activity [53]. The catechol group also acts as the preferred binding site for trace metals leading to significant chelating activity [54]. Phenolic compounds may act as antioxidants by several different mechanisms. Vallejo, Aredes-Fernández and Rodríguez-Vaquero [28] determined the antioxidant capacity of caffeic acid (Figure 2) using three different methods, ferric reducing power (a) and the radical scavenging activity of DPPH and ABTS radicals (b) at different concentrations, the authorsdemostrated that caffeic acid possesses important antioxidant activities, which increase with the concentration.

Inoue et al., [55] reported that plant polyphenols, such as caffeic acids possess both antioxidant as well as prooxidant properties, and it was earlier proposed that the prooxidant action of polyphenols may be an important mechanism of their anticancer and apoptosis inducing-properties [56, 57].

The molecular structure of hydroxycinnamic acids is a key determinant for their radical scavenging and metal chelating properties. It has been proposed that phenolic compounds that are able to inhibit different prooxidants, cooperate with endogenous antioxidants, and are located on the active oxidation sites will be the most efficient in foods [58].

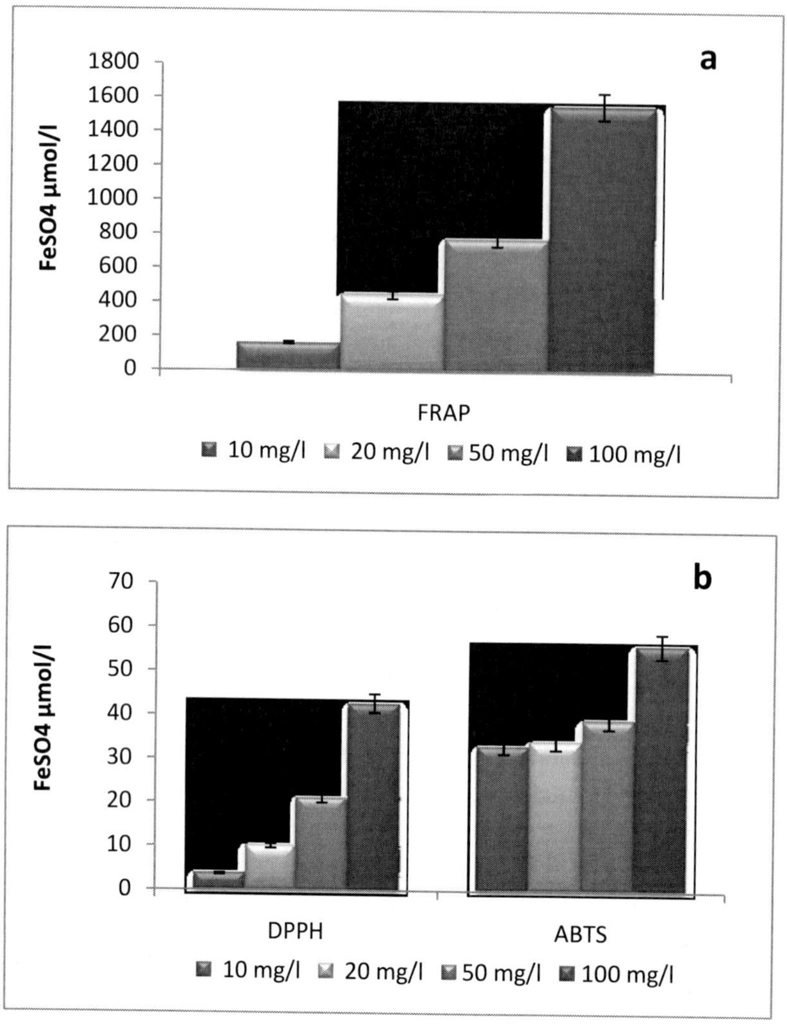

Figure 2. Antioxidant activity of different concentration of caffeic acid determined by Ferric reducing power (a), percentage depuration of DPPH and ABTS radical (b).

Hypertension is an illness that affects million people worldwide. Many phytochemicals including plant polyphenols have gained attention as antihypertensive agents [59, 60], principally, plant secondary metabolites, including flavonoids such as quercetin [61]. Renin–angiotensin–aldosterone system (RAAS) has evolved as a principal clinical target for the treatment of hypertension, myocardial infraction, stroke and renal diseases [62]. Many drugs used in the treatment of hypertension and cardiac disorders elicit their activity through modulation of RAAS enzymes such as renin and angiotensin converting enzyme (ACE) as well as acting as antagonist of aldosterone [63, 64]. Vallejo, Farias, Aredes-Fernandez and Rodríguez-Vaquero [28] study the capacity of caffeic acid to inhibit the ACE enzyme (ACEI); the authors reported that between phenolic acids, caffeic and gallic acid show the highest ACEI activity. With a concentration of 20 mg/l of caffeic acid, the ACE was inhibiting by 82.5%. In other hand, Actis-Goretta et al., [65] demonstrate that phenolic acid commonly found in fruits and vegetables have strong antihypertensive agent in vitro.

A high concentration of caffeic acid was reported in tea or herb infusions around the word [7, 8, 9]. Rodríguez-Vaquero et al., [29] investigated the antihypertensive, antioxidant and antibacterial properties of different Argentinean herbs infusions consumed traditionally; and the modifications for the use of three additives (sugar, sweetener and lemon) commonly used during herbal infusion consumption in Argentine on these activities. The authors demonstrated that ACEI activity of *Ilex paraguaiensis, Lippiaintegrifolia* and *Menthapiperita* was higher than 80%; the lowest activity was shown in presence of *Chenopodioum ambrosioides* and *Minthostachys verticillatta* (between 15 and 37%); the addition of lemon increased the ACEI activity to the herb infusions preparation around 10%; this could be due to the phenolic compounds found in lemon, but the addition of sugar or sweetener did not modify this activity. The results reported by Rodríguez-Vaquero et al., show a high correlation between the phenolic concentration and the antihypertensive activity, and was in agreement with results of Vallejo et al., (2013) [28] that demonstrated a good correlation between the phenolic compounds concentrations in juice fruit and the antihypertensive activity.

So, the intakes of foods with a high amount of polyphenol are recommended to be included in diet and new dietary supplements and functional foods containing polyphenols have been released.

Several experiments demonstrated that caffeic acid inhibited the growth of fungi [28, 66, 67], bacteria [24, 25, 26, 68, 69] and antiviral [70, 71].

Advantages in the Usefulness As Antibacterial Agent

The rapid incidences of drug resistance of microorganisms to antibiotics were reported [73]. When *Staphylococcus aureus* became resistant to penicillin, infections with these strains caused many deaths [74]. The discovery of multidrug-resistant *Mycobacterium tuberculosis*, penicillin resistant *Streptococcus pneumoniae*, methicillin-resistant *S. aureus*, vancomycin-resistant enterococci and vancomycin-resistant *S. aureus* increase the demand for alternative antibiotics to the treatment of these bacteria, and the structure of these new antibiotic drugs must differ from conventional antibiotics.

Caffeic acid is one of the most widely distributed hydroxycinnamate and phenylpropanoid metabolites in plant tissues, it is usually found as various simple derivatives including amides, esters, sugar esters, and glycosides [75]. The antimicrobial activity of these derivatives has attracted much attention and has been studied by many research groups in recent years [24, 25, 26, 75].

Phytopathogenic bacteria are causative infectious agents of plant diseases. The inhibitory effect of caffeic acid against phytopathogenic bacteria was reported by Schaal, and Johnson in 1955 [76] and they reported that the activity of hydroxycinnamic acids is dependent on the positions of the free hydroxy groups; on cis-trans isomerism. Andary (1982) [77] studied that in glycosides, the caffeoyl as well as the non-caffeoyl part may play a role in their antibacterial activity.

The antimicrobial activity of caffeic acid was demonstrated in vitro [78]. The growth of *Xylellafastidiosa* was inhibited by 0.2 mM (0.036 mg/ml) of caffeic acid [79]. The MIC for this phenolic acid was determined to be of 2.5 mg/ml for strains of *Pseudomonas aeruginosa*, *Bacillus cereus* and *Micrococcus luteus*, and 3.7 mg/ml for *Staphylococcus aureus* [74].

*Listeria monocytogenes*is a bacterium responsible for the severe food-borne illness, listeriosis, which has been associated with a mortality rate as high as 30–40% [80]. Garcia et al., [81] reported that most listeriosis cases were asociate with the consumption of contaminated ready-to-eat foods such as dairy products, processed or cured meat and poultry, salads, seafood and uncooked eggs. Rodríguez-Vaquero et al., [25] studied the effect of different concentrations (25, 50, 100, 200 and 500 mg/l) of phenolic acids on *L. monocytogenes* growth, the authors demonstrated that at all concentrations, caffeic acid was more effective than hidroxybenzoic acids (gallic, protocatechuic and vanillic acids) to decrease the final cell density and the μ_{max} of *L. monocytogenes*was totally inhibited only with 500 mg/l of the hydroxycinnamic derivate, caffeic acid. Hydroxycinnamic acids, due to their propenoic side chain, are much less polar than the corresponding hydroxybenzoic acids. Campos et al., [82] reported that in *Oenococcus oeni*, this property might facilitate the transport of these molecules across the cell membrane, which might be related in turn to the stronger inhibitory effect of hydroxycinnamic acids.

Caffeic acid has potential as a food additive to inhibit growth of *Clostridium botulinum* and reduce thermal processing requirements of heat sensitive foods [83]. The caffeic acid phenylethyl esters are demonstrated to have higher antimicrobial activities against *S. aureus*, *B. subtilis* and *Pseudomonas aeruginosa* than 5-chorogenic acid and caffeic acid [84] and the synthesis of caffeic acid amides possessing antimicrobial activities was recently reported by Fu et al., [85].

Rodríguez-Vaquero et al., [24] studied the antibacterial activity of different concentrations of caffeic acid against *Serratia marcescens, Proteus mirabilis, Escherichia coli, Klebsiella pneumonia* and *Flavobacterium sp.* and against ATCC collection bacteria. The authors reported that caffeic acid was effective to inhibit the growth of selected bacteria and the inhibition was related with the concentration. *Klebiella* and *Flavobacterium* was the only resistant to the treatment.

On the basis of previous results among the antibacterial effect of caffeic acid, Rodríguez-Vaquero et al., [86] investigated the antibacterial efficiency of three phenolic compounds combinations on *E. coli* (Table 1) and *L. monocytogenes* (Table 2) viabilityin a fish-meat model system, at 4°C and 20°C. The autors demostrated the use of phenolic compounds as antibacterial agents with refrigerated temperature could be a good combination to prevent fish meat contamination and extend the shelf life of the product. Besides, would provide additional benefits inherent to their natural biological properties and health benefits.

Table 1. Reduction of viable cell numbers of *E. coli* in fish meat added with phenolic cmpounds combinations, at 21 days of storage at 20°C and 4°C

Phenolic compounds Combinations	Log cycles reduction of *E. coli*											
	20 °C						4 °C					
	100 mg/L			200 mg/L			100 mg/L			200 mg/L		
	C	I	*D*	C	I	*D*	C	I	*D*	C	I	*D*
G - P	2.04	-	-	2.58	-	-	5.0	4.92	4.3	9.08	9.0	2,0
G - C	2.73	-	-	3.37	-	-	9.08	9.0	2.1	9.08*	9.0*	1.6
Q - R	4.04	0.05	-	5.51	1.53	13.9	9.08	9.0	1.9	9.08*	9.0*	1.5

C: Log cycles reduction with respect to control. **I:** Log cycle reduction with respect to inoculums. ***D***: Decimal reduction time (days). **(-):** No Inhibition observed. *At 14 d no viable cells are detected.

Table 2. Reduction of viable cell numbers of *L. monocytogenes* in fish meat added with phenolic cmpounds combinations, at 21 days of storage at 20°C and 4°C

Phenolic compounds Combinations	Log cycles reduction of *L. monoytogenes*											
	20 °C						4 °C					
	100 mg/L			200 mg/L			100 mg/L			200 mg/L		
	C	I	*D*	C	I	*D*	C	I	*D*	C	I	*D*
G - P	1.39	-	-	4.87	1.0	21.0	2.6	2.0	9.0	3.96	3.36	4.4
G - C	2.92	-	-	5.83	1.96	8.5	4.65	4.05	5.2	9.6*	9.0*	1.2
Q - R	3.51	-	-	6.67	2.8	6.9	6.15	5.55	4.0	9.6*	9.0*	1.3

C: Log cycles reduction with respect to control. **I:** Log cycle reduction with respect to inoculums. ***D***: Decimal reduction time (days).**(-):** No Inhibition observed. *At 14 d no viable cells are detected.

From Rodríguez-Vaquero et al. (2013).

Caffeic Acid As Antifungal Agent

There are special interests in the study of phenolic acid, especially caffeic acid, for their antifungal activity against different fungi, due that pathogenic and toxinogenic fungi are constantly developing resistance against commonly used fungicides.

This is a critical problem in many areas such as agriculture, human and animal health and safety, in the storage and production of foods and human medicines [87, 88]. Besides their potential to cause yield losses and food decay, many of them produce dangerous secondary metabolites, extremely hazardous to consumers.

The treatment of infections caused by *Fusarium*, *Aspergillus* and *Penicillium* genera is problematic and questionable due to the toxicity and side effects of the medicines used [89, 90, 91, 92]. *Fusarium*, *Aspergillus* and *Penicillium* genera produce the majority of known mycotoxins [93, 94, 95] and they are able to cause very serious human mycoses. So, there is a growing need to research and develop new, environmentally safe substances which will undergo quick and natural degradation in the environment.

Candida albicans, an opportunistic fungal pathogen, is one of the major causes of life threatening, invasive candidiasis with estimated mortality rate of 46-75% [96]. The ability of *C. albicans* to grow as a 'biofilm' and its flexibility in adapting to different environmental conditions contributes to unacceptably high rates of morbidity and mortality, of which ineffectual antifungal therapy is a key factor. Therefore, alternative approaches to prevent biofilm formation or destruction and modulation of established biofilms are highly desirable. Fluconazole is the most commonly used drug to treat *Candida albicans* in the prophylaxis and therapy, however, widespread and repeated use of fluconazole resulted in resistance to or failure of fluconazole therapy [97]. Alternative methods of suppressing pathogenic and toxinogenic fungi, based on the use of natural plant substances, often result in research on the development of highly effective essential oils. Several authors [98, 99, 100], suggested that plant essential oils could be an alternative sources for antifungal drugs, such as phenolic substances (principal antifungal active substances present in essential oils), and the low toxic effect on homeothermic animals was reported by Xu et al.,[101] and Ahmad et al.,[88].

To combat the fluconazole-resistant *C. albicans*, much attention has been paying to the synergism of fluconazole with other agents, as synergists, which can significantly sensitize fungi towards fluconazole [88, 102]. Mukherjee et al., [103] demonstrated that caffeic acid amides, exhibited a synergistic potent activity with fluconazole against fluconazole-resistant *C. albicans*. The mechanism of the antifungal effect thus depends predominantly on the ability to affect the function of cellular lipoprotein membranes, causing an impairment of cellular ionic homeostasis, acidification of vacuolar and cytosolic pH, and even the destruction of structural cellular integrity [99, 103]. Shahzad et al., [105] evaluate the potential role of phenolic compounds, including caffeic acid in the prevention, and possibly management, of chronic oral and systemic fungal infections. Said et al., [106] showed that the presence of caffeic acid (250µg/ml) in a *Neurosporacrassa* culture did not inhibit fungal growth [107].

Conclusion

Polyphenols are secondary plant metabolites and possesseveral beneficial properties. The antimicrobial activity of caffeic acid was extensively studied, so it is possible the use of caffeic acid as antibacterial or antifungal drugs. Caffeic acid could be used as alternative method of suppressing pathogenic and toxinogenic fungi, based on the use of natural plant substances.

REFERENCES

[1] Clifford, MN. Chlorogenic acids and other cinnamates – Nature, occurrence and dietary burden. *Journal of Science and Food Agriculture,* 1999 *79*, 362-372.

[2] Faulds, CB; Williamson, G. The role of Hidroxycinnamates in the plant cell wall, review. *J. Sci. Food Agric.*, 1999 79, 393-395.

[3] Friend, J. Phenolic substances and plant disease. In Annual Proc. Phytochemical Soc. Europe. Eds. C.F. Van Sumere and P.J. Lea. Clarendon Press, Oxford, 1985 367-392.

[4] Nystrom, L; Makinen, M; Lampi, AM; Piironen, V. Antioxidant activity of sterylferulate extracts from rye and wheat bran. *J. Agric. Food Chem.,* 2005 *53(7)*, 2503-2510.

[5] Bravo, L. Polyphenols: Chemistry, dietary sources, metabolism, and nutritional significance. *Nutrition Reviews*, 1998 *56*, 317–333.

[6] Tyszczuk, K; Skalska-Kaminska, A; Wozniak, A. Voltammetric method using a lead film electrode for the determination of caffeic acid in a plant material. *Food Chemistry*, 2011 *125*, 1498.

[7] Sen, S; Sarkar, P. A new diagnostic tool for measuring total antioxidants in tea. *Sensor Letters*, 2013 11, 428-435.

[8] Veljkovic, JN; Pavlovic, AN; Mitic, SS; Tosic, SB; Stojanovic, GS; Kalicanin, BM. Evaluation of individual phenolic compounds and antioxidant properties of black, green, herbal and fruit tea infusions consumed in Serbia: spectrophotometrical and electrochemical approaches. *Journal of Food and Nutrition Research*, 2013 52, 12-24.

[9] Novak, I; Šeruga, M; Komorsky-Lovric, Š. Characterisation of catechins in green and black teas using square wave voltammetry and RP-HPLC-ECD. *Food Chemistry*, 2010 *122*, 1283-1289.

[10] Mudnic, I; Modun, D; Rastija, V; Vukovic, BJ; Brizic, I; Katalinic, V. Antioxidative and vasodilatory effects of phenolic acids in wine. *FoodChemistry,* 2010 *119,* 1205-1210.

[11] Photinon, K; Chalermchart, Y; Khanongnuch, C; Wang, SH; Liu, CC. A thick-film sensor as a novel device for determination of polyphenols and their antioxidant capacity in white wine. *Sensors,* 2010 *10,* 1670-1678.

[12] Rebelo, MJ; Rego, R; Ferreira, M; Oliveira, MC. Comparative study of the antioxidant capacity and polyphenol content of Douro wines by chemical and electrochemical methods. *Food Chemistry*, 2013 141, 566-573.

[13] Šeruga, M; Novak, I; Jakobek, L. Determination of polyphenols content and antioxidant activity of some red wines by differential pulse voltammetry, HPLC and spectrophotometric methods. *Food Chemistry*, 2011 124, 1208-1216.

[14] Sousa, WR; da Rocha, C; Cardoso, CL; Silva, DHS; Zanoni, MVB. Determination of the relative contribution of phenolic antioxidants in orange juice by voltammetric methods. *Journal of Food Composition and Analysis,* 2004 17, 619-633.

[15] NatašaPoklarUlrih. Effects of Caffeic, Ferulic, and p-Coumaric Acids on Lipid Membranes Coffee in Health and Disease Prevention, 2015, *Chapter 89*, 813-821.

[16] Visioli, F; Galli, C. Olives and their production waste products as sources of bioactive compounds. *Current Topics in Nutraceutical Research*, 2003 *1(1)*, 85-88.

[17] Mpofu, A; Sapirstein, HD; Beta, T. Genotype and environmental variation in phenolic content, phenolic acid composition and antioxidant activity of hard spring wheat. *J. Agric. Food Chem.*, 2006 *54*, 1256-1270.

[18] Verma, B; Hucl, P; Chibbar, RN. Phenolic acid composition and antioxidant capacity of acid and alkali hydrolyzed wheat bran fractions. *Food Chem.*, 2009 *116*, 947-954.

[19] Weidner, S; Amarowicz, R; Karamac, M; Dabrowski, G. Phenolic acids in caryopses of two cultivars of wheat, rye and triticale that display different resistance to pre-harvest sprouting. *Eur. Food Res. Technol.*, 1999 *210*, 109-113.

[20] Seal, AN; Haig, T; Pratley, JE. Evaluation of putative allelochemicals in rice root exudates for their role in the suppression of arrowhead root growth. *J. Chem. Ecol.*, 2004 30, 1663-1678.

[21] Cifuentes, A. Recent advances in the application of capillary electromigration methods for food analysis. *Electrophoresis,* 2006 *27*, 283.

[22] Kris-Etherton, PM; Hecker, KD; Bonanome, A; Coval, AE; Binkoski, AE; Hilpert, AE. Bioactive compounds in foods: their role in the prevention of cardiovascular disease and cancer. *American Journal of Medicine*, 2002 *113*, 71.

[23] Van Dam RM, Feskens EJ. Coffee consumption and risk of type 2 diabetes mellitus. *Lancet*, 2002 *360*, 1477-1478.

[24] Rodríguez Vaquero, MJ; Alberto, MR; Manca de Nadra, MC. Influence of phenolic compounds from wimes on the growth of *Listeria monocytogenes. Food Control*, 2007a *18*, 587- 593.

[25] Rodríguez Vaquero, MJ; Alberto, MR; Manca de Nadra, MC. Antibacterial effect of phenolic compounds from different wines. *Food Control,* 2007b *18*, 93-101.

[26] Rodríguez Vaquero, MJ; Manca de Nadra, MC. Growth parameter and viability modifications of *Escherichia coli* by phenolic compounds and argentine wine extracts. *Applied Biochemistry and Biotechnology*, 2008 *151(2),* 342-352.

[27] Rodríguez Vaquero, MJ; Tomassini Seravalle, LR; Manca de Nadra, MC; Strasser de Saad. AM. Antioxidant capacity and antibacterial activity of phenolic compounds from argentinean herbs infusions**.** *Food Control,* 2010 *21*, 779-785.

[28] Vallejo, CV; Rodríguez-Vaquero, MJ; Aredes-Fernandez, PA. Biofilm inhibition of spoilage bacteria by argentinean fruit juices with antihypertensive activity. *Current Pharmaceutical Biotechnology*, 2013. *In press.*

[29] Rodríguez-Vaquero, MJ; Vallejo, CV; Aredes-Fernandez, PA. Biological activity of phenolic compounds from Argentinean herbs infusions. *The Open Conference Proceedings Journal*, 2014 *5*, 1-7.

[30] Hurtado-Fernández, E; Gómez-Romero, M; Carrasco-Pancorbo, A; Fernández-Gutiérrez, A. Application and potential of capillary electroseparation methods to determine antioxidant phenolic compounds from plant food material. *Journal of Pharmaceutical and Biomedical Analysis,* 2010 *53*, 1130-1160.

[31] Da Silva, LF; Stradiotto, NR; Oliveira, HP. Determination of caffeic acid in red wine by voltammetric method. *Electroanalysis,* 2008 *20(11),* 1252.

[32] Briggs, SW; Galanopoulou, AS. Altered GABA signaling in early life epilepsies, *Neural. Plast.*, 2011, 1-16.

[33] Scharfman, HE. The neurobiology of epilepsy, *Curr. Neurol. Neurosci.* Rep., 2007 7, 348-354.

[34] Dua, T; Boer, HM; Prilipko, LL; Saxena, S. Epilepsy care in the world: results of an ILAE/IBE/WHO global campaign against epilepsy survey, *Epilepsia,* 2006 *47*, 1225-1231.
[35] Kaminski, RM; Rogawski, MA; Klitgaard, H. The potential of antiseizure drugs and agents that act on novel molecular targets as antiepileptogenic treatments, *Neurotherapeutics,* 2014 *11*, 385-400.
[36] Löscher, W; Klitgaard, H; Twyman, RE; Schmidt, D. New avenues for anti-epileptic drug discovery and development, *Nat. Rev. Drug Discov.,* 2013*12*, 757-776.
[37] Rodrigues Coelho, V; Gonçalves Vieira, C; Pereira de Souza, L; Moysés, F; Basso, C; KuckMausolffPapke , D; Rocha Pires , T; Rodrigues Siqueira, I; NascimentoPicada, J; Pereira, P. Antiepileptogenic, antioxidant and genotoxic evaluation of rosmarinic acid and its metabolite caffeic acid in mice. *Life Sciences*, 2015 *122* , 65-71.
[38] Mahmoud, NN; Carothers, AM; Grunberger, D; Bilinski, RT; Churchill, MR; Martucci, C; Newmark, HL; Bertagnolli, MM. *Carcinogenesis,* 2000 *21*, 921.
[39] Tapiero, H; Tew, KD; Ba, GN; Mathé, G. Polyphenols: do they play a role in the prevention of human pathologies. *Biomed. Pharmacother,* 2002 *56(4)*, 200-207.
[40] Nagaoka, T; Banskota, AH; Tezuka, Y; Saiki, I; Kadota, S. Selective antiproliferative activity of caffeic acid phenethyl ester analogues on highly liver-metastatic murine colon 26-L5 carcinoma cell line. *Bioorg.Med. Chem.*, 2002 10(10), 3351-3359.
[41] Etzenhouser, B; Hansch, C; Kapur, S; Selassie, CD. *Bioorg. Med. Chem.*, 2001 *9*, 199.
[42] Bengmark, S; Mesa, MD; Gil, A. Plant-derived health: the effects of turmeric and curcuminoids. *Nutr. Hosp.*, 2009 *24(3),* 273-281.
[43] Del Rio, D; Costa, LG; Lean, ME; Crozier, A. Polyphenols and health: what compounds are involved? *Nutr. Metab. Cardiovasc. Dis.*, 2010 *20(1),* 1-6.
[44] Grassi, D; Desideri, G; Croce, G; Tiberti, S; Aggio, A; Ferri, C. Flavonoids, vascular function and cardiovascular protection. Curr. Pharmaceut. *Design*, 2009 *15(10),* 1072-1084.
[45] Soory, M. 2009. Relevance of nutritional antioxidants in metabolic syndrome, ageing and cancer: potential for therapeutic targeting. Infect. Disord. *Drug Targets*, 2009 *9(4),* 400-414.
[46] Marinova, EM; Yanishlieva, NV; Toneva, AG. Antioxidant activity and mechanism of action of ferulic and caffeic acids in different lipid systems. *Rivista Italiana delle Sostanze Grasse*, 2006 *83(1)*, 6-13.
[47] Nardini, M; D□Aquino, M; Tomassi, G; Gentili, V; Di Felice, M; Scaccini, C. Inhibition of human low-density lipoprotein oxidation by caffeic acid and other hydroxycinnamic acid derivatives. *Free RadicalBiology and Medicine*, 1995 *19*, 541-552.
[48] Taubert, D; Breitenbach, T; Lazar, A; Censarek, P; Harlfinger, S; Berkels, R. Reaction rate constants of superoxide scavenging by plant antioxidants. *Free Radical Biology and Medicine*, 2003 *35*, 1599-1607.
[49] Wu, WM; Lu, L; Long, Y; Wang, T; Liu, L; Chen, Q. Free radical scavenging and antioxidative activities of caffeic acid phenethyl ester (CAPE) and its related compounds in solution and membranes: A structure-activity insight. *Food Chemistry*, 2007 *105*, 107-115.
[50] Amorati, R; Pedulli, GF; Cabrini, L; Zambonin, L; Landi, L. Solvent and pH effects on the antioxidant activity of caffeic and other phenolic acids. *J. Agric. Food Chem.*, 2006 54(8), 2932–2937.

[51] Silva, FA; Borges, F; Guimaraes, C; Lima, JL; Matos, C; Reis, S. Phenolic acids and derivatives: studies on the relationship among structure, radical scavenging activity, and physicochemical parameters. *J. Agric. Food Chem.*, 2000 *48 (6),* 2122-2126.

[52] Bors, W; Michel, C; Schikora, S. Interaction of flavonoids with ascorbate and determination of their univalent redox potentials – A pulse-radiolysis study. *Free Radical Biology and Medicine*, 1995 *19*, 45-52.

[53] Laranjinha, J; Lester, P. Redox cycles of caffeic acid with [alpha]-tocopherol and ascorbate. *Methods in Enzymology*, 2001 *335,* 282-295.

[54] Pietta, PG. Flavonoids as antioxidants. *Journal of Natural Products*, 2000 *63*, 1035-1042.

[55] Inoue, M; Suzuki, R; Koide, T; Sakaguchi, N; Ogihara, Y; Yabu, Y. Antioxidant gallic acid induces apoptosis in HL60RG cells. *Biochem.Biophys. Res. Commun.*, 1994 *204*, 898-904.

[56] Azam, S; Hadi, N; Khan, NU; Hadi, SM. Prooxidant property of green tea polyphenols epicatechin and epigallocatechin-3-gallate: implications for anticancer properties. *Toxicol. In Vitro*, 2004 *18(5),* 555-561.

[57] Hadi, SM; Asad, SF; Singh, S; Ahmad, A. Putative mechanism for anticancer and apoptosis-inducing properties of plant-derived polyphenolic compounds. *IUBMB Life,* 2000 *50(3),* 167-171.

[58] Frankel, EN. (1998). Antioxidants. In E. Frankel (Ed.), Lipid oxidation. Dundee, Scotland: The Oily Press.

[59] Li, F; Takahashi, Y; Yamaki, K. Inhibitory effect of catechin-related compounds on renin activity. *Biomed. Res*., 2013*34*, 167-171.

[60] Suzuki, A; Yamamoto, N; Jokura, H; Yamamoto, M; Fujii, A; Tokimitsu, I; Saito, I. Chlorogenic acid attenuates hypertension and improves endothelial function in spontaneously hypertensive rats. *J. Hypertens.,* 2006*6*, 1075-1082.

[61] Perez-Vizcaino, F; Duarte, J; Jimenez, R; Santos-Buelga, C; Osuna, A. 2009.

[62] Ma, TK; Kam, KK; Yan, BP; Lam, YY. Renin–angiotensin–aldosterone system blockade for cardiovascular diseases: current status. *Br. J. Pharmacol.,* 2010*160*, 1273-1292.

[63] McMurray, JJ; Pitt, B; Latini, R; Maggioni, AP; Solomon, SD; Keefe, DL; Ford, J; Verma, A; Lewsey, J. Aliskiren Observation of Heart Failure Treatment (ALOFT) investigators. Effects of the oral direct renin inhibitor aliskiren in patients with symptomatic heart failure. *Circ. Heart Fail.,* 2008 *1*, 17-24.

[64] Calhoun, DA. Use of aldosterone antagonists in resistant hypertension. *Prog. Cardiovasc. Dis*., 2006*6*, 387-396.

[65] Actis-Goretta, L; Ottavian, JI; Fraga, CG. Inhibition of angiotens in converting enzyme activitybyflavanol-richfoods. *J. Agric. Food Chem*. 2006 *54*, 229-234.

[66] Ray, SK; Kumar, S; Sibal, M; Rudra. *Med. Sci. Res*., 1990 341, 18.

[67] Bostock, RM; Wilcox, SM; Wand, G. Adaskaveg. Suppression of Monilinia fruticolacutinase production by peach fruit surface phenolic acids. *Physiol. Mol. Plant Phatol*., 1999 *54*, 37-50.

[68] Velázquez, C; Navarro, M; Acosta, A; Angulo, A; Domínguez, Z; Robles, R; Robles, R; Lugo, E; Goycoolea, FM; Velásquez, EF; Astiazaran, H; Hernández, J. *J. Appl. Microbiol*., 2007 *1747*, 103.

[69] Bowles, BL; Miller, AJ. Caffcic acid activity against *Clostridium botulinum* spores. *J. Food Sci.*, 1994 *59*, 905-908.

[70] Sechi, M; Angotzi, G; Dallocchio, R; Dessi, A; Carta, F; Sannia, L; Mariani, A; Fiori, S; Sanchez, T; Movsessian, L; Plasencia, C; Neamati, N. *Antivir. Chem. Chemother.*, 2004 *67*, 15.

[71] Artico, M; Di Santo, R; Costi, R; Novellino, E; Greco, G; Massa, S; Tramontano, E; Marongiu, ME; De Montis, A; La Colla, PJ. *Med. Chem.*, 1998 *3948*, 41.

[72] Schmidr, LH; Sessler, CL. *Proc. Soc. Exp. Biol. Med.*, 1943 *53*, 353-357.

[73] Gillespie, SH; McHugh, TD. The biological cost of antimicrobial resistance. *Trends Microbiol.*, 1997 *5*, 337-339.

[74] Macheix, JJ; Fleuriet, A; Billot J. (Eds.) Fruit Phenolics, CRC Press, Boca Raton, FL, 1990, pp. 20-34.

[75] Almajano, MP; Carbo□, R; Delgado, ME; Gordon, MH. *J. Food Sci.*, 2007 *72*, 258-263.

[76] Schaal, AL; Johnson, G. The inhibitory effect of phenolic compounds on the groth of Streptomyces scabies as related to the mechanism, of scab resistance. *Phytopathology*, 1955 *45*, 626.

[77] Andary, C. Bulledn de Liaison No. 11 du Groupe Polyphenols,1982, 544-553. Narbonne, France.

[78] Dixon, RA. Natural products and plant disease resistance. *Nature*, 2001 *411*, 843-847.

[79] Maddox, CE; Laur, LM; Tian, L. Antibacterial activity of phenolic compounds against the phytopathogen Xylellafastidiosa. *Curr.Microbiol.*, 2010 *60*, 53-58.

[80] Datta, AR. *Listeria monocytogenes*. In M. D. Miliotis& J. W. Bier (Eds.), International handbook of foodborne pathogens, 2003, pp. 105-121. New York: Marcel Dekker Inc.

[81] Garcia, MT; Canamero, MM; Lucas, R; Omar, NB; Pulido, RP; Galvez, A. Inhibition of *Listeria monocytogenes* by enterocin EJ97 produced by *Enterococcus faecalis* EJ97. *International Journal of Food Microbiology*, 2004 *90*, 161-170.

[82] Campos FM; Couto JA; Hogg TA. Influence of phenolic acids on growth and inactivation of *Oenococcus oeni* and *Lactobacillus hilgardii*. *Journal of Applied Microbiology*, 2003 *94*, 167-174.

[83] Bowles, BL; Miller, AJ. Caffeic Acid Activity Against*Clostridium botulinurn*Spores. *J. FoodSci.*, 1994 *59*, 905-908.

[84] Noriaki, K; Yukari, K; Kazuya, I; Kyo, M; Tokio, F. In vitro antibacterial, antimutagenic and anti-influenza virus activity of caffeic acid phenethyl esters. *Biocontrol Sci.*, 2005 *10*, 155-161.

[85] Fu, J; Cheng, K; Zhang, Z; Fang, R; Zhu, H. Synthesis, structure and structure-activity relationship analysis of caffeic acid amides as potential antimicrobials. *European Journal of Medicinal Chemistry*, 2010 45, 2638–2643.

[86] Rodríguez Vaquero, MJ; Aredes Fernandez, PA; Manca de Nadra, MC.Phenolic compounds from wines as natural preservative of fishmeat. *Food technology and biotechnology*, 2013 *51(3)*, 376-382.

[87] Howard, A; Hoffman, J; Sheth, A. Clinical application of voriconazole concentrations in the treatment of invasive aspergillosis. *Ann. Pharmacother*, 2008 *42(12)*, 1859-1864.

[88] Ahmad, A; Khan, A; Akhtar, F; Yousuf, S; Xess, I; Khan, LA; Manzoor, N. Fungicidal activity of thymol and carvacrol by disrupting ergosterol biosynthesis and membrane integrity against Candida. *Eur. J. Clin.Microbiol. Infect. Dis.*, 2011 *30(1)*, 41-50.

[89] Howard, A; Hoffman, J; Sheth, A. Clinical application of voriconazole concentrations in the treatment of invasive aspergillosis. *Ann.Pharmacother*, 2008, *42(12)*, 1859–1864.
[90] Maertens, J; Meersseman, W; Van Bleyenbergh, P. 2009. New therapies for fungal pneumonia. *Curr. Opin. Infect. Dis.*, 2009 *22(2)*, 183-190.
[91] Mazzei, T; Novelli, A. Pharmacological properties of antifungal drugs with a focus on anidulafungin. *Drugs*, 2009 *69*, 79-90.
[92] Gubbins, PO; Heldenbrand, S. Clinically relevant drug interactions of current antifungal agents. *Mycoses, 201053(2)*, 95–113.
[93] Edwards, SG; O'Callaghan, J; Dobson, ADW. PCR-based detection and quantification of mycotoxigenic fungi. *Mycol. Res.*, 2002 106, 1005-1025.
[94] Niessen, L. PCR-based diagnosis and quantification of mycotoxin producing fungi. *Int. J. Food Microbiol.*, 2007 *119(1-2)*, 38-46.
[95] Palumbo, J; O'Keeffe, T; Abbas, H. Microbial interactions with mycotoxigenic fungi and mycotoxins. *Toxin Rev.*, 2008 *27(3-4)*, 261-285.
[96] Brown GD, Denning DW, Gow NA, Levitz SM, Netea MG, White TC. Hidden killers: human fungal infections. *Sci. Transl. Med.*, 2012 *19(4)*, 113-165.
[97] Gulat, S; Dereli, MD. Flukonazole Dirençli *Candida albicans* İzolatlarında Atım Pompa Ekspresyonlarının Araştırılması. *Mikrobiyol. Bul.*, 2014 *48*, 325.
[98] Singh, P; Kumar, A; Dubey, NK; Gupta, R. Essential oil of aeglemarmelos as a safe plant-based antimicrobial against postharvest microbial infestations and aflatoxin contamination of food commodities. *J. Food Sci.*, 2009 *74(6)*, 302-307.
[99] Zabka, M; Pavela, R; Slezakova, L. 2009. Antifungal effect of Pimentadioica essential oil against dangerous pathogenic and toxinogenic fungi. *Ind. Crop. Prod.*, 2009 *30(2)*, 250-253.
[100] Kumar, A; Shukla, R; Singh, P; Anuradha, Dubey, NK. Efficacy of extract and essential oil of Lantana indica Roxb. against food contaminating moulds and aflatoxin B-production. *Int. J. Food Sci.* Technol., 2010 *45(1)*, 179–185.
[101] Xu, J; Zhou, F; Ji, BP; Pei, RS; Xu, N. The antibacterial mechanism of carvacrol and thymol against Escherichia coli. *Lett. Appl. Microbiol.*, 2008 47(3), 174-179.
[102] Pinto, E; Vale-Silva, L; Cavaleiro, C; Salgueiro, L. Antifungal activity of the clove essential oil from Syzygium aromaticum on *Candida*, *Aspergillus* and dermatophyte species *J. Med. Microbiol.*, 2009, *58*, 1454-1462.
[103] Mukherjee, PK; Sheehan, DJ; Hitchcock, CA; Ghannoum, MA. Combination treatment of invasive fungal infections. *Clin. Microbiol. Rev.*, 2005 *18*, 163-194.
[104] Rao, A., Zhang, Y.Q., Muend, S., Rao, R. Mechanism of antifungal activity of terpenoid phenols resembles calcium stress and inhibition of the TOR pathway. *Antimicrob. Agents Chemother.*, 2010 *54(12)*, 5062-5069.
[105] Shahzad, M; Sherry, L; Rajendran, R; Edwards, CA; Combet, E; Ramage, G. Inhibitory effect of plant (poly)phenolics on growth and biofilm formation by *Candida albicans*. *Proc. Nutr. Soc.* 2014 *73*, (OCE1), E28
[106] Said, S; Neves, FM; Griffiths, AJF. Cinnamic Acid Inhibits The Growth of the Fungus Neurosporacrassa, but is Eliminated as Acetophenone. *Int. Biodeterior. Biodegrad.* 2004 *54*, 1-6.
[107] Neves, FM; Kawano, CY; Said S. Effect of Benzene Compounds from Plants on the Growth and Hyphal Morphology in *Neurosporacrassa*. *Braz. J. Microbiol.* 2005 *36(2)*, 189-194.

In: Caffeic Acid
Editor: Leanna Vaughn

ISBN: 978-1-63483-121-5

Chapter 5

MOLECULAR, PHARMACOLOGICAL AND CHEMICAL BIOLOGY OF CAFFEIC ACID AND DERIVATIVES

Cheorl-Ho Kim*

Molecular and Cellular Glycobiology Laboratory, Department of Biological Sciences, Sung Kyun Kwan University, 300 Chunchun-Dong, Jangan-Gu, Suwon City, Kyunggi-Do 440-746, South Korea

ABSTRACT

Caffeic acid (CA) is one of the most abundant hydroxycinnamic acids in plants, fruits, coffee, and propolis. CA and its derivative CAPE (CA phenethyl ester) have been well characterized on their aspects of chemistry, biochemistry, immunnopharmacology, molecular biology, and cell biology. During the past two decades, remarkable progress in the potential benefits of CA and CAPE has been documented. CA and its natural derivatives are multi-functional and non-toxic stabilizing compounds, which can thus be salubrious for food additives, cosmetic, drugs, pharmaceutics, and overall well-being. Chemically, they easily chelate cations such as Zn^{++}, Mn^{++}, and Fe^{++} and modulate metal enzyme activities such as matrix metalloproteinases. This specific property has been attributed to the anti-angiogenic and tissue remodeling capacities of the compounds. In intracellular signaling pathway, they effectively inhibit the phosphorylation of kinases such as mitogen-activated protein kinase (MAPK) and transcription factors such as nuclear factor-κB (NF-κB), thus contributing to the down-regulation of specific genes and cellular phenotype changes. The cellular regulatory functions have stimulated the researchers to study the pharmacological activities of CA and CAPE. However, there has been no definitive conclusion about how to regulate the cell functions when they initially meet the cells, more specifically how to interact with the cellular receptors or counter-parted molecules on the cell surface.

CA, as a leader compound, can potentially be applied for new functional and chemotherapeutic agents using fused chemi-biological synthesis in oxidative stress-related diseases such as inflammation, cell transformation, and degenerative pathology. It is also a candidate for cooperative adjuvant therapy in various human diseases, such as

* Corresponding author: Email: chkimbio@skku.edu.Office Tel: +82-31-290-7002. FAX:82-31-290-7015,www.glycobio.org, http://bio.skku.edu/glter/wiki/Cheorl%20Ho%20Kim.

degenerative diseases, inflammatory diseases, and transformed cancer. It is assumed that in contrast to conventional medicines, CA does not have side effects. Understanding the mechanism of the interaction of CA with biological targets at the molecular level will allow researchers to explain its positive effects on human. Chemobiological fusion science, using CA as the chemical scaffold, will construct new drugs for human diseases because the creative drug design of small functional compounds is timely and requisite for various human diseases. Medical scientists know that the advent of Aspirin as an active ingredient of acetylsalicylic acid enhanced human health and welfare since the twentieth century. As acetylsalicylic acid is also a small leader compound, CA should be classified into that category. That is the reason why a naturally occurring CA is useful for drug design by scaffold structural variation and chemobiological fusion. CA should be used as a medium and substrate for the creation and development of new chemotherapeutic agents for many human diseases, several of which are fundamentally based on different origins. One should note that the naturally occurring CA derivative, CAPE, is relatively more active against specific inhibition to signaling pathways, such as intracellular phosphorylation and transcription, than the original CA. To date, the chemical syntheses of ester and amide derivatives has progressed. The present-day studies on CA and CAPE have thus, been analyzed for a future direction in their application to health fields. At present, a comprehensive series of schololarly monographs, which describes CA and related derivatives, are needed for drug design and pharmaceutical application. The purpose of this chapter is to provide a comprehensive review, based on recent reports, on the structural basis and pharmacological activities of CA and its natural derivatives. This chapter, "**Molecular, pharmacological and chemical biology of Caffeic acid and derivatives**," is composed of two topics: [1] General and biochemical basis of CA; and [2] Biological and pharmacological functions in the health sciences.

1. General and Biochemical Basis of CA

1.1. Finding, Preparation, and Importance of CA

CA was originally discovered from plant sources, such as coffee phenolics, insect colonies, such as propolis from honey bee hives. Plant phenolics are secondary metabolites synthesized during metabolic process of plants and several compounds, such as phenolic acids, anthocyanins, and flavonoids, and are well reported with regard to plant self-defense capacities or their innate immune system from the invasion. Among them, a major class of phenolic compounds is hydroxycinnamic acids (HCs), of which a major representative is CA [(3,4-dihydroxycinnamic acid; 3-(3,4-dihydroxyphenyl)-2-propenoic acid)]. CA, as a HC derivative, has been found as one of an abundant plant-derived polyphenol compounds with two phenolic hydroxyl groups (-OH). Furthermore, CA, as a secondary metabolite, is widely distributed in plants and is a widespread phenolic acid that occurs naturally in many agricultural products such as fruits, vegetables, wine, olive oil, and coffee [1].

As a plant constituent, CA is a non-flavonoid catecholic compound present in many plants [2]. CA is an important phenolic acid present in medicinal plants, vegetables, bee propolis, and beverages such as wine, tea, coffee, and apple juice [3, 4]. Ubiquitous phenolic compounds, such as HCs (3-phenyl-2-propanoic acids), are esterified to produce conjugated forms in plants. Thus, CA is generally found as an esterified form with quinic acid and chlorogenic acid (CGA). CA frequently occurs from a metabolic enzyme reaction in the plant phenylpropanoid pathway because HC is a major class of phenolic compounds. Yet, CA is the major metabolite produced by the hydrolysis of CGA in coffee, fruits, vegetables, and grains.

CA is commonly produced by extraction from plant sources, such as coffee beans. CGA and CAPE are the most commonly occurring natural derivatives of CA. Natural and synthetic derivatives of CA have been reported in various studies, in comparison with the effect of its parent compound, CA. Several monomeric HCs, including CA, sinapic acid, ferulic acid, and *p*-coumaric acid have antioxidant activities regarding *in vitro* and *in vivo* human low density lipoprotein oxidation (LDL) [5, 6]. CA and its phenethyl ester (CAPE) have attracted increasing attention for their various pharmaceutical properties and health-enhancing effects. In general, CAPE was extracted from honeybee propolis [3] and has been chemically synthesized by the esterification of CA [7]. Nowadays, the large-scale production of drugs or drug precursors, via microbial approaches, provides a promising alternative to chemical synthesis and extraction from plant sources. Plant polyphenols are economically important because of their contributions to the taste, flavor, and color of food and drink. In plants, phenols are important for protection from herbivores and deleterious flowering, pollination, and symbioses.

CA has potentially ameliorating capacities for food and drug uses, as it has been extensively studied for therapeutic pharmacology and antioxidative protection, with a focus on the health benefits of polyphenols resources such as coffee, crucifers, and the propolis of honey bees. Dietary consumption of those polyphenols is linked with lowered risks in various human cancers, rheumatoid arthritis, vascular stroke, dementia, and atherosclerosis with promising immunomodulatory, neuroprotective, anti-bacterial, anti-inflammatory, anti-type II diabetic, and prevention or attenuation of cardiomyopathy, anti-oxidative, anti-infectious, anti-tumor, and anti-metastatic effects [8, 9, 10, 11]. Thus, it is not surprising that CA and its derivatives are attractive for their broad biological activities for future use.

1. 2. Physicochemical Properties of CA

Physicochemically, three fundamental properties are chemically known in phenol compounds. First, phenols are intrinsically reactive with various chemicals, such as hydrogen bonding is formed in all phenols. For example, the binding interactions between plant phenols and peptide links of proteins are known to induce a "tanning reaction." Second, cationic metal-chelating property is induced by *o*-dihydroxyl (cathechol) phenolics. These metal chelates may be important in biological systems. Considering that most of the cellular catalytic enzymes have metal chelates for the cationic and divalent metal ions, such as Zn^{++}, Mn^{++}, Mg^{++}, and Fe^{++} located as cofactors at the adjacent region of the catalytic site, the metal-chelating property is crucial for the cellular functions operated by these catalytic environments. The matrix metalloproteinases (MMP) have a well studied mechanism; two oxygen atoms are known to effectively chelate the tri-coordinated Zn^{++} with the two nearly identical bonds. Hence, the major forms of the enzyme inhibitors are reported to contain the hydroxamic acid group as the Zinc chelator and functional group. Third, phenols are very susceptible to oxidation with the adjacent compounds in tissues and cells. The oxidative potentials are catalyzed by specific enzyme, phenolases, as well reported studies on plants attest to the cascade course of monophenols to diphenols, diphenols to quinines, and these products to polymeric colored materials, the phlobaphenes of the polyphenol structure.

CA is widely recognized for its antioxidant activity because of its direct scavenging capacity of deleterious reactive species by hydrogen donation or electron transference [12]; its chelating capacity of pro-oxidant transition metals (cationic Zn^{++}, Cu^{++}, and Fe^{++}) [13]; its cytoprotective response by activation of the redox-sensitive transcription factors [14, 15]; and its regenerating capacity of the reduced endogenous antioxidants [16], CA is widely recognized for its antioxidant activity. The final destination of polyphenol formation, as the end products, represents the brown colored polyphenol forms in fruits and plants. These physicochemical properties give the phenolic compounds their versatile functional capacities with taste, flavor, color and defense-based storage from herbivores and pathogenic microbes in plants. Hence, to mention this plant immunity, if any plant defense mechanism is working in plant immunity, these chemical and enzymological reactions coupled together should be considered an innate immune system of the plant. Practically, humulone and anthocyanin are examples that meet the above criteria for self-defense. Antioxidative stress and wound healing properties in plants are also known to exist in more structurally diverse polyphenols, such as curcumin, epigallocatechin gallate (EGCG), and resveratrol.

1.3. Structural Basis of CA

Phenolics are classified by the number of carbon atoms in the skeleton. Plant phytochemicals with a polyphenol structure, which are different and diverse with more than about 8,000 derivatives, have been chemically studied, using the identified natural antioxidants. CA is classified as a phenolic compound, which is an aromatic compound with hydroxyl substituents and residues. The simplest and starting compound of CA, as the parent and monomeric type, is phenol; and most of phenol compounds are polyphenolic.

The vicinal hydroxyl groups, on an aromatic residue in CA, exert reactive oxygen species (ROS)-scavenging capacity because of its ionizing and electron transfer capacity [17]. CA is not related to caffeine on a structural basis. The phenolic hydroxyl group of CA, particularly the ortho-hydroxy phenol group, has been hypothesized as making a more stable oxidation intermediate, the ortho-hydroxyphenoxyl radical, which can quench free radicals [18]. For example, a major plant flavonoid, quercetin, is known to carry five hydroxyl groups.

The dimeric structure form of CA has been studied for their function-structure basis and several structurally different compounds have been identified, such as quinines, lignans, xanthones, and coumarins. In addition, the multimeric structural form of CA has been well-studied pharmacologically, such as three structurally different compounds, the lignins of the plant cell wall, black melanin pigments, and tannins of woody plants. Thus, CA is an intermediate of the biosynthesis of lignin or the related polymers in plants.

Phenolic units are also encountered among nitrogen compounds such as tyrosine. In nature, the HC or CA is found in the combined forms and hence the polyphenols are frequently originated from the hydrocinnamic family, such as CGA and CA, as the main initial forms. Cinnamic acids in plants generally exist in a mono- or multiesterized form with D-quinic acid, such as CA, ferulic acid, and *p*-coumaric acid. For example, CGA is the ester of D-(-)-quinic acid and CA, and the 5-O-caffeoylquinic acid is the most representative and abundant form. In this case, CA is associated with the cyclohexane carboxylic acids. For example, CA is linked with quinic acid to form CGA. The CGAs are further converted to free CA or ferulic acid during digestion and adsorption in the colon, although the deesterification is not yet well studied. CA, and its derivatives such as CGA and CA phenyl ester, are the major phenylpropanoid compounds in plants and act as a substrate for polyphenols oxidases or peroxidases [19].

The simple phenols are phenol (C_6H_6O), catechol ($C_6H_4(OH)_2$ or 1,2-dihydroxybenzene), and phloroglucinol (1,3,5-trihydroxybenzene). Depending on the side chain carbons, if the carbon number in side chain is gradually increased, the differentially structured phenolic compounds are newly named as the subclasses (**Figure 1**), such as salicylic acid (C6-C1), p-hydroxyphyneylacetic acid (C6-C2), and HC such as CA (C6-C3). CA (**Figure 2**) is a HC, enzymatically produced by the smooth endoplasmic reticulum of many plant species as a secondary metabolite of the shikimate pathway [20, 21]. CA exhibits the chemical framework of phenylpropanoid (C6-C3) units with a 3,4-dihydroxylated aromatic ring linked to a carboxylic acid moiety by a trans-ethylene tether. The number of carbon atoms in the basic skeleton of CA differentiates the phenolic compounds into each unique nomenclature.

In short, a coumarin compound has a benzene ring closure and lactonization of the HC. If it is a dimerized phenylpropanoid, such as HC or CA, it is then a lignin $(C6\text{-}C3)_n$, such as podophyllotoxin, which has anti-tumor potential. The HC or CA is then further used for flavonoid (C6-C3-C6) biosynthesis via a basic flavone compound in plants. According to the oxidative capacity of the pyran ring of flavonoids, some different species can be synthesized, such as anthocyanin (C6-C3-C6), malvin, and flavonol (C6-C3-C6), as well as quercetin. Additionally, functional groups, such as hydroxyl and methoxyl, make subtypes of flavonoids such as genistein, which is known to be a phytoestrogenic compound.

1.4. Chemico-Physical Properties and Biosynthesis of CA

The phenolic compounds are chemicophysically classified as lipophilic (e.g., leaf waxes and bud feathers) and thus extracted by methanol or chloroform. Also, the phenol compounds are water soluble due to the high hydroxyl g-hydroxyl groups. For biosynthesis of the phenolic compounds, two different synthetic pathways of the shikimate pathway and the polyketide pathway are used. Cinnamic acid in the shikimate pathway is initially used and oxidized to hydroxylated forms, whereas the polyketide pathway produces hydroxylated phenols in the 1, 3, 5-positions, such as phloroglucinol (**Figure 1**).

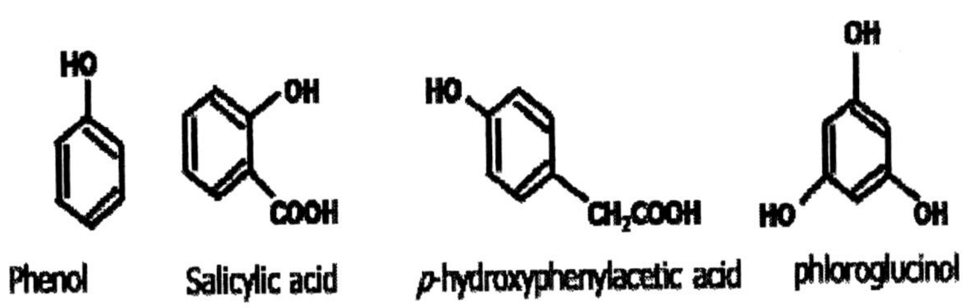

Figure 1. Molecular structures of phenol, salicylic acid, *p*-hydroxyphenylacetic acid and phloroglucinol.

The polyketide pathway produces hydroxylated phenols in the 1, 3, 5-positions, such as phloroglucinol (**Figure 1**).

Shikimic acid was originally isolated from the Japanese Shikimi tree fruits by a Dutch professor Eykmann at the University of Tokyo in 19th century and the name Shikimic acid pathway was given by Davis and Sprinson in the 1950s. The Shikimic acid pathway is divided into 2 different subpathways at chorismate. One is the prephenate pathway and the other is isochorismate pathway. Isochorismate is important for other natural products. Plants produce C6-C3 compounds, phenylpropanoids which are derived from the phenylalanine. Phenylalanine is a Shikimic acid pathway metabolite and the high production of phenylalanine in plants is closely associated with photosynthesis.

CA acts as an electrophile with a nucleophilic property and the presence of nucleophilic (catechol) and electrophilic moieties (Michael acceptor) in CA collaboratively activate the intracellular signal transduction and related phosphorylation enzymes [22]. Because of those chemical properties, recently, CA has been used as a template for the synthesis of new chemical derivatives with therapeutic potentials in human diseases associated with oxidative stress. Since CA has high chemical versatility and modifiability, as a synthetic strategy for the therapeutic application, these advantages are also applicably transferred to esters, amides, and hybrids. For example, CA derivatives and their stabilizing properties have been applied for cosmetic efficiency. CA contains hydroxyl groups in the 3- and 4-positions of cinnamic acid and also has a free radical scavenging and metal chelating properties [23, 24]. The applications for the synthesis of CA-derived derivatives with new functionalities (using a chemical scaffold) is conjectured as possibly creating a new generation effective for therapeutic agents against inflammatory, diabetic, neuropathic, vascular, cancer, and viral diseases [25]. At pH5.8, CA has a substantial negative charge because the pKa of an ionizable group is 4.36 [26]. The negatively charged CA may have the power to interact weakly with cellular membranes because of the electrostatic repulsion with the negatively charged outer surface of the phospholipid bilayer [27]. This can explain the low partitioning of CA into the lipid portion of vesicles, liposomes, and water-lipid interfaces [28, 29].

CA is one of the pivotal intermediates of the plant phenylpropanoid pathway and for the plant biosynthesis of CA. Plants use a phenylalanine residue as the starting precursor for the serial process of the deamination of the phenylalanine to cinnamic acid by phenylalanine ammonia lyase (PAL), cinnamic acid to p-coumaric acid by cinnamate-4-hydroxylase (C4H), and p-coumaric acid to CA by 4-coumarate 3-hydroxylase (C3H).

The phenylpropanoid pathway is present in all plants and cinnamic acid is converted into *p*-coumaric acid, CA, ferulic acid, feruloyl CoA and coniferyl alcohol. The core reaction of the phenylpropanoid pathways produces the lignin as the fortifying compound of the plant cell wall so that tress can sand highly. Lignin is a polymer of coniferyl and cinapyl alcohols and polymerization is a radical reaction catalyzed by peroxidase. The polymerization reaction is nonstereo- and non-regiospecific. Lignins are the dimers of phenylpropanoids formed by oxidation. Using simple carbon sources, plants metabolize them via the pentose phosphate pathway or glycolysis to synthesize the CA and its derivatives, using enzymes such as phosphoenolpyruvate synthase, transketolase, chorismate mutase-prephenate dehydrogenase, 3-deoxy-D-arabino-heptulosonate-7-phosphate synthase, hydroxyphenylacetate 3-hydroxylase, tyrosine ammonia-lyase or phenylalanine ammonia-lyase. Because plants use the TAL enzyme, the C4H is a key enzyme for CA synthesis in plants (**Figure 3**). C3H and C4H are the plant-specific cytochrome P450-dependent monooxygenases [30]. Cinnamic acid is converted into CA via *p*-coumaric acid, by a two-step sequential hydroxylation at the 4- and 3- position of the benzyl ring [31, 32, 33]. Broad substrate-specific PAL enzymes can also directly convert tyrosine residue to *p*-coumaric acid because of tyrosine's 4-hydroxyl group and this process is simple for production of CA [34, 35]. Chemical or enzymatic hydrolysis of caffeoylquinic acid derivatives, which is produced in plants, is also employed to produce CA [36, 37]. The hydroxylation of 4- and 3-ring carbons of cinnamic acid derivatives during monolignol biosynthesis is a key step that determine the structure and properties of lignin. Among them, CA is suggested as being potentially the highest for antioxidant activity because of the presence of the o-dihydroxy group in the phenolic ring. In plants, for conversion of a simple cinnamic acid derivative, CA, the conversion of p-coumaric acid (4-hydroxycinnamic acid) to CA is catalyzed by two hydroxylation enzymes, such as copper-containing polyphenol oxidases (monophenol monooxygenase (EC 1.14.18.1) and P450-dependent hydroxylases [38]. Cinnamic acid 4-hydroxylase (C4H; CYP73A) catalyzes the aromatic ring-4 hydroxylation of cinnamic acid into *p*-coumaric acid [39], the most direct precursor for monolignols. Then, *p*-coumaric acid 3-hydroxylase (C3H; CYP98A) [40] hydroxylates *p*-coumaric acid at the ring-3 position to CA. In vitro conversion of *p*-coumaric acid to CA has been reported using the red grape (*Vitis vinifera*) berries skins of Cabernet Sauvignon as the catalyzing resource [41]. The enzymes in the grape, monophenol- or flavonoid 30-monooxygenase (EC 1.14.18.1 or EC 1.14.13.21), have been proposed. In the results, *p*-coumaric acid was apparently converted to CA by o-hydroxylation catalytic activity. The CA hydroxylation reaction is: *p*-Coumaric acid + O_2+ AH_2 ⇒ CA + H_2O + A, where AH_2 indicates electron donors such as NADH or *L*-dopa.

CA has practical pharmaceutical applications. CA traditionally is obtained by direct extraction from plants, requiring low yields, high temperature, petroleum and solvent extraction, high energy-consumption, and environmental disadvantages [42]. To meet the phenylpropanoids market needs, genetically engineered microorganisms are constructed [43, 44, 45, 46]. For the microbial biosynthesis of CA, genetically modified microbial cells have been developed, such as *Escherichia, Streptomycess*, and *Actinomycetes*. The microbial conversion process provides an alternative approach to CA production. Sachan et al. (2006) reported the co-production of CA and p-hydroxybenzoic acid in *Streptomyces caeruleus* by feeding *p*-coumaric acid [47]. Recently, to overcome the limited production yield of CA in plants, biotechnological approaches have been made to produce CA in microorganisms. The conversion of tyrosine to CA and ferulic acid in *E. coli* was achieved by the co-expression of the two enzymes from *S. espanaensis* and an *O*-methyltransferase from *Arabidopsis thaliana* [48]. Recently, genes and enzymes involved in CA biosynthesis were also reported in the actinomycete *Saccharothrix espanaensis* [49]. Isolated TAL and C3H genes for CA biosynthesis, in the actinomycete *S. espanaensis* and expressed in *Streptomyces fradiae* strains and CA, have been produced in *E. coli*. Other bacterial TAL (Rhodobacter capsulatus and Rhodotorula glutinis) were also used for the production of *p*-coumaric acid and hydroxyphenylacetate 3-hydroxylase (4HPA3H) from *E. coli* or *Pseudomonas aeruginosa*, and cytochrome P450 CYP199A2 from *Rhodopseudomonas palustris* was shown to convert p-coumaric acid to CA with a high yield.

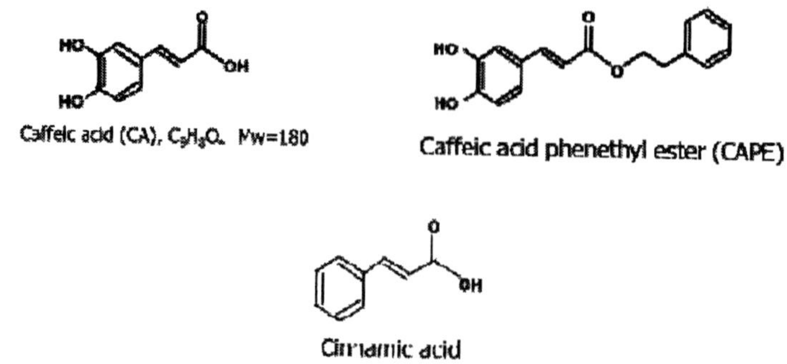

Figure 2. Molecular structures of CA, CAPE and cinnamic acid.

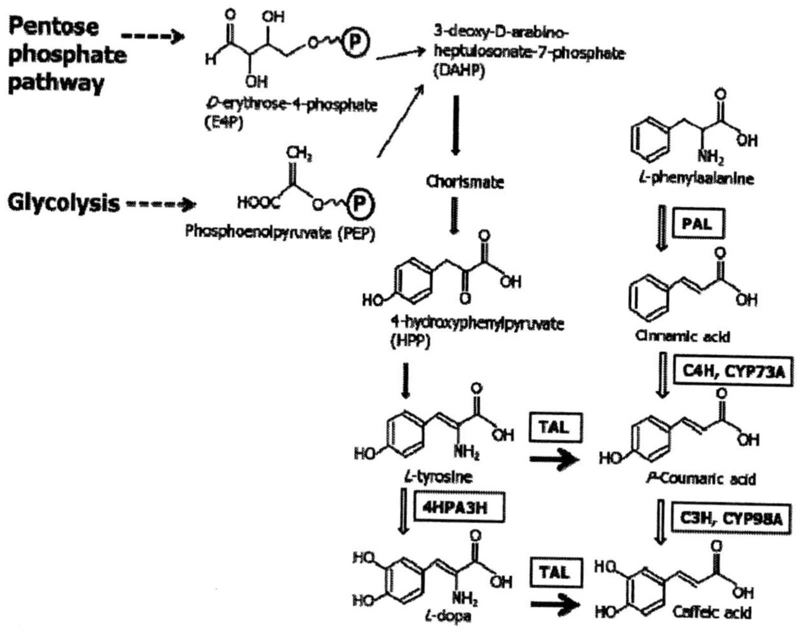

Figure 3. CA biosynthetic pathway in planst and recombinant microbes. C3H, p-coumaric acid 3-hydroxylase; C4H, p-coumaric acid by cinnamate-4-hydroxylase; PAL, phenylalanine ammonia lyase; TAL, tyrosine ammonia lyase; 4HPA3H, hydroxyphenylacetate 3-hydroxylase; 4-hydroxyphenylacetate 3-hydroxylase.

The TAL converts *L*-dopa to CA, while 3-hydroxylase converts *p*-coumaric acid into CA. Lin et al. (2012) recently first identified that an *E. coli* native hydroxylase complex, 4HPA3H, can convert *p*-coumaric acid to CA [50]. This enzymatic step is comparable to a plant system, as a membrane-associated cytochrome P450 enzyme, C3H catalyzed [51]. The two bacterial TAL enzymes possess activities towards both tyrosine and *L*-dopa. Using these enzymes, the novel conversion pathway for the production of CA has been constructed in the recombinant *E. coli* strain. Recently, Rodrigues et al. (2015a) reported the direct production of CA from tyrosine or p-coumaric acid with a higher yield than ever before. They used TAL from R. glutini for tyrosine to p-coumaric acid conversion and C3H from *S. espanaensis* or CYP199A2 from *R. palustris* for *p*-coumaric acid to CA conversion, respectively [52].

Because CA is recognized for its therapeutic properties, CA is viable for metabolic engineering and synthetic chemical biology. The engineered production of CA has already been conducted via metabolic engineering technology in *E. coli*. CA is a phenylpropanoic acid and the phenylpropanoid pathway can be further applicable to produce valuable products, like flavonoids, stilbenoids, isoflavonoids, and curcuminoids. For example, Rodrigues et al. (2015b) reported an engineered artificial pathway in *E. coli* to produce natural curcuminoids through CA, such as *Arabidopsis thaliana* 4-coumaroyl-CoA ligase and *Curcuma longa* diketide-CoA synthase and curcumin synthase. Curcuminoids as phenylpropanoids have pharmaceutical potential [53]. Curcuminoids were produced from tyrosine by the CA pathway using TAL from *Rhodotorula glutinis* and 4-coumarate 3-hydroxylase from *S. espanaensis*. Curcumin derivatives, such as methoxycurcumin and demethoxycurcumin, were also produced. Feruloyl-CoA was also produced from caffeoyl-CoA by caffeoyl-CoA 3-*O*-methyltransferase from *Medicago sativa*.

1.5. Natural Derivatives of CA

As a naturally occurring CA derivative, chicoric acid has been known as a CA diester of tartaric acid in plants such as chicory (*Cichorium intybus*, *Cichorium endivia* or *Cichorium pumilum*), Echinacea (*Echinacea purpurea*), the peanut (*Arachis hypogaea*), and dandelion (*Taraxacum officinale*) [54]. Chicoric acid has been shown to exhibit anti-inflammatory [55], antiviral [56] and antidiabetic activities [57, 58, 59]. In addition, chicoric acid has insulin-stimulating properties, in vitro and in vivo, acting as an insulin sensitizer. This pharmacological activity of chicoric acid has been suggested to have applications and therapeutic potential for diabetes and metabolic syndromes such as obesity. Another type of naturally occurring CA derivative, tricaffeoylaldaric acid, has also been isolated as a potential antidiabetic agent with a strong antioxidant activity [60, 61]. From the roots and leaves of plant yacón (*Smallanthus sonchifolius*), 2,4- or 3,5-dicaffeoylaltraric acid, 2,5-dicaffeoylaltraric acid, and 2,3,5- or 2,4,5-tricaffeoylaltraric acid were isolated as diester and trimester forms of CA with aldaric acid. Because of its strong inhibition on hyperglycemia and potential antidiabetic activity, di- and tricaffeoylaldaric acid have been considered in possible uses as a so-called nutraceutical source.

On the other hand, there already is a recognized CA derivative. That is CAPE, a naturally occurring CA derivative, which is a well-known compound. CAPE is a diphenolic compound that has the empirical formula $C_{17}H_{16}O_4$ with a molecular weight of 284.3. The chemical name of CAPE is: (E)-3-(3,4-dihydroxyphenyl)-2-propionic acid, 2-phenylethyl 3-(3,4-dihydroxyphenyl)-2-propenoate (**Figure 2**). CAPE is a white, fine crystalline powder, insoluble in water but soluble in ethanol, methanol, acetone, and dimethyl sulfoxide (DMSO). CAPE was first identified as a component of propolis in 1987 [62]. CAPE can be chemically synthesized from CA and phenethyl alcohols with a molar conversion value of 96% [63]. CAPE has been widely and safely used worldwide for many years [64]. CAPE has also multiple biological activities and many up-to-date studies on CAPE have uncovered a wide range of CAPE bioactivities, namely antibacterial [65], antiviral [66], antioxidant [67], anti-inflammatory [68, 69], immunomodulatory [70], and anticancer effects [71, 72]. In addition, CAPE has cytoprotective activities, protecting against chemotherapy-induced adverse drug reactions (ADRs) including several antibiotics (streptomycin, vancomycin, isoniazid, ethambutol) and chemotherapeutic agents (mitomycin, doxorubicin, cisplatin, methotrexate). Moreover, CAPE has drawn attention because of its demonstrated therapeutic effects, including its potential as an anti-5-lipoxygenase (5-LP), anti-diabetic, and liver-protective agent, as well as an anti-tumor drug for human breast cancer treatment [72, 73, 74].

1.6. CA-Drug and CA-Protein Hybrids as Multi-Target Drugs and Adsorption-Transporter

In the view of phenolic compound drug creation, the development of multi-target drugs (MTD) for human diseases is recently receiving increased interest in medicinal chemoenzymatic fusion science [75, 76, 77, 78]. For example, in a recent anti-influenza viral drug pharmaceutical application, therapeutic drugs have been developed, such as Tamiflu as an oseltamivir carboxylate and Relenza (zanamivir). However, because the drugs are single-targeted toward the sialidase or *N*-acetylneuraminidase inhibitors, they are limited in efficacy and thus the MTD is an evolving concept for complicated influenza treatment and prevention [79]. The concept if a MTD CA-based hybrid drug has been proposed as being chemoenzymatically synthesized as anti-influenza agent with anti-influenza and anti-inflammatory activities. The CA-based hybrid drugs have been suggested as inhibiting the expression of proinflammatory cytokines in activated innate immune cells such as neutrophils, dendrocytes, and macrophages. More specifically, doxorubicin (DOX)-cinnamic acid hybrids, which were enzymochemically synthesized, exhibited a lower cardiotoxicity than the original DOX [80]. In a case of a Tacrin-CA hybrid, it showed a higher acetylcholinesterase (AChE) inhibitory activity than the starting drug, tacrine, with increased selectivity for AChE over butyrylcholinesterase (BChE) [81, 82]. Moreover, the Tacrin-CA hybrid drug has an improved antioxidant and Alzheimer's disease (AD)-therapeutic activity. These are just some examples of natural cinnamic acid or CA scaffolds as the templates to construct new MTD candidates. Thus, considering the small molecular weight and dihydroxyl functional group of a CA scaffold, further potentially interesting new drug candidates can be created through the fusion science of medicinal chemistry and enzymology.

On the other hand, from another view of phenolic compound-protein interaction, the concept of a drug-protein hybrid is based on a mechanism in the delivery, transport, and adsorption of organic medicinal drugs in the human body. The best known example is the heme-transferring protein, albumin, in plasma serum. After CA ingestion, CA-protein hybrids are specifically formed during fast adsorption, delivery, and transport of CA in the intestinal epithelial cells to the serum of a vascular vessel. Polyphenols such as CA and ferulic acid, permeate the gastrointestinal and colon barrier, and then are absorbed in the plasma [83]. The binding of an organic chemical drug to a protein in the intestine and serum should play an important role in controlling the adsorption, transport, and functionality of the drug in vivo [84]. Actually, dietary polyphenolic compounds, including CA and rosmarinic acid, were reported to be absorbed by the intestine through an active transport system [85]. CA and p-coumaric acid are also absorbed in the gastrointestinal tract (GIT) through monocarboxylic acid transporters [86, 87] and transepithelial flux [88]. Once absorbed, the cinnamic acids, including CA, undergo extensive metabolic transformations in the GIT, liver, and kidneys through CA-protein interaction, even though it is not yet well elucidated [89]. For plasma delivery and transport of CA, the human serum albumin has been suggested as playing a key role for the CA–protein interaction because it is the most abundant protein in human plasma and is particularly important in the binding and transport of many low-molecular-weight compounds in blood [90, 91]. Human serum albumin contains two binding sites for small organic molecules from three homologous helical domains [92]. Thus, serum albumin can bind some endogenous ligands, including as non-esterified fatty acids, bilirubin, heme, thyroxine, bile acids, and any other hydrophobic compounds, hence indicating an interaction with hydrophobic chemicals [93]. For example, multiple ligands interacted with albumin, such as phenolic compounds from dietary sources including curcumin, flavonoids, flavanol, flavonol, flavone, isoflavone, flavanones, and anthocyanidins, as well as long-chain fatty acids, steroids, warfarin, tryptophan, ketoprofen, propranolol, and diazepamare [94]. The interactions of catechins with bovine albumin have previously been reported, such as [(–)-epigallocatechin-3-gallate, (–)-epigallocatechin, (–)-epicatechin-3-gallate], flavones (kaempferol, kaempferol-3-glucoside, quercetin, naringenin), and HCs (rosmarinic acid, CA, *p*-coumaric acid) [95, 96]. Resveratrol binds to human serum albumin, although the binding is irreversibly regulated by stearic acid [97]. The specific binding of CA with human serum albumin has previously reported as being Kd=6 μM [90].

On the other hand, once CA is administered, it is extensively metabolized by phase II enzymes and transformed into sulfated or glucuronidated conjugates as the predominant forms in plasma. In kidney tissue, CA was reported to inhibit the human organic anion transporters OAT-1 and OAT-3, which are expressed in the basolateral membrane of proximal epithelial cells and function in the tubular uptake of drugs from the blood [99]. The human OAT proteins are also known to interact with flavonoids and cinnamic acids [100, 101]. In the renal proximal tubule, when CA is ingested and transported, OAT proteins contribute to the CA-interacted regulation of the blood concentration of drugs by the tubular secretion of the drugs. For example, the adsorbed CA inhibits the transport of endogenous compounds, such as p-aminohippuric acid transport, with a significant inhibition (IC_{50}=5.22 μM) by OAT-1 [102]. Accordingly, administration of CA, which inhibits the renal transporter proteins with the consumed drugs, possibly triggers and allows the drug-drug interaction. Thus, the resulting CA-drug interaction may help to prevent the nephrotoxicity induced by the anion transporter proteins-mediated uptake of the consumed drugs. The inhibition of the renal transporters by CA is favorable and beneficial for renal protection, although the mechanisms of CA-drug interaction have not been elucidated at the molecular level.

1.7. CA Polymerization: New Function in Immune-Stimulation and Hemostasis

The biological and pharmacological functions of polyphenolic compounds are diverse. In general, low-molecular weight phenolic compounds, such as CA, are known to have diverse biological and pharmacological activities. However, the polymerization reaction is frequently mediated by a plant endogenous polyphenol oxidase and peroxidases. They often enzymatically synthesize the CA unit-based polymers such as lignin, especially leading to immune-stimulating activity. Monomeric phenolic compounds are easily converted to high-molecular weight polyphenol polymers by oxidative enzymes such as polyphenol oxidase and peroxidase. For example, the enzymatical polymerization of phenylpropanoic acids, such as

CA and ferulic acid, give immunomodulating activity. The enzyme catalysis uses the phenolic compounds as the donor substrates, resulting in diverse polymerization. The phenol oxidase reaction with CA as the substrate results in polymerization of oxidation products. To date, plant phenol oxidases have been used to form, such as a dimer from CA with a caffeicin-like structure, as well as trimeric and tetrameric derivatives. The phenol oxidases oxidize phenolic compounds ranging from simple phenols to multimeric and polyphenols with an ortho-diphenolic structure. It is suggested from *in vivo* reactions in plants that phenol oxidases may contribute to the antioxidant mechanism of the organisms, such as fungus and plants; and this might be of future value in food processing and biotechnology applications [103].

High molecular weight polymerized polyphenols have immunomodulating and stimulating activities as additional action mechanisms. Lignin-like polymerized polyphenols strongly activate lymphocytes and induce cytokine synthesis. High-molecular weight phenolic compounds, such as Oolong tea polymerized polyphenols, which are catechin polymers, have anti-postprandial hypertriglyceridemia [104]. Lignin, naturally occurring high-molecular polyphenols, has the following bioactivities: anti-tumor, anti-influenza virus, anti-herpes simplex virus, anti-human immunodeficiency virus, and T-cell immunomodulating [105, 106, 107]. Phenolic polymer compounds have been reported to inhibit over-activated immune systems by inhibiting mitogen-activated protein kinase (MAPK) and nuclear factor-κB (NF-κB) activities [108]. In plants, natural polyphenols are enzymatically polymerized and the resulting polyphenol polymers exhibit anti-tumor and immunoenhancing activities [109, 110, 111, 112].

The natural and synthetic lignins, polyphenol polymers polymerized from CA (pCA) but not monomers, induced the proliferation of splenocytes in a dose-dependent manner, as previously reported [113]. Polymerized polyphenols can induce cytokine production from murine splenocytes. In particular, cell proliferation and IFN-γ and granulocyte macrophage-colony stimulating factor (GM-CSF) production were strongly induced by the polymerized polyphenols from murine splenocytes in the presence of T cells [89], thus suggesting indicators of the bioactivity of polymerized polyphenols. In the case of the mitogenic activity of natural lignin and the polymerized pCA as a synthetic lignin, the T lymphocytes are the primary target cell for the DNA synthesis activity [113]. In the study, as the representative T cell surface receptors of CD4 and CD8, the polymerized pCA-induced IFN-γ and GM-CSF production were strongly suppressed by neutralizing the anti-CD4 mAb, but not by anti-CD8 mAb in murine splenocytes. Thus, the polymerized pCA was shown to directly modulate the CD4 through binding to the CD4 receptor. The polymerized pCA was shown to react with CD4; however, it did not bind to a soluble CD8a protein. In addition, an anti-CD4 mAb binding to a CD4 protein was inhibited by the polymerized pCA, while monomeric CA did not interfere with the binding of an anti-CD4 mAb to CD4. When the reactivity of leukocytes to the polymerized pCA was examined, splenocytes and resident peritoneal cavity cells, treated with polymerized polyphenols, released cytokines such as IFN-γ [114]. Thus, polymerization can confer immunoactivating properties to phenylpropanoic acids (such as pCA and the polymerized pCA) bind specifically to CD4 molecules, and exhibit immunoenhancing activity on murine splenocytes, so indicating that polymerization is required for these activities and leads to subsequent cell–cell interaction, and hence resulting in the activation of the immune system.

For a case of the carboxylic acid-based scaffold of CA, chemo-enzymatically prepared oligomers of 4-HCs were also reported to display anticoagulant properties [115]. The CA oligomers were synthesized via a chemoenzymatic oxidative coupling of the monomeric units using horseradish peroxidase as the enzyme. Oligomers, from tetramers to pentadecamers, were formed by a peroxidase-catalyzed oxidative coupling of CA and sulfated using a triethylamine-sulfur trioxide complex. The 4-HCs were composed of inter-monomer linkages similar to the natural lignins. Both the sulfated and unsulfated derivatives of the 4-HCs display anticoagulant activity with the activated thromboplastin and prothrombin time. Hydroxylated and sulfated CA trimmers, tetramers, and oligomers exhibited potent thrombin and factor Xa inhibitory and anticoagulant activity [115]. Thus, the inhibition of a factor Xa and thrombin was the first example in the inhibitors of coagulation and is an example of a non-heparin structure having the anticoagulation activity of the CA oligomers. They are applicable for the treatment of aberrant coagulation or inflammatory disorders with excessive neutrophil elastase activity and oxidative damage, as related with lung disorders such as emphysema, acute lung injury, and chronic obstructive pulmonary disease. This anticoagulant activity can be replaced for the heparin with the prolongation of clotting time, expressed as the prothrombin time and activated partial thromboplastin time. This activity is based on the direct inhibition of clotting factors by CA oligomers. Recently, Saluja et al. (2014) reported that sulfated CA dehydropolymer (CDSO3) is a novel potent triple-action inhibitor of elastolysis, oxidation, and inflammation *in vitro* as a potential anti-emphysema agent [116]. CDSO3 at dose of 30 to 100 μg/kg attenuated the development of emphysema affected by barrier damage, such as luminal hemorrhage and protein leak, elastolytic, oxidative, and inflammatory activities, luminal neutrophil infiltration, and tissue myeloperoxidase activity, as well as functional impairment of exercise endurance and airspace enlargement.

REFERENCES

[1] Shahidi, F. & Naczk, M. (1995). Food Phenolics. Sources, Chemistry, Effects, Applications. Technomic Publishing Company, Inc., Lancaster, PA.

[2] Friedman, M. (1997). Chemistry, biochemistry, and dietary role of potato polyphenols. A review. *Journal of Agricultural and Food Chemistry*, *45*, 1523–1540

[3] Grunberger D., Banerjee, R., Eisinger, K., Oltz, E. M., Efros, L., Caldwell, M., Estevez, V. & Nakanishi, K. (1988). Preferential cyt,otoxicity on tumor cells by caffeic acid phenethyl ester isolated from propolis. *Experientia*, *44*, 230–232.

[4] Naczk, M. & Shahidi, F. (2004). Extraction and analysis of phenolics in food. *J Chromatogr A*, 1054, 95–111.

[5] Meyer, A. S., Donovan, J. L., Pearson, D. A., Waterhouse, A. L. & Frankel, E. N. (1998). Fruit hydroxycinnamic acids inhibit human low-density lipoprotein oxidation in vitro. *J Agric Food Chem*, *46*, 1783–1787.

[6] Andreasen, M. F., Landbo, A. K., Christensen, L. P., Hansen, A. & Meyer, A. S. (2001). Antioxidant effects of phenolic rye (Secale cereale L.) extracts, monomeric hydroxycinnamates, and ferulic acid dehydrodimers on human low-density lipoproteins. *J Agric Food Chem*, *49*, 4090–4096.

[7] Nagaoka, T., Banskota, A. H., Tezuka, Y., Saiki, I. & Kadota, S. (2002). Selective antiproliferative activity of caffeic acid phenethyl ester analogues on highly liver-metastatic murine colon 26-L5 carcinoma cell line. *Bioorg. Med. Chem., 10*, 3351–3359.

[8] Johnson, A. A., Marchand, C. & Pommier, Y. (2004). HIV-1 integrase inhibitors: a decade of research and two drugs in clinical trial. *Curr Top Med Chem, 4*, 1059–1077.

[9] Nardini, M., D'Aquino, M., Tomassi, G., Gentili, V., Di Felice, M. & Scaccini, C. (1995). Inhibition of human low-density lipoprotein oxidation by caffeic acid and other hydroxycinnamic acid derivatives. *Free Radic Biol Med, 19*, 541–552.

[10] Okutan, H., Ozcelik, N., Yilmaz, H. R. & Uz, E. (2005). Effects of caffeic acid phenethyl ester on lipid peroxidation and antioxidant enzymes in diabetic rat heart. *Clin Biochem, 38*, 191– 196.

[11] Yamada, Y., Yasui, H. & Sakurai, H. (2006). Suppressive effect of caffeic acid and its derivatives on the generation of UVA-induced reactive oxygen species in the skin of hairless mice and pharmacokinetic analysis on organ distribution of caffeic acid in ddY mice. *Photochem Photobiol, 82*, 1668–1676.

[12] Sroka, Z. & Cisowski, W. (2003). Hydrogen peroxide scavenging, antioxidant and anti-radical activity of some phenolic acids. *Food Chem Toxicol, 41*, 753-758.

[13] Borges, F., Guimaraes, C., Lima, J. L., Pinto, I. & Reis, S. (2005). Potentiometric studies on the complexation of copper(II) by phenolic acids as discrete ligand models of humic substances. *Talanta, 66*, 70-73.

[14] Zhang, M., Zhou, J., Wang, L., Li, B., Guo, J., Guan, X., Han, Q. & Zhang, H. (2014). Caffeic acid reduces cutaneous tumor necrosis factor alpha (TNF-alpha), IL-6 and IL-1beta levels and ameliorates skin edema in acute and chronic model of cutaneous inflammation in mice. *Biol Pharm Bull, 37*, 347-354.

[15] Gopalakrishnan, A. & Tony Kong, A. N. (2008). Anticarcinogenesis by dietary phytochemicals: cytoprotection by Nrf2 in normal cells and cytotoxicity by modulation of transcription factors NFkappa B and AP-1 in abnormal cancercells. *Food Chem Toxicol, 46*, 1257-1270.

[16] Yeh, C. T., Ching, L. C. & Yen, G. C. (2009). Inducing gene expression of cardiac antioxidant enzymes by dietary phenolic acids in rats. *J Nutr Biochem, 20*, 163-171.

[17] Rice-Evans, C. A., Miller, N. J. & Paganga, G. (1996). Structure-antioxidant activity relationships of flavonoids and phenolic acids. Free Radic. *Biol. Med., 20*, 933–956.

[18] Devipriya, N., Sudheer, A. R. & Menon, V. P. (2008). Caffeic acid protects human peripheral blood lymphocytes against gamma radiation-induced cellular damage. *Journal of Biochemical and Molecular Toxicology, 22*, 175–186.

[19] Bassil, D., Makris, D. P. & Kefalas, P. (2005). Oxidation of caffeic acid in the presence of Lcysteine: isolation of 2-S-cysteinyl caffeic acid and evaluation of its antioxidant properties, *Food Res. Int., 38*, 395–402.

[20] Razzaghi, N., Garrido, J., Khazraei, H., Borges, F. & Firuzi, O. (2013) Antioxidant properties of hydroxycinnamic acids: a review of structure- activity relationships. *Curr Med Chem, 20*, 4436-4450.

[21] Benfeito, S., Oliveira, C., Soares, P., Fernandes, C., Silva, T., Teixeira, J. & Borges, F. (2013). Antioxidant therapy: still in search of the 'magic bullet'. *Mitochondrion, 13*, 427-435.

[22] Friedman, M. (1997). Chemistry, biochemistry, and dietary role of potato polyphenols. A review. *Journal of Agricultural and Food Chemistry*, *45*, 1523–1540.

[23] Huang, M. T., Smart, R. C., Wong, C. Q. & Conney, A. H. (1988). Inhibitory effect of curcumin, chlorogenic acid, caffeic acid, and ferulic acid on tumor promotion in mouse skin by 12-O-tetradecanoylphorbol-13acetate. *Cancer Research*, *48*, 5941–5946.

[24] Rice-Evans, C. A., Miller, N. J. & Paganga, G. (1997). Antioxidant properties of phenolic compounds. *Trends in Plant Science*, *2*, 152–159.

[25] Touaibia, M., Jean-Francois, J. & Doiron, J. (2011). Caffeic Acid, a versatile pharmacophore: an overview. *Mini Rev Med Chem*, *11*, 695-713.

[26] Silva, F. A., Borges, F., Guimaraes, C., Lima, J. L. F. C., Matos, C. & Reis, S. (2000). Phenolic acids and derivatives: Studies on the relationship among structure, radical scavenging activity, and physicochemical parameters. *Journal of Agricultural and Food Chemistry*, *48*, 2122–2126.

[27] Kristinova, V., Mozuraityte, R., Storro, I. & Rustad, T. (2009). Antioxidant activity of phenolic acids in lipid oxidation catalyzed by different prooxidants. *Journal of Agricultural and Food Chemistry*, *57*, 10377–10385.

[28] Medina, I., Gallardo, J. M., Gonzalez, M. J., Lois, S. & Hedges, N. (2007). Effect of molecular structure of phenolic families as hydroxycinnamic acids and catechins on their antioxidant effectiveness in minced fish muscle. *Journal of Agricultural and Food Chemistry*, *55*, 3889–3895.

[29] Nakayama, T., Ono, K. & Hashimoto, K. (1998). Affinity of antioxidative polyphenols for lipid bilayers evaluated with a liposome system. *Bioscience, Biotechnology, and Biochemistry*, *62*, 1005–1007.

[30] Kim, Y. H., Kwon, T., Yang, H. J., Kim, W., Youn, H., Lee, J. Y. & Youn, B. (2011). Gene engineering, purification, crystallization and preliminary X-ray diffraction of cytochrome P450 p-coumarate-3-hydroxylase (C3H), the Arabidopsis membrane protein. *Protein Expr Purif*, *79*, 149-155.

[31] Schoch, G., Goepfert, S., Morant, M., Hehn, A., Meyer, D., Ullmann, P. & Werck-Reichhart, D. (2001). CYP98A3 from Arabidopsis thaliana is a 3′-hydroxylase of phenolic esters, a missing link in the phenylpropanoid pathway. *J Biol Chem*, *276*, 36566–36574.

[32] Mir Derikvand, M., Sierra, J. B., Ruel, K., Pollet, B., Do, C. T., Thevenin, J., Buffard, D., Jouanin, L. & Lapierre, C. (2008). Redirection of the phenylpropanoid pathway to feruloyl malate in Arabidopsis mutants deficient for cinnamoyl-CoA reductase1, *Planta*, *227*(5), 943–956.

[33] Kojima, M. & Takeuchi, W. (1989). Detection and characterization of p-coumaric acid hydroxylase in mung bean, Vigna mungo, seedlings. *J Biochem*, *105*, 265-270.

[34] Miyahisa, I., Kaneko, M., Funa, N., Kawasaki, H., Kojima, H., Ohnishi, Y. & Horinouchi, S. (2005). Efficient production of (2S)-flavanones by *Escherichia coli* containing an artificial biosynthetic gene cluster. *Appl Microbiol Biotechnol*, *68*, 498–504.

[35] Hwang, E. I., Kaneko, M., Ohnishi, Y. & Horinouchi, S. (2003). Production of plant-specific fla-vanones by Escherichia coli containing an artificial gene cluster. *Appl EnvironMicrobiol*, *69*, 2699-706.

[36] Wang, J., Lu, D. Q., Zhao, H., Ling, X. Q., Jiang, B. & Ouyang, P. K. (2009). Application of response surface methodology optimization for the production of caffeic acid from tobacco waste. *Af J Biotechnol*, *8*, 1416-1424.

[37] Yoshimoto, M., Kurata-Azuma, R., Fujii, M., Hou, D. X., Ikeda, K., Yoshidome, T. & Osako, M. (2005). Enzymatic production of caffeic acid by koji from plant resources containing caffeoylquinic acid derivatives. *Biosci Biotechnol Biochem*, *69*, 1777-1781.

[38] Strack, D. (2001). Enzymes involved in hydroxycinnamate metabolism. In: Methods in Enzymology. Academic Press Inc, San Diego,70–81

[39] Russell, D. W. (1971). The metabolism of aromatic compounds in higher plants. *J Biol Chem* , *246*, 3870-3878.

[40] Bourgaud, F., Hehn, A., Larbat, R., Doerper, S., Gontier, E., Kellner, S. & Matern, U. (2006). Biosynthesis of coumarins in plants: a major pathway still to be unravelled for cytochrome P450 enzymes. *Phytochem Rev.*, *5*, 293-308.

[41] Arnous, A. & Meyer, A. S. (2009). Grape skins (Vitis vinifera L.) catalyze the in vitro enzymatic hydroxylation of p-coumaric acid to caffeic acid. *Biotechnol Lett.*, *31*(12), 1953-1960.

[42] Xing, Y., Peng, H. Y., Zhang, M. X., Li, X., Zeng, W. W. & Yang, X. E. (2012). Caffeic acid product from the highly copper-tolerant plant Elsholtzia splendens post-phytoremediation: its extraction, purification, an identification. *J Zhejiang Univ Sci B*, *13*, 487–493.

[43] Furuya, T., Arai, Y. & Kino, K. (2012). Biotechnological production of caffeic acid by bacterial cytochrome P450 CYP199A2. *Appl Environ Microbiol*, *78*, 6087–6094.

[44] Furuya, T. & Kino, K. (2013). Catalytic activity of the two-component flavin-dependent monooxygenase from Pseudomonas aeruginosa toward cinnamic acid derivatives. *Appl Microbiol Biotechnol*, *98*, 1145–1154.

[45] Huang, Q., Lin, Y. & Yan, Y. (2013). Caffeic acid production enhancement by engineering a phenylalanine over-producing Escherichia coli strain. *Biotechnol Bioeng*, *110*, 3188–3196.

[46] Watts, K. T., Lee, P. C. & Schmidt-Dannert, C. (2004). Exploring recombinant flavonoid biosynthesis in metabolically engineered Escherichia coli. *ChemBioChem*, *5*, 500–507.

[47] Sachan, A., Ghosh, S., Sen, S. K. & Mitra, A. (2006). Co-production of caffeic acid and phydroxybenzoic acid from p-coumaric acid by Streptomyces caeruleus MTCC 6638. *Appl Microbiol Biotechnol*, *71*, 720-727.

[48] Choi, O., Wu, C. Z., Kang, S Y., Ahn, J. S., Uhm, T. B. & Hong, Y. S. (2011). Biosynthesis of plantspecific phenylpropanoids by construction of an artificial biosynthetic pathway in *Escherichia coli*. *J Ind Microbiol Biotechno*, *38*, 1657-1665.

[49] Berner, M., Krug, D., Bihlmaier, C., Vente, A., Muller, R. & Bechthold, A. (2006). Genes and enzymes involved in caffeic acid biosynthesis in the actinomycete Saccharothrix espanaensis. *J Bacteriol*, *188*, 2666-2673.

[50] Lin, Y. & Yan, Y. (2012). Biosynthesis of caffeic acid in Escherichia coli using its endogenous hydroxylase complex. *Microb Cell Fact.*, *11*, 42.

[51] Chen, H. C., Li, Q., Shuford, C. M., Liu, J., Muddiman, D. C., Sederoff, R. R. & Chiang, V. L. (2011). Membrane protein complexes catalyze both 4- and 3-hydroxylation of cinnamic acid derivatives in monolignol biosynthesis. *Proc Natl Acad Sci U S A.*, *108*(52, 1253-1258.

[52] Rodrigues, J. L., Araújo, R. G., Prather, K. L., Kluskens, L. D. & Rodrigues, L. R. (2015). Heterologous production of caffeic acid from tyrosine in Escherichia coli. *Enzyme Microb Technol.*, *71*, 36-44.

[53] Rodrigues, J. L., Araújo, R. G., Prather, K. L., Kluskens, L. D. & Rodrigues, L. R. (2015). Production of curcuminoids from tyrosine by a metabolically engineered Escherichia coli using caffeic acid as an intermediate. *Biotechnol J.*, *10*(4), 599-609.

[54] Scarpati, M. L. & Oriente, G. (1958). Chicoric acid (dicaffeyltartic acid): its isolation from chicory (Chicorium intybus) and synthesis. *Tetrahedron*, *4*, 43-48.

[55] Park, C. M., Jin, K. S., Lee, Y. W. & Song, Y. S. (2011). Luteolin and chicoric acid synergistically inhibited inflammatory responses via inactivation of PI3K-Akt pathway and impairment of NF-kappaB translocation in LPS stimulated RAW 264.7 cells. *Eur J Pharmacol*, *660*, 454-459.

[56] Reinke, R. A., Lee, D. J., McDougall, B. R., et al. L-chicoric acid inhibits human immunodeficiency virus type 1 integration in vivo and is a noncompetitive but reversible inhibitor of HIV-1 integrase in vitro. *Virology*, 2004, *326*, 203-19

[57] Azay-Milhau, J., Ferrare, K., Leroy, J., King, P. J., Victoria, J., Mao, Y., Lei, X., Reinecke, M. G. & Robinson, W. E. Jr. (2013). Antihyperglycemic effect of a natural chicoric acid extract of chicory (Cichorium intybus L.): a comparative in vitro study with the effects of caffeic and ferulic acids. *J Ethnopharmacol*, *150*, 755-760.

[58] Muthusamy, V. S., Saravanababu, C., Ramanathan, M., Bharathi Raja, R., Sudhagar, S., Anand, S. & Lakshmi, B. S. (2010). Inhibition of protein tyrosine phosphatase 1B and regulation of insulin signalling markers by caffeoyl derivatives of chicory (Cichorium intybus) salad leaves. *Br J Nutr*, *104*, 813-823.

[59] Tousch, D., Lajoix, A. D., Hosy, E., Azay-Milhau, J., Ferrare, K., Jahannault, C., Cros, G. & Petit, P. (2008). Chicoric acid, a new compound able to enhance insulin release and glucose uptake. *Biochem Biophys Res Commun*, *377*, 131-135.

[60] Takenaka, M., Yan, X., Ono, H., Yoshida, M., Nagata, T. & Nakanishi, T. (2003). Caffeic acid derivatives in the roots of yacon (Smallanthus sonchifolius). *J Agric Food Chem*, *51*, 793-796.

[61] Genta, S. B., Cabrera, W. M., Mercado, M. I., Grau, A., Catalán, C. A. & Sánchez, S. S. (2010). Hypoglycemic activity of leaf organic extracts from Smallanthus sonchifolius: Constituents of the most active fractions. *Chem Biol Interact.*, *185*(2), 143-152.

[62] Bankova, V., Dyulgerov A., Popov S. & Marekov, N. (1987). A GC/MS study of the propolis phenolic constituents. *Z Naturforsch*, *42*, 147-151.

[63] Chen, H. C., Chen, J. H., Chang, C. & Shieh, C. J. (2011). Optimization of ultra sound accelerated synthesis of enzymatic caffeic acid phenethyl ester by response surface methodology. *Ultrason Sonochem*, *18*, 455–459.

[64] Wagh, V. D. (2013). Propolis: a wonder bees product and its pharmacological potentials. *Adv Pharmacol Sci*, 2013, 308249.

[65] Abdel-Latif, M. M., Windle, H. J., Homasany, B. S., Sabra, K. & Kelleher, D. (2005). Caffeic acid phenethyl ester modulates *Helicobacter pylori*-induced nuclear factor-kappa B and activator protein-1 expression in gastric epithelial cells. *Br J Pharmacol*, *146*, 1139-1147.

[66] Shen, H., Yamashita, A., Nakakoshi, M., Yokoe, H., Sudo, M., Kasai, H., Tanaka, T., Fujimoto, Y., Ikeda, M., Kato, N., Sakamoto, N., Shindo, H., Maekawa, S., Enomoto,

N., Tsubuki, M. & Moriishi, K. (2013). Inhibitory effects of caffeic acid phenethyl ester derivatives on replication of hepatitis C virus. *PLoS One*, *8*, e82299.

[67] Bai, H., Liu, R., Chen, H. L., et al. Enhanced antioxidant effect of caffeic acid phenethyl ester and trolox in combination against radiation inducedoxidative stress. *Chem Biol Interact*, 2014, *207*,7-15.

[68] Hassan, N. A., El-Bassossy, H. M., Mahmoud, M. F. & Fahmy, A. (2014). Caffeic acid phenethyl ester, a 5-lipoxygenase enzyme inhibitor, alleviates diabetic atherosclerotic manifestations: effect on vascular reactivity and stiffness. *Chem Biol Interact*, *213*, 28-36.

[69] Zhao, W. X., Wang, L., Yang, J. L., Li, L. Z., Xu, W. M. & Li, T. (2014). Caffeic acid phenethyl ester attenuates pro-inflammatory and fibrogenic phenotypes of LPS-stimulated hepatic stellate cells through the inhibition of NF-kappaB signaling. *Int J Mol Med*, *33*, 687-694.

[70] Nam, J. H., Shin, D. H., Zheng, H. Kang, J. S., Kim, W. K. & Kim, S. J. (2009). Inhibition of store-operated Ca2+ entry channels and K+ channels by caffeic acid phenethylester in T lymphocytes. *Eur J Pharmacol*, *612*, 153-160.

[71] Omene, C., Kalac, M., Wu, J., Marchi, E., Frenkel. K. & O'Connor, O. A. (2013). Propolis and its active component, caffeic acid phenethyl ester (CAPE), modulate breast cancer therapeutic targets via an epigenetically mediated mechanism of action. *J Cancer Sci Ther*, *5*, 334-342.

[72] Wu, J., Omene, C., Karkoszka, J., Bosland, M., Eckard, J., Klein, C. B. & Frenkel, K. (2011). Caffeic acid phenethyl ester (CAPE), derived from a honeybee product propolis, exhibits a diversity of antitumor effects in pre-clinical models of human breast cancer. *Cancer Lett*, *308*, 43-53.

[73] Morin, P. J., Ferguson, D., Leblanc, L. M., Hébert, M. J., Paré, A. F., Jean-François, J., Surette, M. E., Touaibia, M. & Cuperlovic-Culf, M. (2013). NMR metabolomics analysis of the effects of 5-lipoxygenase inhibitors on metabolism in glioblastomas, *J. Proteome Research.*, *12*(5), 2165–2176.

[74] Celik, S., Erdogan, S. & Tuzcu, M. (2009). Caffeic acid phenethyl ester (CAPE) exhibits significant potential as an antidiabetic and liver-protective agent in streptozotocin-induced diabetic rats. *Pharmacol Res*, *60*, 270-276.

[75] Csermely, P., Agoston, V. & Pongor, S. (2005). The efficiency of multi-target drugs: the network approach might help drug design. *Trends Pharmacol Sci*, *26*, 178-182.

[76] Geldenhuys, W. J. & Van der Schyf, C. J. (2013). Rationally designed multi-targeted agents against neurodegenerative diseases. *Curr Med Chem*, *20*, 1662-1672.

[77] Aprile, G., Macerelli, M. & Giuliani, F. (2013). Regorafenib for gastrointestinal malignancies : from precl9inical data to clinical results of a novel multi-target inhibitor. *BioDrugs*, *27*, 213-224.

[78] Youdim, M. B. & Oh, Y. J. (2013). Promise of neurorestoration and mitochondrial biogenesis in Parkinson's Disease with multi target drugs: an alternative to stem cell therapy. *Exp Neurobiol*, *22*, 167-172.

[79] Baz, M., Abed, Y., Papenburg, J., Bouhy, X., Hamelin, M. E. & Boivin, G. (2009). Emergence of oseltamivir-resistant pandemic H1N1 virus during prophylaxis. *N Engl J Med*, *361*, 2296-2297.

[80] Chegaev, K., Riganti, C., Rolando, B., Lazzarato, L., Gazzano, E., Guglielmo, S., Ghigo, D., Fruttero, R. & Gasco, A. (2013). Doxorubicin-antioxidant co-drugs. *Bioorg Med Chem Lett*, *23*, 5307-5310.

[81] Chao, X., He, X., Yang, Y., Zhou, X., Jin, M., Liu, S., Cheng, Z., Liu, P., Wang, Y., Yu, J., Tan, Y., Huang, Y., Qin, J., Rapposelli, S. & Pi, R. (2012). Design, synthesis and pharmacological evaluation of novel tacrine-caffeic acid hybrids as multi-targeted compounds against Alzheimer's disease. *Bioorg Med Chem Lett*, *22*, 6498-6502.

[82] Chen, X. & Decker, M. (2013). Multi-target compounds acting in the central nervous system designed from natural products. *Curr Med Chem*, *20*,1673-1685.

[83] Renouf, M., Guy, P. A., Marmet, C., Fraerinf, A. L., Longet, K. & Moulin, J. (2010). Measurement of caffeic and ferulic acid equivalents in plasma after coffee consumption: Small intestine and colon are key sites for coffee metabolism. *Molecular Nutrition & Food Research.*, *54*, 760–766.

[84] Tsuruda, P. R., Yung, J., Martin, W. J., Chang, R., Mai, N. & Smith, J. A. M. (2010). Influence of ligand binding kinetics on functional inhibition of human recombinant serotonin and norepinephrine transporters. *Journal of Pharmacological and Toxicological Methods*, *61*, 192–204.

[85] Konishi, Y., Hitomi, Y., Yoshida, M. & Yoshioka, E. (2005). Pharmacokinetic study of caffeic and rosmarinic acids in rats after oral administration. *Journal of Agricultural and Food Chemistry*, *53*, 4740-–4746.

[86] Konishi, Y. & Kobayashi, S. (2004). Transepithelial transport of chlorogenic acid, caffeic acid, and their colonic metabolites in intestinal caco-2 cell monolayers. *J Agric Food Chem*, *52*, 2518-2526.

[87] Konishi, Y., Hitomi, Y. & Yoshioka, E. (2004). Intestinal absorption of p-coumaric and gallic acids in rats after oral administration. *J Agric Food Chem*, *52*, 2527-2532.

[88] Konishi, Y., Kobayashi, S. & Shimizu, M. (2003). Transepithelial transport of p-coumaric acid and gallic acid in Caco-2 cell monolayers. *Biosci Biotechnol Biochem* *67*, 2317-2324.

[89] El-Seedi, H. R., El-Said, A. M., Khalifa, S. A., Göransson, U., Bohlin, L., Borg-Karlson, A. K. & Verpoorte, R. (2012). Biosynthesis, natural sources, dietary intake, pharmacokinetic properties, and biological activities of hydroxycinnamic acids. *J Agric Food Chem*, *60*, 10877-10895.

[90] Gabriella, F., Alessandra, M., Viviana, T., Maria, M., Mauro, F. & Paolo, A. (2012). Human serum albumin: from bench to bedside. *Molecular Aspects of Medicine*, *33*, 209–290.

[91] Yue, Y. Y., Liu, J. M., Fan, J. & Yao, X. J. (2011). Binding studies of phloridzin with human serum albumin and its effect on the conformation of protein. *Journal of Pharmaceutical and Biomedical Analysis*, *56*, 336–342.

[92] Sudlow, G., Birkett, D. J. & Wade D. N. (1975). The characterization of two specific binding sites on the human serum albumin. *Molecular Pharmacology*, *11*, 824–832.

[93] Varshney, A., Sen, P., Ahmad, E., Rehan, M., Subbaro, N. & Khan, R. H. (2010). Ligand binding strategies to human serum albumin: How can the cargo be utilized? *Chirality*, *22*, 77–87.

[94] Pal, S. & Saha, C. (2014). A review on structure-affinity relationship of dietary flavonoids with serum albumins. *Journal of Biomolecular Structure and Dynamics*, *32*, 1132–1147.

[95] Skrt, M., Benedik, E., Podlipnik, C. & Ulrih, N. P. (2012). Interactions of different polyphenols with bovine serum albumin using fluorescence quenching and molecular docking. *Food Chemistry*, *135*, 2418–2424.

[96] Sinisi, V., Forzato, C., Cefarin, N., Navarini, L. & Berti, F. (2015). Interaction of chlorogenic acids and quinides from coffee with human serum albumin. *Food Chem.*, *168*, 332-340.

[97] Pantusa, M., Sportelli, L. & Bartucci, R. (2012). Influence of stearic acids on resveratrol–HSA interaction. *European Biophysics Journal.*, *41*, 969–977.

[98] Min, J., Meng-Xia, X., Dong, Z., Yuan, L., Xiao-Yu, L. & Xing, C. (2004). Spectroscopic studies on the interaction of cinnamic acid and its hydroxyl derivatives with human serum albumin. *Journal of Molecular Structure.*, *692*, 71–80.

[99] Uwai, Y., Ozeki, Y., Isaka, T., Honjo, H. & Iwamoto, K. (2011). Inhibitory effect of caffeic acid on human organic anion transporters hOAT1 and hOAT3: a novel candidate for food-drug interaction. *Drug Metab Pharmacokinet.*, *26*(5), 486-493.

[100] Whitley, A. C., Sweet, D. H. & Walle, T. (2005). The dietary polyphenol ellagic acid is a potent inhibitor of hOAT1. *Drug Metab Dispos.*, *33*(8), 1097-1100.

[101] Hong, S. S., Seo, K., Lim, S. C. & Han, H. K. (2007). Interaction characteristics of flavonoids with human organic anion transporter 1 (hOAT1) and 3 (hOAT3). *Pharmacol Res.*, *56*(6), 468-473.

[102] Wong, C. C., Barron, D., Orfila, C., Dionisi, F., Krajcsi, P. & Williamson, G. (2011). Interaction of hydroxycinnamic acids and their conjugates with organic anion transporters and ATP-binding cassette transporters. *Mol. Nutr. Food Res.*, *55*, 979–988.

[103] Koclar Avci, G., Coruh, N., Bolukbasi, U. & Ogel, Z. B. (2013). Oxidation of phenolic compounds by the bifunctional catalase-phenol oxidase (CATPO) from Scytalidium thermophilum. *Appl Microbiol Biotechnol.*, *97*(2), 661-672.

[104] Toyoda-Ono, Y., Yoshimura, M., Nakai, M., Fukui, Y., Asami, S., Shibata, H., Kiso, Y. & Ikeda, I. (2007). Suppression of postprandial hypertriglyceridemia in rats and mice by oolong tea polymerised polyphenols. *Biotechnol Biochem Biosci*, *71*, 971–976.

[105] Nagata, K., Sakagami, H., Harada, H., Nonoyama, M., Ishihama, A. & Konno, K. (1990). Inhibition of influenza virus infection by pine cone antitumor substances. *Antiviral Res*, *13*, 11–21.

[106] Lai, P. K., Donovan, J., Takayama, H., Sakagami, H., Tanaka, A., Konno, K. & Nonoyama, M. (1990). Modification of human immunodeficiency viral replication by pine cone extracts. *AIDS Res Hum Retroviruses*, *6*, 205–217.

[107] Fukuchi, K., Sakagami, H., Ikeda, M., Kawazoe Y., Oh-Hara, T., Konno, K., Ichikawa, S., Hata, N., Kondo, H. & Nonoyama, M. (1989). Inhibition of herpes simplex virus infection by pine cone antitumor substances. *Anticancer Res*, *9*, 313–317.

[108] Rasheed, Z., Akhtar, N., Anbazhagan, A. N., Ramamurthy, S., Shukla, M. & Haqqi, T. M. (2009). Polyphenol-rich pomegranate fruit extract (POMx) suppresses PMACI-induced expression of pro-inflammatory cytokines by inhibiting the activation of MAP Kinases and NF-kappaB in human KU812 cells. *J Inflamm (Lond)*, *6*, 1.

[109] Yamanaka, D., Tamiya, Y., Motoi, M., Ishibashi, K.I., Miura, N. N., Adachi, Y. & Ohno, N. (2012). The Effect of Enzymatically Polymerised Polyphenols on CD4 Binding and Cytokine Production in Murine Splenocytes. *PLoS ONE*, *7*(4), e36025.

[110] Ohno, N., Furukawa, M., Miura, N. N., Adachi, Y., Motoi, M. & Yadomae, T. (2001). Antitumor beta glucan from the cultured fruit body of Agaricus blazei. *Biol Pharm Bull*, *24*, 820–828.
[111] Liu, Y., Fukuwatari, Y., Okumura, K., Takeda, K., Ishibashi, K. I., Furukawa, M., Ohno, N., Mori, K., Gao, M. & Motoi, M. (2008). Immunomodulating activity of Agaricus brasiliensis KA21 in mice and in human volunteers. *Evid Based Complement Alternat Med*, *5*, 205–219.
[112] Yamanaka, D., Motoi, M., Ishibashi, K., Miura, N. N., Adachi, Y. & Ohno, N. (2012). Effect of Agaricus brasiliensis-derived cold water extract on Toll-like receptor 2-dependent cytokine production in vitro. *Immunopharmacol Immunotoxicol.*, *34*(4), 561-570.
[113] Kurakata, Y., Sakagami, H., Oh-Hara, T., Kawazoe, Y., Asano, K., Fujinaga, S., Takeda, M. & Sato, T. (1990). Mitogenic activity of natural and synthetic lignins against cultured splenocytes. *In Vivo*, *4*, 377–38
[114] Yamanaka, D., Motoi, M., Ishibashi, K., Miura, N. N., Adachi, Y. & Ohno, N. (2013). Modulation of interferon-γ synthesis by the effects of lignin-like enzymatically polymerized polyphenols on antigen-presenting cell activation and the subsequent cell-to-cell interactions. *Food Chem.*, *141*(4), 4073-4080.
[115] Monien, B. H., Henry, B. L., Raghuraman, A., Hindle, M. & Desai, U. R. (2006). Novel chemo-enzymatic oligomers of cinnamic acids as direct and indirect inhibitors of coagulation proteinases. *Bioorg Med Chem.*, (*23*), 7988-7998.
[116] Saluja, B., Li, H., Desai, U. R., Voelkel, N. F. & Sakagami, M. (214). Sulfated caffeic acid dehydropolymer attenuates elastase and cigarette smoke extract-induced emphysema in rats: sustained activity and a need of pulmonary delivery. *Lung*, *192*(4), 481-492.

2. BIOLOGICAL AND PHARMACOLOGICAL FUNCTIONS IN HEALTHY SCIENCE

CA (3,4-dihydroxycinnamic acid) is a natural phenolic compound initially found in many plants such as the carrot, tomato, strawberry, and blueberry [1]. In view of the contents of CA, the best sources are the coffee bean and propolis. First, with regard to CA-rich coffee, coffee is one of the most popular beverages all over the world. According to human studies, the daily coffee consumption is related to a lowered risk of chronic liver diseases, including chronic hepatitis C virus, hepatic fibrosis, and hepatocellular carcinoma (HCC) [2, 3, 4]. Although caffeine is a well-known component of coffee, it constitutes only 1–2.5% of dry coffee beans, while polyphenols such as CGA and are about 10% [5]. Coffee consumption results in an increased CA content in human total plasma [6, 7]. A 200 ml cup of coffee is estimated to provide about 35 to 175 mg of CA. In the daily consumption of coffee, CA intake alone was estimated to be about 206 mg/day adult in western countries, and the major source was coffee, which supplied 92% of CA [8]. It means that a dose of 0.08% CA in human daily food consumption (500 g/dry weight) is about 400 mg. Studies have noted a high variability in the polyphenol intake of adults in a Bavarian subgroup ranging from 6 to 987 mg/day in Germany [9]. The CA concentration in human total plasma at approximately 1 h after coffee consumption is approximately 80–116 nM, which gradually decreases (approximately 50 nM at 8 h) and reduces to almost zero at 12 h [10]. Hence, it is conjectured that the intake of CA from coffee may be helpful in the increased CA concentration in human plasma and coffee intake may regulate many human diseases including pathogenic, inflammatory, and transformed diseases. CA exists as a chemically bound form of CGA, which is an ester of CA and quinic acids and is one of the major polyphenols in coffee. CGA is heat-unstable and, following its absorption into the body, it metabolically decomposes to CA and quinic acids. Thus, coffee consumption increases CA concentration in human total plasma, whereas CGA is undetectable [6]. Daily coffee consumption having a high polyphenol content of coffee is reported to lower the risk for various cancers [11, 12, 13, 14]. The risk for type II diabetes is also decreased by daily coffee consumption [5] with antibacterial [15, 16], cancer preventive [17, 18], and antioxidants in the GI tract [19].

Second, traditionally a CA-rich propolis ethnomedicine has long been used in many cultures for time immemorial throughout the world. CA is mainly enriched in the propolis and propolis has been used to treat inflammatory diseases since at least the Neolithic Era. Propolis is a naturopathic substance produced by honey bees (*Apis mellifera* L.) that is rich in polyphenolic compounds. Propolis, a honey bee product, has been used by folk medicine in various capacities: neuroprotective, hepatoprotective, cardioprotective, anti-inflammatory (treatment of abscesses caused by bacterial infection, parasites, or foreign substances), antibacterial, antiviral, antioxidant, and anticancer. For the collection and selection of the active components in propolis, the bees accumulate natural protective substances into a beehive for nectar and pollen, forming a resin complex extracted from tree buds, leaves, branches, and bark with their salivary secretions and wax. Propolis has been utilized as a type of apitherapy from ancient times, for example, the ancient Egyptian priests used propolis obtained from honey bees to embalm mummies. The ancient Greeks used propolis for suppuration and the ancient Romans used propolis to treat the injuries and bacterial infection during their many imperial wars. The etymology of propolis is, of course, from the Greek, with pro meaning "before" and polis, "city," that is, propolis was placed by the bees at the entrance of the hive to prevent the intrusions of predators. Later, the Persians, the Romans, and the Incas used it for their ethnomedicine. The renowned ancient Greek physician Galen (c. 130–230 C.E.) mentioned propolis in his extensive oeuvre. The Persian polymath and philosopher-scientist Ibn Sina (980-1037 C.E.), known as Avicenna in the west, summarized its properties, "Propolis has the quality to remove arrowheads and thorns, … easy to use, and highly efficient" in the eleventh-century. Interestingly, Avicenna was the first known person to systematically describe coffee. Inca empire used it against febrile infections and fever. The Roman legions regularly used propolis to heal wounds. In France during the eighteenth and nineteenth centuries, propolis was used in the treatment of wounds. During the Boer War (1899–1902) in South Africa, the use of propolis reached its peak, for anti-infection and healing of soldiers. In the twenty-first century, in Germany, Japan, and Russia propolis is a significant and popular remedy [20]. In recent years, many studies have focused on the chemical composition and biological effects of the substances in propolis. This composition varies according to its origin, the bee species, and time of harvest. The main antioxidant groups of propolis were CA, CAPE, quercetin, kaempferol, galangal, and cinnamic acid ester.

CA is a phenylpropanoic acid that is derived from the phenylpropanoid pathway in plants. CA is the major dietary HC and abundant in natural food materials [1]. The CA scaffold is abundant in nature and extremely versatile and exists in a number of biologically active compounds. Hence, CA is regarded as a potent candidate for the treatment of many diseases with concomitant human health benefits [21], thus there is an increased interest in CA, owing to its important properties because of a wide variety of pharmacological activities, including hypothalamic protection [22], anti-fibrillation [23], anti-coagulatory [24], anti-inflammatory [25], prooxidant [26], antioxidant [27, 28, 29, 30, 31, 32], neuroprotective [33], analgesic [34], immunomodulatory [35, 36], virucidal [37], anti-viral [37], anti-herpes simplex virus (HSV) [38], anti-human immunodeficiency virus (HIV) [39, 40, 41, 42], anti-depressive [43], anti-cancer [44, 45], anti-arthritis [46], anticarcinogenic [47], anti-glycative [48], free radical scavenger [30], anti-diabetic agent [49], antithrombosis [50], LDL-protective [51] antihypertensive [52], antifibrosis [53], free-radical scavenging [54], metal chelating [55], enzymatic activity modulating [56], signal transduction pathway regulation [57], and anti-inflammatory effects [27, 30, 34, 58, 59]. CA has recently been shown to act as a carcinogenic inhibitor [60] and antimelanogenic agent [61].

In short sum, this section described and summarized several significant activities of CA and its derivatives including anti-oxidative, anti-inflammatory, anti-cancer, anti-viral, anti-diabetic, anti-cardiovascular, neuroprotective, and skin-aging protection (**Figure 4**).

2.1. Anti-Oxidant Activity

A major site of attack by free radicals in an organism is the cellular membrane because cell membranes are the primary site of lipid oxidation because of the abundance of phospholipids containing polyunsaturated fatty acids. The polyunsaturated fatty acid side chains of membrane phospholipids are susceptible to modification by reactive oxygen intermediates [62]. Oxidation of unsaturated lipids in cellular membranes causes structural changes for fluidity, membrane damage, and cell death. The exposure of biological membranes to reactive oxygen species (ROS) creates an oxidized phospholipid species, which contribute to the development of chronic inflammatory diseases and metabolic disorders. The phospholipase (PL) A2-arachidonic acid/cyclooxygenase (COX)-2 axis-specific oxidation produces a series of endogenous inflammatory eicosanoid mediators as second messengers, while an oxidative radicals-mediated oxidation reaction is at radom.

In mitochondrial ATP synthetic oxidoreduction, 2 mols of water (H_2O) are produced with 1 mol of oxygen (O_2), formed using 4 electrons and 4 protons. However, oxidative radicals are produced in unusual cell states such as overloaded oxidative phosphorylation. For example, if 1 molecule of O_2 accepts only 1 electron, then a free radial is additively generated, like ROS or an active oxygen superoxide anion (O_2^-). Because the remaining 3 electrons and 4 protons automatically react with the resulting anion (O_2^-) to produce a water (H_2O) molecule, additional byproducts are automatically generated, such as hydrogen peroxide (H_2O_2), hydroxyl radical (OH^-), and peroxynitrite ($ONOO^-$). Thus, these byproducts are highly reactive and reproduce secondary radicals, such as hydroxyl radicals (OH^-), with free metal ions (iron or cupper) by a Fenton reaction with perhydrol by a cascade reaction. However, normal cells protect themselves from those radicals by using their intrinsic antioxidative system. In abnormal cell states, secondary radicals, such as ROS, stimulate the peroxidation of membrane arachidonic acid, inducing cell senescence, ADRs, and various human diseases, such as preeclampsia, cancer, myocardial ischemia, and autoimmune diseases, immune disorders, vascular cardiovascular disease, ischemic injury, diabetes mellitus, neurodegenerative disorders, AD, Parkinson's disease, rheumatoid arthritis, and other degenerative diseases [63]. Thus, antioxidants are important for normal homeostasis.

The antioxidant properties of dietary polyphenols from plant sources and propolis are well-known. Some polyphenols exhibit strong radical-trapping capability as potent antioxidants, which can prevent or ameliorate diseases initiated by oxidative stress [64]. An antioxidation event is a chemical reaction, and efficient antioxidants are more easily oxidized than other biomolecules via radical intermediates during the antioxidation process [3]. The mechanism of the interaction of phenolic compounds with biological membranes, including membrane lipids, has not yet been fully explained. It is assumed that the phenolic compounds may bind both electrostatically with the polar groups of membrane phospholipids and hydrophobically with their alkyl chains. Hence, the extent of antioxidative levels might be basically dependent on the chemical structure of the phenolic compounds. From a structural basis, it is considered that the phenolic compounds with multiple hydroxyl groups can interact with the hydrophilic part of the plasma membrane, and the hydrophobic portions of the phenolic compounds potentiate to permeate deeper into the membrane lipid bilayer, so significantly changing its mosaic fluidity.

Thus, understanding the mechanism of the interaction of CA with the biological membrane and oxidative radicals at the molecular level will result in further applications of CA. The discussion in this chapter is concerned with both the antioxidant activity of CA and its effect on the properties of lipid and biological membranes. This chapter discusses the antioxidant activity of CA and it derivatives.

2.1.1. CA Has Multiple Antioxidative Activities

CA as a phenolic compound has been reported to have a potent antioxidative activity by free radical scavenging and metal chelating properties. More fundamentally, CA has: 1) non-enzymatic antioxidant activity such as scavenging free radicals; and 2) enzymatic antioxidant activity such as increasing the protein level of antioxidant enzymes. Oral administration of CA resulted in the presence of its intact form in the liver without transport through GIT [65]. The inhibitory effects of CA on iron-induced and gamma-radiation induced oxidative DNA damage have also been reported [66]. In a 2,2-diphenyl-1-picrylhydrazyl (DPPH) free radical scavenging assay, CA scavenged DPPH free radicals as efficiently as a known antioxidant, butylated hydroxyanisole [67, 68, 69, 70]. Besides their well-known antioxidant activity, CA is a selective inhibitor for 5-LP and leukotriene biosynthesis with ID_{50} of 3.7 X 10(-6) M and CA inhibits 5-LP in a non-competitive manner, whereas CA did not inhibit prostaglandin synthase activity [71]. However, CA inhibited oxidative enzyme activities such as COX, glutathione S-transferase (GST), and xanthine oxidase (XO) [72, 73, 74, 75, 76, 77]. Because CA has vicinal hydroxyl groups on an aromatic residue, which scavenge ROS, it exhibits anti-mutagenic, anti-carcinogenic, and anti-oxidant activities *in vitro* [1]. The beneficial effects of CA were observed, as the *in vitro* uptake of CA is reported to be as fast occurring, by Caco-2 human intestinal cell models [78]. CA exhibited radical oxidative reactivity to generate the final oxidation products in biological systems of living cells. The Fe^{++} ion-catalyzed radical oxidation products of CA, but not CA, showed an XO-inhibitory activity [74]. The XO regulation can prevent hyperuricemia and gout because XO is a well-known physiological source of superoxide anions in eukaryotic cells and a key enzyme in the catabolism of purine-based nucleic acids, and produces uric acid from xanthine and hypoxanthine in body [79]. In animal tissue, the XO is highly expressed in intestinal mucosa and thus the hypoxanthine-XO system is believed to be a causative factor for the free radical production after hypoxia and reperfusion of the gut. Furthermore, if the XO is inhibited, the tissue can be protected from oxidation damage. Interestingly, it was reported that the oxidation product of CA inhibited XO activity, whereas CA itself did not [76]. The pharmacological activity of the orally administered CA was evident after intestinal absorption and interaction with the target tissues. In an *in vivo* intestinal ischemia–reperfusion model, by which ROS are directly linked with the intestinal injury, during reperfusion, oxygen inflowing with normal nutritive blood induced XO and produced superoxide anions. Interestingly, CA protected the intestinal ischemia–reperfusion injury by its strong antioxidant activity [80].

On the other hand, it was recently reported that CA directly bind hemoglobin (Hb), causing the CA-bound Hb formation by electron donation from CA to liganded O_2 in Hb, and CA effectively inhibited hemin dissociation from Hb [81]. In general, exogenously added CA has difficulty incorporating into the phospholipids of plasma membranes; rather it inhibits iron-promoted oxidation, by which the vascular Hb-origined heme releases the iron ions. The CA inhibited Hb-mediated lipid oxidation beyond free radical scavenging and metal chelating capacities. The fact that the CA-bound Hb hybrid inhibited Hb-mediated lipid oxidation is notable, because the lipid oxidation by Hb is a health issue. Hb-associated reactions can cause degenerative diseases, including Hb autooxidation, hemin dissociation, ferryl Hb radical formation, and iron release upon destruction of the porphyrin ring [82]. Hb and myoglobin (Mb) facilitates lipid oxidation while Hb is more effective at enhancing lipid oxidation compared to Mb, and thus Hb is regarded as a catalyst of lipid oxidation. The Hb-mediated lipid oxidation involves formation of peroxyl and alkoxyl radicals via decomposition of lipid hydroperoxides by hemin [83]. Alternatively, iron dissociated from hemin can catalyze the generation of hydroxyl radicals. Similarly, CA has also its therapeutic effects on Mb-catalyzed LDL oxidation and hepatotoxicity [84]. For novel mechanisms associated with CA inhibiting Hb-mediated lipid oxidation, CA inhibited hemin loss from Hb by directly binding CA to Hb, not to albumin. CA-Hb binding increased hemin-globin linkage by conformational change in Hb. Previously, quercetin and rutin were reported to bind and cause conformational changes in Hb [82]. Hence, a mechanism by which CA and other phenols inhibit hemin-Hb dissociation is hypothesized. Another mechanism of the inhibition of lipid oxidation by CA was the α-tocopherol regeneration and free radical scavenging [85].

2.1.2. CAPE, a CA Derivative, Has Multiple Antioxidative Activities

As a potential antioxidative agent, CAPE, which contains catechol, is also known to exhibit a strong antioxidant and cytoprotective activity *in vitro* and *in vivo* [86, 87]. CAPE is known as a beneficial dietary supplement to improve human health and ameliorate diseases that originated from ROS. Many studies on CAPE have examined the antioxidant and cytoprotective activities. Recently, the pleiotropic action mechanisms of CAPE for therapeutic advantages has been described, suggesting a translational application of CA for human chronic diseases. CAPE is a hydroxyl derivative of cinnamic acid. The presence of CH_2=CH-COOH group in cinnamic acids supports a superior antioxidant potential compared to other phenolic acids. The steric hindrance of the phenolic hydroxyls by a neighboring inert group, such as methyl, was known to enhance and strengthen its antioxidant activity [88]. Diphenolics act as antioxidants via inhibiting both the propagation and formation reactions of free radicals and also, they have the capacity to chelate the transition metal or to inhibit the enzymes implicated in the initiation reactions [89].

Recently, it was reported that CAPE inhibits the production of superoxide anion during β-mercaptoethanol autoxidation and quench a DPPH radical. CAPE also inhibited XO and COX-2 activities likely CA [90]. Moreover, CAPE exhibits potent antilipoperoxidative cytoprotective and antigenotoxic potentials against oxidative damage [91]. CAPE activated the expression of the antioxidant gene heme oxygenase-1 (HO-1) and the intercellular adhesion molecule 1 (ICAM-1) gene. CAPE induced alleviation of malonedialdehyde (MDA) levels with concurrent elevation in superoxide dismutase (SOD), catalase (CAT), glutathione peroxidase (GPx) activities, and reduced glutathione (GSH) levels from old rats as the experimental animal. CAPE protects against lipid peroxidation in the streptozotocin (STZ)-induced diabetic rat model, where STZ is an inducer of diabetes and producer of oxidative stress in multiple target tissues. CAPE reversed the oxidative stress in cardiac tissues of STZ-induced diabetic rats [92]. CAPE protected against radiation-induced cardiac tissue damage in rats by reduction of the MDA, adenosine deaminase (ADA), XO, and SOD activities [93]. More specifically, CAPE is a selective extracellular signal-regulated kinase (ERK) inhibitor, forming the CAPE-ERK2 complex in cellular signaling [94] and CAPE decreased expression of lymphocyte function-associated antigen-1 (LFA-1) and ICAM-1, expressed by the endothelial cells in a vascular vessel.

2.2. Anti-Inflammatory Activity

Inflammation is the central self-defense system in response to tissue injury in host against microbial invasion or allergens [95]. In an initial inflammatory reaction, neutrophils and monocytes migrate into the injured tissue by an increased interaction with the ICAM-1 and substantial E/P-selectins on endothelial cells. The macrophages exhibit an immediate defense against foreign agents, depending on the inflammatory stimulus. During inflammatory reaction in tissues, TNF-α and IL-1β expressed from the innate immune cells, such as neutrophils and macrophages, mediate immune responses by activating the NF-κB signaling pathway for chronic inflammation and inflammatory diseases. Activated macrophages and mast cells further express pro-inflammatory mediators, such as prostaglanoids like arachidonic acid metabolites, and these are harmful because they induce ROS. The expression of pro-inflammatory molecules is regulated by several signaling pathways and their downstream transcription factors. The inflammatory cells lead to various inflammatory diseases such as arthritis, cancer, and atherosclerosis. Thus, to prevent and treat inflammatory diseases, it has been suggested that a search be conducted for non-toxic and nutraceutical anti-inflammatory agents from natural resources in the current drug pharmacology.

CA is considered an antioxidant, detoxicant, or anti-inflammatory agent, as CA has a preventive and therapeutic capacity in *in vivo* inflammatory injury and *in vitro* inflammation of cells. Traditionally, CA has been used in folk medicine and traditional medicine as a healthy drink and drug to improve health and prevent inflammatory diseases. The antioxidant activity and anti-inflammatory properties of CA are linked together, as suggested from the known reports. Recently, it has been scientifically known that CA possesses numerous biological activities through fundamental modulation of inflammatory pathways, such as anti-oxidative, anti-cancer, anti-invasion, anti-angiogenesis, and anti-metastasis of tumor cells [44,

96]. For nutraceutical aspects, although a high concentration of CA exerted a cytotoxic action on macrophage Raw 264.7 cells; however, low doses of CA showed anti-inflammatory activity. CA also exhibited anti-inflammatory and anti-coagulatory activities in cardiac tissue of diabetic mice [24], and inhibitory capacity on the generation of ultraviolet A (UVA)-induced ROS in the mouse skin [65]. The mechanisms underlying the anti-inflammatory effects of CA in acute and chronic skin inflammation in mice have been attributed to the inhibitory effects of CA on the TNF-α production, the TNF-α-induced IL-6 and IL-1β production, and TNF-α-induced NF-κB activation in human keratinocytes [97]. The study of Chao et al. (2009) revealed that CA has anti-inflammatory protection by down-regulated TNF-α, IL-6, and IL-1β mRNA, and protein expression in cardiac tissue of diabetic mice [24]. Furthermore, the study of Song et al. (2008) indicated that CA had anti-inflammatory effects by reducing myeloperoxidase (MPO) activity and PLA2 activity in skin-incised mice [98].

The *in vitro* anti-inflammatory activity of CA had been reported previously in monocytic THP-1 cells by measuring the expression levels of cytokines, including IL-1*β*, and adhesion molecule ICAM-1, and TNF- *α* [99]. In addition, CA exhibited the high-density lipoprotein (HDL)-mediated cholesterol efflux in macrophages, reducing the cholesterol in vascular vessel, the inflammatory reaction in vascular endothelial cells, and eventually arteriosclerosis [100]. CA also diminished TNF-α-induced vascular inflammation in human umbilical vein endothelial cells [101].

2.2.1. Suppression of Inflammatory Responses via MAPK-JNK1/2-NF-κB Pathway

Inflammation is a complexed reaction through the activation of many intracellular signaling pathways, such as MAPKs or ERK-mediated intracellular cascade and transcription factors, notably NF-κB [102]. The MAPKs are a group of serine and threonine kinases. They regulate gene expression by modulating transcription factors such as NF-κB. They have also been implicated in inflammatory and cell death pathways. MAPKs function in the cellular response during growth, differentiation, cell survival/apoptosis, cellular response to cytokines, stress and inflammation, hence, the MAPK targets NF-κB. Three different MAPKs are mainly studied: p38 MAPK, ERK, and c-Jun NH2-terminal kinase (JNK) [103]. The inflammatory signaling pathways and down stream phosphorylation is forwarded to the transcription factor NF-κB [104]. The NF-κB transcription factors have been well characterized in inflammation responses. The functional NF-κB protein is a hetero-dimer composed mainly of p65 and p50 subunits. NF-κB has beneficial properties due to p50 homodimers and it activation is known to prevent cholestatic and chronic liver injury. Under normal conditions, NF-κB is localized on the cytoplasm in an inactive state and bound to inhibitory protein κB (IκB). IκB phosphorylation is mediated by the inhibitory protein kappa B kinase (IKK), a complex that contains two subunits, IKKα (IKK1) and IKKβ (IKK2), with a NF-κB essential modulator, IKKγ [105]. The degradation of IκBα releases NF-κB and translocation to the nucleus for transcriptional expression of cytokines, chemokines, anti-apoptotic, and cell growth factors. The NF-κB signal pathway, which is activated by ROS-dependent mechanisms, also plays an important role in the gene expression of a large number of proinflammatory cytokines [106, 107]. Because the immune and inflammatory responses are directly regulated by NF-κB, NF-κB-inhibiting agents are candidates for being anti-inflammatory agents. Thus, NF-κB regulation is a target for the prevention and treatment of inflammatory diseases. In this way, anti-inflammatory drugs or nutraceuticals might involve strategies aimed at interrupting the oxidant and inflammatory cascades. Specifically, there is an interest in the discovery of natural products inhibiting pro-inflammatory mediators, thereby serving as beneficial therapeutic agents. Natural phenolic products, such as EGCG and HC derivatives from tea leaves and plants on macrophages/monocytes-triggered inflammatory processes, can contribute to preventive or therapeutic drugs for inflammatory diseases [108, 109]. HC and its derivatives exert anti-inflammation, anti-cancer, and hypocholesterolemic activity through the suppression of NF-κB [101, 110, 111]. More specifically, the antioxidant and anti-inflammatory properties of CA with MAPKs/ERK/NF-κB signaling pathway regulation are potentially valuable for human health [112].

In the anti-inflammatory activity of CA in Raw 264.7 macrophages, CA was reported to inhibit production of nitric oxide (NO), a strong pro-inflammatory mediator and lipopolysaccharide (LPS)-induced intracellular signaling pathways triggered during inflammation without cytotoxicity [75]. The signaling inhibition of CA suppressed the targets: p38 MAPK, JNK1/2, NF-κB, and ERK1/2. With a strong anti-inflammatory potential [101], a transcription of TNF- α, COX-2, and inducible NO synthase (iNOS) was suppressed by CA [111]. CA also attenuates a Toll-like receptor (TLR)-4-mediated inflammatory reactions, including NO and prostaglandin E2 (PGE2) production as proinflammatory mediators in LPS-treated RAW 264.7 cells and ethanol-induced gastritis [112]. CA inhibited LPS-induced NO and PGE2 production by Raw 264.7 macrophages, acting at the transcriptional level. It has been suggested that the anti-inflammatory effect of CA, concerning the NF-κB and inflammatory mediators' expression, is mediation by down-regulating the NF-κB and p38 MAPK-JNK1/2. However, CA treatment decreases the LPS-mediated phosphorylation of p38 MAPK and JNK1/2. In a likewise fashion, CA derivative, CAPE exhibits the anti-inflammatory effect as a specific inhibitor of NF-κB and protein kinase B (Akt) phosphorylation in T cells [113]. In transcription factor regulation, CA directly targeted the transcription factor NF-κB and other related compounds of captopril, curcumin, pyrrolidine dithiocarbamate, resveratrol, silymarin, and thalidomide, as the inhibitors targeted the NF-κB. These NF-κB inhibitors are also known to exhibit antinecrotic, anticholestatic, antifibrotic. and anticancer activities. CA inhibited NO production by down-regulating p38 MAPK-JNK1/2-NF-κB pathways and IκBα degradation in the LPS-induced pro-inflammatory signaling cascades, so contributing to the anti-inflammatory properties. In addition, CA showed no hepatotoxicity using the same anti-inflammatory concentrations. CA suppressed the nuclear translocation of AP-1 and NF-κB family proteins and the related upstream signaling cascade, composed of IL-1 receptor-associated kinase 1 (IRAK1), IRAK4, TGF-β-activated kinase 1 (TAK1), MAPK kinase 4/7 (MKK4/7), and JNK [127]. Either CA can suppress the activation of another transcription factor, NFAT [128]. Some other inflammatory factors of Fyn, 5-LP, MMP, and ectonucleotidase were also targeted by CA [112, 114, 115, 116]. CA as an inhibitor of the 5-LP activity is also known to prevent acute and chronic inflammation, as CA protected the CCl_4-induced liver hepatitis and cirrhosis [117, 118], because the leukotrienes of the 5-LP product are deleterious to the liver. Thus, CA had a protective effect on hepatic inflammation by scavenging ROS and inhibiting 5-LP and NF-κB activities [119, 120, 121, 122], hence confirming that CA is a phenolic compound with an anti-inflammatory activity.

It was shown that caffeoylquinic acid derivatives inhibited LPS-induced Raw 264.7 macrophage inflammation by suppressing NO/iNOS and PGE2/COX-2 pathways through inhibiting NF-κB activity [123]. Tri-caffeoylquinic acid also inhibited TNF-α-stimulated inflammatory mediator production in keratinocytes, deactivating the Akt and NF-κB pathways provoked by ROS as an inflammatory skin disease-blocker [124]. Thus, CA should be considered as a crucial anti-inflammatory agent; moreover, CA is extensively used as a dietary resource for treating various inflammations, and a potential product for anti-inflammatory drugs.

2.2.2. Phagocytic Activation via TLR-AKT Signaling Pathway in Innate Immunity

Monocytes are a subset of circulating leukocytes, terminally differentiated into tissue macrophages and dendritic cells. Neutrophils are the most abundant white blood cells in circulation and act as the first defense cells against microbial pathogens with a restricted set of pro-inflammatory functions [125]. Monocytes are responsible for host defense events in inflammation, phagocytosis, antigen processing and presentation, cytokines production and T-cells activation by its receptors as surface monocyte markers. Many receptors responsible for phagocytosis are known and the receptors, such as mannose receptor, mannose-binding protein, and TLRs, recognize the surface patterns of pathogens. For example, TLRs recognize foreign antigens and induce host immune and inflammatory responses [126]. Monocyte receptor TLR-2 specifically binds gram-positive bacteria and zymosan, and TLR-4 binds gram-negative bacterial endotoxin, LPS for host defense events in inflammation and phagocytosis. Some receptors such as the Fc-γ receptors (CD16) and complement receptors CR1 (CD35) and Mac-1 (CD11b/CD18) recognize components derived from the host cells [127], while antigen-presenting cells exhibit leukocyte antigen (HLA)-DR for antigen presentation to T cells, and co-stimulatory CD80 (B7-1) for T-cells activation [126, 128]. The phagocytic behaviors of neutrophils and macrophages are controlled by the PI3K/AKT and MAPK signaling pathways [129]. Recently, CA has received attention for its effects on the leukocyte immune response, such as immunomodulation and activation of neutrophils and monocytes. The action mechanisms of CA in its biological activity are specifically interested in human immune cells such as monocytes, neutrophils, and dendrites. For example, the question how does CA act on human monocytes, neutrophils, or dendrites is raised to answer its activity on innate immunity. As a CA derivative, CAPE has recently been studied for its immunosuppressive capacities on T lymphocytes and macrophages [130]. CAPE exerted inhibitory effects on cytokine (IL-12, IL-10, IL-5 and IFN-γ) production in human dendritic cells, and T cell proliferation by inhibiting NF-κB and Akt signaling pathways [131, 132]. Additionally, CAPE blocked TLR-4 activation for dermal anti-inflammation, release of inflammatory mediators, and NF-κB activity [133]. Thus, information on CA-regulated immune cells and immunomodulatory action of monocytes is of interest for its meritorious health benefits and alleviation of human diseases.

On the other hand, for the regulation of leukocyte infiltration into the inflammatory sites such as tissue injury, CA inhibited inflammatory leukocyte rolling and adhesion in cerebral venules with the decreased ROS production in human umbilical vein endothelial cells (HUVEC), and decreased plasma macrophage inflammatory protein-2, and monocyte chemoattractant protein-1 levels. CA reduced the E/P-selectin and ICAM-1 expression on endothelium and CD11b/CD18 on leukocytes. CA further suppressed CD18 phosphorylation events in the Src/PI3K/Akt pathway in leukocytes [134]. In monocytes, CA inhibited the expression levels of cell surface receptors including TLR-2, TLR-4, HLA-DR, and CD80 in activated monocytes of innate immunity [35, 135]. Considering TLRs in recognition of pathogenic microbes and HLA-DR in antigen presentation to T cells, CA has an inhibitory activity on the immune cells related to innate immunity. Furthermore, CA and cinnamic acid inhibited proinflammatory TNF-α and anti-inflammatory IL-10 cytokine production in monocytes in a similar fashion [136]. Since TNF-α may activate monocytes, whereas IL-10 may suppress them, it is conjectured that CA-suppressed expression of TLR-4 possibly inhibits the TNF-α release, and CA double-blocked expression of TLR-2/TLR-4 may inhibit IL-10 production in monocytes. Thus, it is suggested that CA probably interact with the two TLR-2/4 receptors in monocytes, yet the molecular interaction is unknown. Although CA stimulated the monocyte functions, the expression and cytokine production were downregulated, suggesting an anti-inflammatory action of CA. Thus it is hypothesized that CA inhibit or stimulate the monocyte activity with a modulatory effect by a mechanism that TLRs are direct targets of CA for immunomodulatory action in human monocytes.

Interestingly, in the recent study [137], CA accelerated phagocytic uptake of gram-negative bacterium, *E. coli* by differentiated HL-60 cells and neutrophils, blocking *E. coli* survival within the differentiated HL-60 cells. Because there is no available neutrophil cell line, and owing to the inherent differences between neutrophils and differentiated HL-60 cells, human promyelocytic HL-60 cells are used as a model of differentiation. Additionally, HL-60 cells are differentiated into macrophage/neutrophil-like cells by retinoic acid (RA), DMSO or ganglioside GM3 [138, 139, 140]. Because of the very short lifespan of neutrophils, the differentiated HL-60 cells are also used as a neutrophil model [141, 142]. To elucidate the CA effect on the neutrophil phagocytic uptake, CA was examined for its immunoregulatory effect on neutrophils. CA has obviously increased the neutrophil cell phagocytosis, suggesting that CA has immunoregulatory potential in human neutrophils. The enhancement of neutrophil phagocytic capacity by CA is consistent with its effect on macrophages. Furthermore, CA enhanced the expression of cell surface receptor, Mac-1 (CD11b/CD18) for host cell-derived components and activated Akt/MAPK signaling pathway in *E. coli*-captured neutrophils. In the events of phagocytic uptake of *E. coli* by the differentiated HL-60 cells or neutrophils, as well as bacterial survival, Mac-1 expression, and Akt activation of neutrophils, CA was a key effector for the phagocytic uptake. CA enhanced inflammatory responses by improved phagocytic capacity, inhibited bacterial survival within cells, and increased neutrophil Mac-1 expression [140]. CA also exhibited its effect on neutrophil elastase activity, destroying bacteria and host tissue injured during inflammation [143]. Thus, there is a reasonable expectation of CA immunoregulatory activities, based on the effects of CA on neutrophils, monocytes, and dendrites in the human body. The neutrophil phagocytosis is reported to be mechanistically mediated by Akt/ERK activation, where Akt pathway induces the Mac-1 activity [134]. CA is suggested to activate both the Akt signaling pathway and Mac-1 expression in neutrophils, indicating that CA is important for phagocytic uptake. Thus, CA-activated Akt signaling pathways are the basic reason for CA-induced modulation of neutrophil phagocytic uptake.

2.2.3. Suppression of Leukocyte- and Platelet-Endothelial Interaction during Inflammation

As mentioned previously, neutrophils are key players in the recognition and elimination of pathogenic invaders, but they are a two-edged sword under microenvironmental situations of tissues in organisms. In a specific condition, the dysregulated function of neutrophils induces tissue damage and lesions, and triggers a pathophysiological status of various inflammatory diseases [144]. Activated neutrophils, stimulated by various stimulants, induce vascular cell-mediated inflammation and tissue injury through the production of ROS metabolites and cytotoxic NADPH oxidoreductase or MPO. Thus, to clear the inflammation ROS scavengers and antioxidative substances are required to elicit and reduce the oxidative insult, inflammation, and tissue injury. Although neutrophil NAPDH oxidase or MPO is used as the magnitude of neutrophil activation status, the inhibition of the MPO activity is indicative of the anti-inflammatory status in the chronic or acute inflammation response. Vascular inflammation recruits leukocytes to the endothelium area by rolling and homing for vascular remodeling. The leukocytes migrated to endothelium are then stimulated by pro-inflammatory cytokines to induce expression of the inflammatory genes [145]. Leukocyte-endothelium adhesion is mediated by the receptor-ligand interaction between the leukocyte molecules such as integrins, selectins, specific ligands of Sialyl-Lewis X and their endothelial counterpart receptors [146]. Vascular cell adhesion molecule 1 (VCAM-1) and ICAM-1 induce inflammation through the binding of lymphocyte function-associated antigen 1 (LFA-1, αL/β2 integrin) with ICAM-1 on endothelial cells [147]. Thus, the leukocyte-endothelium interaction is unique in vascular inflammation process, as escalated by E-selectin for monocytes to rolling or homing along the endothelium.

Interestingly, CA effectively blocks the leukocyte rolling/homing recruitment to the activated endothelium by suppressing endothelial expression of VCAM-1, ICAM-1, and E-selectin and monocyte expression of β1, β2, and α4 integrins during the inflammation reaction. CA eliminated THP-1 monocyte adhesion on endothelium with decreased inflammatory IL-8 production for ICAM-1 and β2 integrin [148]. CA diminished the leukocyte−endothelium interaction by eliminating the binding of VLA-4-VCAM-1 and LFA-1-ICAM-1. ICAM-1 expression is reported to be increased, depending on β2 integrin, and the ICAM-I expression is sharply increased when the cells are treated with IL-8. However, a recent result showed that CA diminished TLR-4 expression, although the endothelial IL-8 expression induces TLR4 expression. CA reduced expression of endothelial cell adhesion molecules in HUVEC and monocyte integrins in THP-1 monocytic cells in the inflammation condition. In addition, by blocking inflammatory IL-8-mediated NF-κB signaling pathway, CA disrupted monocyte trafficking, rolling and homing to endothelium [148]. The inflammatory status elevates the TLR-4 expression of endothelial cells, leading to synthesis of inflammatory cytokines, such as TNF-α and IL-1. CA diminished the inflammatory TLR4 signaling for IL-8 production and VCAM-1, ICAM-1, and E-selectin expression in endothelial cells. IL-8 expressed via inflammation upregulates the downstream of NF-κB signaling pathway in the leukocyte-endothelium interaction system. TLR4 is also known to be induced during adipocyte differentiation via NF-κB [149]. Thus, CA has a therapeutic potential in preventing inflammation-related human diseases.

In a recent study [150], CA inhibited collagen-induced platelet aggregation and the production of thromboxane A2 (TXA2) via inhibition of platelet COX-1. In addition, CA inhibited collagen-elevated [Ca^{2+}] mobilization, while increasing the aggregation-inhibiting cAMP and cGMP, with a subsequent increase in the inositol 1,4,5-trisphosphate receptor (IP3R) phosphorylation. This indicates the inhibition of IP3-mediated Ca^{2+} release from internal stores in endoplasmic reticulum is regulated by the IP3R on collagen-activated platelets. Thus, CA is potentially beneficial to elevate the cAMP-dependent protein phosphorylation in collagen-induced platelet aggregation, and disturbs the platelet aggregation in vascular endothelium. In a similar fashion, as in the CA-blocked leukocyte-endothelial interaction, CA prevents thrombotic and atherosclerotic diseases by blocking the platelet-endothelial interaction.

2.2.4. Inhibition of In Vivo Skin Inflammation

The skin is an initial defense tissue with epidermal cells such as keratinocytes and Langerhans cells [151]. Several reports showed that CA has anti-inflammatory and anti-allergic activities in *in vitro* and *in vivo* animal skin models [152, 153]. The anti-inflammatory effects of CA in skin inflammation have also been reported using acute and chronic skin inflammation in *in vivo* mice and *in vitro* human keratinocytes [97]. From the above current reports on anti-skin inflammation, CA was apparently effective. CA was previously reported to be a protective agent against UVB-induced skin damage like skin inflammation [154]; however, the action mechanism(s) underlying the anti-inflammatory effects of CA in skin inflammation are only partially explained. Zhang et al. (2014) demonstrated the capacity and mechanisms of CA on acute and chronic skin inflammation in *in vivo* mice and in keratinocyte cells *in vitro* [97]. Additionally, because 12-O-tetradecanoyl-phorbol-13-acetate (TPA)-induced mouse ear inflammation model mimics the skin inflammatory diseases [155], this TPA-induced mouse ear inflammation model is often used as an *in vivo* skin inflammation model system. Thus, this system has been used for both acute and chronic effects of CA and CAPE [124, 156]. Then, it was shown that CA reduced the TPA-induced increase in ear thickness and weight in both acute and chronic irritant contact dermatitis of the animal model [97]. TPA induced a neutrophil NAPDH-oxidase, MPO activity [155], while CA markedly inhibited the MPO activation [97, 157] in mouse skin. Tropical treatment with TPA increased the levels of TNF-α, IL-6, and IL-1β in ear skin biopsies in TPA-induced mouse ear inflammation model, while CA inhibited the expression of the pro-inflammatory cytokines TNF-α, IL-6, and IL-1β, leading to the anti-inflammatory response of CA in acute skin inflammation. CA can also be regarded as a potent inhibitor of tumor promotion, since the pro-inflammatory cytokines TNF-α, IL-6, and IL-1β also act as messengers in the ROS production, inflammation, and hyperplasia in the cancer promotion process of carcinogenesis. Khan et al. (2012) showed that CA reduced TNF-α expression in a mouse skin model [157]. Local treatment with CA on the ear skin epidermis targets keratinocytes since most of epidermal cells (about 85%) consist of keratinocytes. CA also inhibited TPA-induced TNF-α production in human keratinocytes. IL-6 induces the growth and differentiation of dermal and epidermal cells, and acts as a chemotactic factor for T cells. IL-1β, a pro-inflammatory cytokine, activates neutrophils, monocytes, eosinophils, and basophils. Treatment of keratinocytes with TNF-α induces IL-6 and IL-1β expression for the skin inflammation [151]. However, CA blocked their expression in human keratinocytes. As TNF-α provokes inflammation reaction by NF-κB signaling pathway, the TNF-α-induced NF-κB activation provokes expression of inflammatory cytokines of IL-6 and IL-1β with endothelial adhesion receptor proteins.

It was shown that CA blocked the TNF-α-induced endothelial adhesion to HUVECs by NF-κB deactivation in human keratinocytes [97, 101]. CA also modulated the transcription of NF-κB in cutaneous inflammation. CA has anti-inflammatory activities in both acute and chronic irritant contact dermatitis, by blocking the expression of TNF-α, IL-6, and IL-1β. Thus, it is suggested that CA targets inflammatory mediators and modulates the expression of the TNF-α, IL-6, and IL-1β at both the transcription and post-transcription levels. specifically in the keratinocytes of the acute irritant contact dermatitis. CA might be a good candidate, for the prevention and treatment of inflammatory skin diseases in both acute and chronic contact dermatitis models, because CA inhibited TPA-induced skin edema, decreasing the edema-derived thickness and tissue weight swelling and neutrophil NAPDH-oxidase, MPO activity, and inflammatory damage in mice exposed by chronic TPA [97]. CA can be used for a therapeutic agent against inflammatory skin edema.

2.2.5. Suppression of Inflammation-Mediated Nociceptive Pain

In a recent study [34], CA inhibited acetic acid-induced writhing and formalin-induced pain as a pain model in mice. Writhing is specifically derived from acute pain because of tissue damage and sensitization of nociceptors by inflammatory mediators [158]. In the late phase pain model, CA administration (5–100 mg/kg, p.o.) also diminished the inflammation-associated nociceptive pain in a dose-dependent manner in mice and rats. For inflammation-associated nociceptive inhibition, several derivatives of CA such as CGA and CAPE also exhibited similar anti-inflammatory, anti-pain and anti-nociceptive activity [159, 160]. In the experimental pain model, CA inhibited painful writhes by inhibiting the mediators of inflammation and pain. Thus, CA reduced nociceptive behavior in the late phase, rather than the early phase of pain, suggesting the CA reduces the central mediators associated with peripheral inflammation. This pain-controlling analgesic activity of CA against inflammatory pain and mediators has also been observed using LPS and carrageenan-induced hyperalgesia models in rats. The inhibition of hyperalgesia by CA in these models indicates its interference in sequential release of inflammatory mediators peripherally. The oral administration of CA (200 mg/kg, p.o.) showed analgesic activity and inhibited neutrophil NAPDH oxidase MPO, MDA, and nitrite generation in the inflamed paw with the reduced neutrophil infiltration [161]. Thus, CA exhibits peripheral analgesic effect in mice and rats. Since the MPO levels indicate neutrophil recruitment and activation at inflammatory site, neutrophils are involved in mediation of hypernociception. CA inhibits the local oxidative tissue damage and neutrophil-mediated and free radical-mediated hypernociception [162]. CA reduced the nitrite levels for its analgesic activity by iNOS inhibition. In addition, a derivative CAPE suppressed the expression levels of NF-κB and 5-LP in this pain model [162, 163]. As NF-κB plays a pivotal role in regulating pro-inflammatory cytokine gene expression [164], the nerve injury-derived pain is reported to be regulated for the pain genesis and persistence [165]. CA inhibited LPS- and carrageenan-induced hyperalgesia by inhibition of NF-κB regulated transcription of pain mediators and release of inflammatory mediators that trigger the pain reaction. Thus it is concluded that CA is analgesic by the way of reducing neutrophil, free radical, and NO-mediated hypernociception, as confirmed from the reduction in MPO, MDA, and nitrite levels.

2.2.6. Protection of Allergic Inflammatory Asthma and Atopy

Asthma and atopy reactions are representatives of chronic inflammatory and allergic diseases of the airways and skin by aberrant T-helper 2 (TH2) immune activation. An activation of TH2 cells induces the secretion of IgE through an IL-4 and IL-5-stumulated eosinophilic reaction. The asthma reaction is accelerated by IL-13 for airway hyper-responsiveness [166]. At present, potent drugs against asthma are limited to exceptional cases of an immune suppressor glucocorticoid or a bronchodilator. Because of their side effects, such as retarded osteogenesis in young people, CA and its derivatives have been suggested for the drugs. It is not exactly known how CA or CAPE affects asthma. CAPE regulates the cytokine expression of peripheral blood mononuclear cells (PBMCs), reduced the expression of IFN-γ during the stimulation with the TGF-β1, and decreased the expression of IL-2, IL-4, and IFN-γ in mice [167, 168]. CAPE also suppressed mitogen-induced T-cell proliferation by an inhibition of NF-κB activity [169, 170]. In the Th1/Th2 ratio, CAPE accelerated the shift change from Th2 to Th1 type [171]. Thus, the regulatory capacity of CAPE on Th1/Th2 phenotype shift of PBMC is significant. Since Th1 cells produce IFN-γ mainly and asthma disease is caused by Th2 cells, asthmatic patients produce lower IFN-γ levels compared with the normal persons [172]. Hence, the immunoregulatory effect of CAPE can be managed by regulatory T-like cells with the inhibition of Th1/Th2 cytokine secretion [173]. Hence, it is hypothesized that CA or CAPE activates the phenotype functions of the regulatory T-like cells.

2.3. Anticancer Activity

Plant-derived products have been used for preventing and treating various diseases in humans since they maintain health and cure diseases without toxicity. Dietary polyphenolic acids have inhibitory properties against invasion and metastatic behaviors including adhesion, migration, and angiogenesis in cancer cells *in vitro* or *in vivo*. Among them, CA, the major subgroup of phenolic compounds, has various pharmacological properties [174]. CA is predicted as an anticancer agent from the fact that CA is a well-known major phenolic compound in coffee. CA is a main metabolite of CGA in coffee, which lowers the risk of human cancers, as reported for human breast cancer, brain tumor, melanoma skin cancer, liver, colorectal, mouth and throat cancer, and skin cancer [175, 176, 177, 178, 179, 180].

As carcinogenesis is a multi-step process in which tumor-causing metabolites and free radicals species lead to cancer, however, the underlying molecular mechanisms are not completely understood, such as the suppression of carcinogenesis, development, and proliferation of cancers by CA. Hence the anti-tumor mechanism of CA in the various cancer cell lines still needs to be investigated. The molecular mechanisms for cancer occurrence are, for example, oxidative stress and inflammation, as these factors are directly associated with tumor promotion [181]. CA and its derivative CAPE, as the inhibitors of the lipoxygenation and inflammation, have been characterized as having antitumor activity [182, 183]. The working molecular mechanisms for the antitumor capacity of CA has been explained by the efficient inhibition of cancer inducing factor-induced transcription factor NF-κB activation [184, 121], and an internal signal transduction pathway such as ERK-1/2 [185]. As aforementioned, CA or CAPE is well known as a potent and specific inhibitor of the transcription factor NF-κB [121].

As a CA derivative, it was reported that CAPE also has a cytotoxic activity against human tumor cells but not for normal cells [186, 187]. Although the action mechanisms for anti-tumor activity of CAPE have not yet been well studied, there are indications that CAPE selectively inhibits the translocation of transcription factors NF-κB and AP-1 to nuclear site in the hepatocarcinoma cells [188]. CAPE inhibited proliferation, and induced apoptosis of tumor cells or transformed cells, including SW480 colorectal tumor cells [188], HCT116 human colorectal cancer cells [189, 190], ME180 human cervical cancer cells [191], C6 glioma cells [186], MCF-7 breast cancer cells [122], transformed rat embryonic fibroblast cells [192], and transformed rat fibroblast cells [193]. For anticancer activity, CAPE inhibits histone deacetylases (HDACs) and regulated epigenetic gene expression [194, 195], while CAPE inhibited HDAC in human breast cancer MCF-7 cells (estrogen and progesterone receptor positive, ER^+/PR^+), and MDA-231 cells (triple negative breast cancer cells, ER-/PR-/Her2-). CAPE also inhibited the tumor growth of breast cancer MCF-7 and MDA-231 xenografts in nude mice [195]. It is thus suggested that CAPE modulates the HR expression [decrease in HR^+ positive cells, and induction of HR expression in HR- negative cells, decrease in the expression of epidermal growth factor receptor (EGFR) in HR- cells, and decrease in the expression of p-Her2 in $Her2^+$ cells]. Especially, the fact that CAPE decreased the EGFR expression indicates a future therapeutic possibility for $Her2^+$ patients with an anti-Her2 antibody therapy. Thus, CA or CAPE is considered to be a naturally occurring, safe, nontoxic, and affordable form of epigenetic therapy that can be used in combination with conventional anticancer drugs.

2.3.1. Prevention of Skin Cancer Carcinogenesis

CA has the potential to inhibit skin cancer development. CA was highly effective in decreasing UV-induced skin carcinogenesis, as the Dong group [185] recently reported that CA inhibits the colony formation of human skin cancer cells and EGF-induced neoplastic HaCaT cells. The local administration of mouse skin with CA reduced skin tumor formation, inhibiting the Ki-67 expression as a proliferation protein [196]. The MAPK pathway of the Ras–Raf–MEK-ERK1/2 pathway is the common link in human cancer, regulating proliferation and senescence [198]. *In vitro* human skin cancer cells and HaCaT cells there can be a solar UV induced phosphorylation of ERKs' downstream signals, RSK2, Elk1, and c-Myc. In an *in vivo* skin cancer xenograft model, CA suppressed the tumor growth and solar UV–induced skin carcinogenesis by a reduction of the MAPK signaling, and inhibition of ERK1/2 and its downstream signals, RSK2, Elk1, and c-Myc. Thus, CA can be regarded as a chemopreventive agent. An interesting point is that CA directly interacted with ERK1/2 and as a resultant inhibited ERK1/2 activities *in vitro* to prevent the solar UV-promoted skin cancer development. The crystallographic analysis of CA-ERK2 complex was successfully made to elucidate the binding sites at amino acid residues Q105, D106, and M108. This means that CA directly targets ERK,s as confirmed from the ERK2/CA cocrystal structure, hence binding to the ATP-binding cleft through the formation of hydrogen bonds between hydroxyl groups and amino acids located at the hinge loop [185]. In addition, the genetical disruption of the ERK2 gene in A431 cells lost the binding capacity with its ligand, CA, so indicating CA targets ERK1/2 at molecular level. This is the first evidence that CA directly targets and interacts with ERKs at the molecular level, suppressing the solar UV-induced ERKs' phosphorylation and downstream signaling in primary mouse epidermal keratinocytes and HaCaT cells. This research group first reported the cocrystal structure of CA with ERK2, and showed that CA directly inhibited solar UV induced activation of ERK1/2 and suppressed ERKs signaling.

On the other hand, CA administration, with UV-exposed skin, decreased the skin tumor formation in mice. In parallel, CA inhibited the *in vitro* formation of human skin cancer cell colony and EGF-induced neoplastic transformation of HaCaT cells. This same group identified ERKs as a target of CA by *in silico* screening. CA inhibited solar UV-induced ERK1/2 signaling *in vitro* and knockdown of ERK2 decreases the anticancer effects of CA. CA also suppressed solar UV-induced skin carcinogenesis in SKH-1 hairless mice *in vivo*.

The detailed mechanisms underlying the anti-tumor properties of CA and CAPE have been explained by the controlled signaling capacity of PKC, FAK, Cdc42, Ras, Rac1, ERK, JNK, p38, PI3K, p70S6K, PAK1, Akt, RhoA, and RhoB with the downregulated transcription capacity of AP-1, NF-κB, and STAT3. Additionally, the controlled regulation of several factors, in many different tumor cells or animal models, are raised for CA and CAPE, including matrix metalloproteinase (MMP)-1/-2/-3/-9/-14, MT-1 MMP, uPA/uPAR, actin fiber, F-actin, integrin, NCAM, ICAM-1, CXCR4, ENPP2, RON, b-FGF, COX-2, CCL2, IL-6/-8, TIMP-1/2, HER2, ROS, HGF, VEGF/EGF/EGFR, HO-1, SPARC, VCAM-1, HIF-1a, ADAM17, HLJ1, E-cadherin, PAI-1, maspin, β-catenin, filopodia, and tensin. Thus, it is a reasonably proven hypothesis that dietary consumption of CA in coffee and its derivative CAPE in phenolics is helpful for tumor prevention and carcinogenesis.

2.3.2. Prooxidant Roles in the Cancer Prevention, Anticancer and Cancer-Apoptosis

Recent studies have concluded that plant polyphenolic compounds are antioxidants to protect cells and lipid peroxidation. However, it should also be noted that polyphenols are known to enhance oxidative DNA damage in cancer cells [198]. Considering the fact that flavonoids auto-oxidize in aqueous medium and may form highly reactive free radicals, polyphenols may act as substrate for peroxidases and other metalloenzymes, yielding quinonetype prooxidants [199]. More specifically, it should be mentioned that the polyphenols can function as prooxidants, particularly in the presence of transition metal ions such as iron and copper [200]. The prooxidant capacity of phenolics causes the cellular ROS generation, inducing DNA fragmentation and apoptosis in cancer cells. Thus, this fact indicates that the prooxidant properties of plant phenolics are comparable with apoptosis-based antitumor potentials. When considered for cancer cells, cancer cells are especially rich in acidic molecules such as sialic acids or sulfated glycans in that condition. Hence, phytochemicals cannot function as antioxidants, and instead they function as prooxidants in cancer circumstances. Generally, antioxidant enzymes protect cells attacked by tumor-promoting agents and thus cellular transformation is accelerated by the decreased antioxidative enzymes, such as SOD, CAT, and GPx, thereby increasing the cell sensitivity to prooxidants [201]. This indicates that phenolics may deplete intracellular antioxidants, thus inducing cancer cell death. Likely, CA decreases the activities of antioxidative enzymes such as SOD, CAT, and GPx in cancer cells. For example, Indap et al. (2006) reported that CA and cinnamoyl compounds decreased GSH levels in human malignant cancer cells [202]. In that study, they showed that CA induces growth inhibition, alteration in mitochondrial membrane potential, lipid peroxidation, DNA damage, and apoptosis in HT-1080 fibrosarcoma cells. The oxidative mechanism for the antiproliferative effect of CA can be considered as an anticancer agent. For the intracellular oxidative reaction, the phenolic CA is oxidized by peroxidase/ROS to phenoxyl radicals that cooxidize GSH to form a thiyl radical (GS-). This radical may then react with GSH to form a disulfide radical anion and this is why GSH levels are decreased in CA-treated cancer cells, as reported in previous studies on anti-cancer activity of CA.

Similarly to the prooxidant property of CA, Prasad et al. (2011) reported the anticancer and cytotoxic effect of CA in the same fibrosarcoma HT-1080 cells [44], showing the disrupted mitochondrial dehydrogenase activity raised by its prooxidant property. With regard to the disrupted mitochondrial potentials, it has been explained that CA inhibited the proliferation of breast cancer T47D cells, by inhibiting the mitochondrial activity of the breast

T47D cancer cells [203] and causing a rapidly increased intracellular ROS in HT-1080 cells [202]. Thus, it is hypothesized that the increases in the ROS levels in CA-treated cancer cells are basically due to its prooxidant property. The prooxidant activity of CA has also been reconfirmed by Sakihama et al. (2002) [204]. Hence, a question arises, as follows:
how does CA generate and increase the ROS for the CA-induced cytotoxicity in cancer cells? The conjectured mechanisms are as follows. A semiquinone is oxidatively generated by CA and the resulting semiquinone reacts with oxygen to form a superoxide anion O_2^{-}, and then the O_2^{-} radicals autocatalytically oxidize the next adjacent phenolic compound to regenerate the semiquinone and H_2O_2 in a cascade manner. Then, peroxidases of the cancer cells produce phenoxyl radicals through an electron oxidation of phenols and the newly produced phenoxyl radicals oxidize NADH to radical NAD^{+}. Finally, the NAD^{+} radicals reduce oxygen O_2 to oxygen radial O_2^{-} by an electron transfer.

For the CA-mediated antitumor and cancer cell death activity, DNA is also thought to be a key molecule and, as an initial target, subject to DNA damage [205]. Previously, polyphenolics have been reported to induce DNA damage in cervical cancer cells by ROS [198]. This might be the reason for the increased oxidative DNA observed in CA-treated cancer cells. Changes in the mitochondrial membrane potential are frequently observed in CA-treated tumor cells. As mitochondria are an important organelle in cell death and apoptosis, the CA-induced change in mitochondrial membrane potential and cytochrome C is the crucial indicator of cancer cell death [206]. Previously, Galati et al. (2002) [207] showed that CA and other dietary phenolic compounds induce mitochondrial collapse by alteration of the mitochondrial membrane potential, as confirmed by the cytochrome C release and signals. CA increased apoptosis in HT-1080 cells with increased ROS levels and DNA damage. Similarly, CAPE induced lipid peroxidation in plasma membrane and Fas-mediated apoptosis in human breast cancer MCF-7 cells [208]. Recently, the mechanism of CAPE-induced apoptosis in human myeloid leukemia U937 cells has been elucidated by Jin et al. (2008) [209]. The CAPE treatment of U937 cells decreased cell viability with DNA fragmentation as the typical ladder profile of oligonucleosomal fragments and nuclear condensation as a typical phenotype of apoptosis. Apoptotic changes in release of cytochrome C, reduction of Bcl-2 expression, increase of BAX expression, activation/cleavage of caspase-3, and activation/cleavage of PARP were also accompanied, but not by Fas protein as the death signaling mediator or by the phospho-eIF2α and CHOP as mediators in an ER-mediated apoptosis in the CAPE-treated U937 cells. Thus, CAPE induces the mitochondria-mediated apoptosis but not death receptors- or ER-mediated apoptosis in human myeloid leukemia U937 cells. CAPE is involved in the inhibition of proliferation and induction of apoptosis on HL-60 human leukemic cells [210, 211]. The structural basis of the prooxidant activity of CA on DNA damage and lipid peroxidation has been suggested in the presence of Cu(II) ions [212]. CA also has the ameliorated activity on oxidative damage and tumor promotion in Fe-NTA-induced renal injuries and hyperproliferation in rat kidneys [213]. In the antioxidative capacity, CA protects from the depletion in the antioxidant enzymes through its antioxidant and free radical scavenging activities. CA recovered the depleted glutathione reductase, quinine reductase, GPx, GSH, and GST in animals.

On the other hand, many polyphenols such as curcumin, EGCG and resveratrol, gain their beneficial properties in cancer prevention through the activation of the Nrf2/Keap1 pathway [214, 215, 216]. Recently, Sirota et al. (2015) reported that the Nrf2/EpRE pathway activation of CA, involving the enhanced Nrf2 translocation and transcription into the nucleus, contribute to the resistance toward tumor therapy through induction of phase II enzymes [217]. CA acts as an electrophile with its nucleophilic properties. The induction is enhanced, via the double bond known as the Michael acceptor of CA and the hydrogen peroxide generated by the catechol group. It has been shown that CA activated the Nrf2 and phase II enzymes [217]. Structurally, the electrophilic moiety in CA directs the oxidation of the Keap1 protein, and the nucleophilic moiety (the catechol/quinone moiety) provides scavenging activity, although the nucleophilic moiety cannot direct the Nrf2 induction. The Nrf2 interacts with basic leucine zipper transcription factors such as the Maf and Jun family members and forms the Nrf2–EpRE complex with the electrophile response element (EpRE or antioxidant response element-ARE) [218]. The Nrf2–EpRE targets the genes for the phase II enzymes for antioxidant responses, cytoprotection or detoxification. The following phase II enzymes then protect the cells from cellular stress, such as HO-1, ferritin, NAD(P)H quinone oxidoreductase (NQO1), thioredoxin and glutathione metabolizing GST, γ-glutamyl transpeptidase (GGT), glutathione reductase (GSR), GPx, and glutamyl cysteine ligase (GCL) [219]. To date, several compounds are reported as Nrf2–EpRE inducers, including redox-active bisphenols, quinones, phenylenediamines, isothiocyanates, dimercaptans, hydroperoxides (H_2O_2), metals, polyphenols (CA, curcumin, EGCG and resveratrol) and Michael acceptors [220]. Nrf2 is important for cellular homeostasis, oxidative stress, inflammation, and carcinogenesis [221], however, Nrf2 activation functions as a double-edged sword. For example, the abnormal expression or nuclear translocation of Nrf2 provokes detoxification of anti-cancer drugs, using the phase II detoxifying enzymes in cancer cells [222], as reported in the study that GSTP1-producing cancer cells are resistant to cisplatin treatment [223]. The GSTs detoxify chemotherapeutic drugs by the conjugation of GSH to the electrophilic site of the drugs [224]. If GST is thus overexpressed in cancer cells, then the cells become resistant to the drugs [225, 226], hence indicating that the Nrf2-targeted induction of phase II enzymes, such as GSTs, give cancer cells resistance against radiation and chemotherapy. Hence, CA or coffee can be preventive against cancers; however, it might decrease the chemotherapeutic capacity by phase II gene induction.

2.3.3. Antiangiogenic Activity

2.3.3.1. VEGFR Inhibition

Recently, CA was reported to directly target the proteins and molecular pathways responsible for the invasion and metastasis of cancer cells [227]. Metastasis is a multistep process accompanying invasion, migration, adhesion, infiltration, colonization, and subsequent formation of new capillaries and stable angiogenesis; invasive tumor cells overcome the barriers by attachment to basement membrane or extracellular matrices (ECM), degradation of the ECM, and migration into the new vascular area [228]. Hence, the drug inhibition of tumor invasion involves various signaling pathway molecules as targets and these molecules are controlled by agents for chemoprevention [229]. A crucial issue is identification of the protein targets, including epithelial and mesenchymal markers, and the molecular pathways for invasion and metastasis of cancer cells. The growth and metastasis of human solid tumors and the diseases progress such as diabetic retinopathy, rheumatoid arthritis, inflammatory psoriasis, and any other related pathological conditions are regulated by the balance between angiogenic stimulators and inhibitors released in the angiogenic–pathological microenvironment tissues. Vascular endothelial growth factor (VEGF), as an angiogenic factor, is a potent endothelial-specific mitogen that activates to grow endothelial cells in pathophysiological angiogenesis. Recently, Chung et al. (2013) demonstrated that CA and CAPE inhibit tumor growth, invasion, and metastasis with the molecular action mechanism underlying VEGF-mediated angiogenesis blocking [230]. CAPE suppressed VEGF-induced proliferation, tube formation, migration, formation of actin stress fibers, and loss of VEGF-cadherin at cell–cell contacts in endothelial cells. CAPE explained the inhibition of VEGF-mediated VEGF receptor-2 (VEGFR-2) signal pathway, and inhibiting VEGF-enhanced tyrosine phosphorylation of VEGFR-2 in the endothelial cells. CAPE effectively blocked VEGF-stimulated neovascularization, vascular permeability in mouse skin capillaries, and formation of new blood vessels of HUVECs in VEGF-treated Matrigel plugs in C57BL/6 mice, in vivo [231, 232]. CAPE inhibited the growth and neovascularization of primary tumor cells in C57BL/6 and BALB/c mice inoculated with Lewis lung carcinoma, colon carcinoma, and melanoma cells. Hence, it was evident that CAPE is a negative agent against VEGF-induced angiogenesis and is a suppressor of VEGFR-2 activation pathway. Thus, the precise molecular mechanism underlying the inhibitory effect of CAPE on VEGF/VEGFR-2-angiogenesis axis related to tumor progression is evident both *in vitro* and *in vivo*. Suppression of the VEGF-response VEGFR-2 activation and the neovascularization of primary tumor cells in tumor-bearing mice has shown that CAPE inhibits intracellular signal pathways line on PI-3K/AKT, ERK, VE-cadherin, β-catenin and FAK.

The expression of VEGF in many tumors induces microvascular permeability, which leads to tumor angiogenesis. VEGF has also been observed to induce microvascular hyperpermeability in many pathological angiogenesis [233]. CAPE also inhibits VEGF-mediated vascular permeability in various pathological angiogenesis, as evidenced by the Miles assay. For the molecular interaction between CAPE and its targets, three possibilities have been suggested: (1) direct binding of CAPE and the extracellular domain of VEGFR-2, inhibiting VEGF-VEGFR-2 binding; (2) direct binding of CAPE and VEGF, inhibiting VEGF-VEGFR-2; and (3) direct binding of APE and the intracellular region of VEGFR-2. Additionally, CAPE may target directly ERK, VE-cadherin, and FAK phosphorylation. It was known that CAPE has cytotoxic effects against various metastatic tumor cells, such as B16F10 melanoma and CT26 colon carcinoma cells [231, 234]. Chung et al. (2004) also confirmed the *in vivo* cytotoxic capacity of CAPE against murine Lewis lung carcinoma (LLC), B16F10, and CT26 cells in mice [188]. CAPE inhibited the proliferation of VEGF-stimulated endothelial cells as a metastatic potential with anti-angiogenic activities, both in vitro and *in vivo*. Moreover, CAPE inhibited the growth and angiogenesis of primary tumors in C57BL/6 and BALB/c mice inoculated with Lewis lung carcinoma, colon carcinoma, and melanoma cells, which secrete VEGF. Based on cell toxicity of CAPE against B16F10 and CT26 tumor cell lines and endothelial cells, and anti-angiogenic activities of CAPE *in vitro* and *in vivo*, CAPE can be regarded as a dual inhibitor of tumor growth and VEGF-stimulated endothelial cell activation. Thus, based on the above studies, it is reasonable to deduce that CAPE exerts anti-angiogenic property through the signaling pathways of endothelial cells, and might be a therapeutic agent for anti-angiogenesis.

2.3.3.2. Matrix Metalloproteinase-2/9 Inhibition

Invasion and metastasis are fundamental properties of malignant cancer cells. A number of proteolytic enzymes participate in these processes, which involve degradation of environmental barriers such as the ECM and the basement membrane. Tumor invasion and metastasis are a multi-step process covering cell division and proliferation, ECM proteolysis, cell migration to vascular circulation system, and remigration and growth of tumors at the metastatic sites [229]. MMPs are important proteolytic enzymes for angiogenesis, tumor metastasis, progression, and invasion of tumors. For example, MMP-9 expression is associated with the capsular infiltration of HCC and growth of small HCC [188]. Thus, the angiogenic MMP-2/9 regulate tumorigenesis and metastasis, and the MMP-2/9 inhibitors are antitumor candidates. The MMP family plays a key role in the proteolysis of the components of the ECM. MMP-2 and MMP-9 (of the MMP family) are known to degrade the major constituent of the basement membrane type IV collagen during cancer invasion and metastasis. The expression of MMP-9 is directly associated with the progression and invasion of cancer [235]. Although development of MMP inhibitors has been tried to prevent tumor metastasis, broad-spectrum MMP inhibitors may have side effects such as arthralgia, myalgia, stiffness, and edema. Thus, specific and selective MMP inhibitors are required. Recently, Chung et al. (2004) showed that CAPE suppresses the MMP-9 expression and demonstrated that CAPE inhibits the tumor growth, invasion, and metastasis in vivo, through the suppression of MMP-9 enzyme activity and NF-κB activation [188]. CA and CAPE have dual inhibitory agents on MMP-9 enzyme activity and transcription for tumor regression, without cellular cytotoxicity as side effects. For the MMP-2/9-inhibitory agent, CA as an anti-tumor agent, with tumor regression and anti-metastasis, was initially isolated from the methanol extracts of plant *Euonymus alatus* (Thunb.) Also known as a Siebold, or winged euonymus, its derivative, CAPE, was synthesized [236]. The methanol extract of winged euonymus was not toxic to mammalian cells. CA and CAPE selectively inhibited MMP-2 and -9, but not -1, -3, -7, or cathepsin K. The treatment of HepG2 cells with CA (100 μg/mL) and CAPE (5 μg/mL) suppressed phorbol 12-myristate 13-acetate (PMA)-induced MMP-9 expression by inhibiting the NF-κB activity, but not AP-1. CA and CAPE suppressed the growth of HepG2 tumor xenografts in nude mice *in vivo*. The subcutaneous and oral administrations of CA and CAPE significantly reduced liver metastasis, corroborating the therapeutic potential and suggesting that the anti-metastatic and anti-tumor effects of CA and CAPE originated from the dual inhibition of NF-κB, as well as MMP-9 activity.

On the other hand, MMP-9 stimulators, such as cytokines and PMA, control the MMP-9 expression by the activation of transcription factors of AP-1 and NF-κB through downstream signaling of the Ras/Raf/ERK, JNK, and PI-3K/Akt pathway, since the promoter region of MMP-9 has AP-1 and NF-κB binding sites as the cis region [237]. NF-κB regulates the expression of a number of genes for anti-apoptosis genes, such as TRAF, bcl-2, cyclin D1, c-Myc, and cIAP, as involved in tumorigenesis. NF-κB is a key factor involved in the activation of genes that encode inflammatory cytokines such as the TNF-α and IL-1β. Thus, NF-κB induces the activation of MMP-9 and COX-2. Inversely, agents to inhibit NF-κB activity can be used to suppress tumorigenesis and metastasis with therapeutic potentials. To date, many plants-derived anticancer drugs, including green tea polyphenols of EGCG, resveratrol, limonene, and organosulfur compounds, have been known to cause inhibition of the MMP-2/9 and MMP-12 activities. Interestingly, as CA and CAPE have been reported to have anti-metastatic activity, CA and CAPE directly inhibit the MMP-9 gene expression and enzyme activity [238]. In that study, it was shown that CA and CAPE play an important role in metastasis and the invasion of cancer. Although many potent anti-cancer drugs can be clinically applicable, many other agents have cellular cytotoxicity and side effects. Thus, anti-cancer drugs such as CA and CAPE without cytotoxicity and side effects are desirable to meet the above criteria.

2.3.4. Inhibition of Tumor Initiation

Concerning the inhibitory capacity of CA on tumor initiation, Khan et al. (2012) observed that CA attenuates TPA-induced tumor promotional triggers by the inhibition of oxidative stress and pro-inflammatory cytokine production [157]. In addition, it was also reported that CA inhibits skin tumor promotion induced by TPA in mouse skin [239]. CA and other polyphenolics (such as ellagic acid, CGA, and ferulic acid) inhibited the initiation phase of 4-nitroquinoline-1-oxide (4-NQO)-induced tongue carcinogenesis and proliferation of the tongue squamous epithelium in rats in a similar fashion [182]. The oral administration of CA and the above three phenolic compounds decreased the incidences of tongue neoplasms (squamous cell papilloma and carcinoma) and preneoplastic lesions (hyperplasia and dysplasia) with decreased proliferation rates. Thus, CA, together with ellagic acid, CGA, and ferulic acid, were ascertained as inhibiting tongue carcinogenesis induced by 4-NQO when treated with the carcinogen, thus suggesting cancer chemoprevention in the tongue in addition to the skin, lung, liver, and esophagus.

2.3.5. Protection against drug Adverse Reactions

The side reactions and effects of the anticancer drugs, observed during chemopreventive and chemotherapies, are major limiting factors for their future use in humans. Interestingly, CAPE protected against chemotherapy-induced ADRs including several antibiotics (streptomycin, vancomycin, isoniazid, ethambutol) and chemotherapeutic agents (mitomycin, doxorubicin, cisplatin, cyclosporine, gentamycin, and methotrexate). To minimize these side effects, CA and CAPE are therefore used as the protecting drugs. It was shown that CAPE treatment attenuated streptomycin-induced injury and apoptosis in the inner ear hair cells [240]. Similarly, it was also indicated that CAPE protects against vancomycin-induced kidney dysfunction through the elevation in MDA and NO levels [241]. Recently, Sahin et al. (2013) demonstrated that CAPE decreases the oxidative stress in the retina and optic nerve in isoniazid and ethambutol-treated rats [242]. CAPE protects against mitomycin-induced clastogenesis and diminished the frequency of chromosomal aberrations and micronuclei induced by mitomycin [243]. CAPE induced the restored mitotic activity in the bone marrow cells of mice challenged with mitomycin; this has been attributed to antioxidant effects.

Another anti-tumor agent, DOX, is one of the most important chemotherapies against solid tumors. DOX or adriamycin is an anthracycline antibiotic and cytotoxic/antineoplastic agent. It is commonly used in uterine, ovarian, breast and lung cancers, Hodgkin's disease, and soft tissue sarcomas. DOX binds to the DNA and inhibits the topoisomerase, thereby preventing the DNA double helix formation and stopping the replication. Oxidative stress and free radical production are characteristics of DOX, along with antineoplastic effects and toxicity. Despite its high antineoplastic effect, DOX treatment has been restricted because of widespread toxic effects targeting the heart, kidneys, blood, and testes [244]. The problem of DOX use is the toxicity in the target tissues of patients. However, as reported, CAPE protected rat renal tissues against DOX induced toxic damages, inhibiting free radical generation [245]. Thus, as an effective cytoprotective agent against DOX-induced cardiotoxicity in rats, CAPE attenuates DOX-induced neuronal injury through its antioxidant properties. CAPE activates the antioxidant defense system via reducing ROS and increasing antioxidant enzyme activity.

On the other hand, cisplatin is an anticancer alkylating agent, effective for germ cell tumors. The clinical problem of cisplatin administration is nephrotoxicity, inducing renal failures. However, CAPE attenuated the cisplatin-induced nephrotoxicity and hepatotoxicity in rats by compensating for the alterations in blood urea nitrogen, creatinine, NO, CAT, SOD, GPx, and MPO [246, 247]. In this study, the protective effect of CAPE on cisplatin hepatotoxicity and nephrotoxicity was examined using New Zealand rabbits. Although cisplatin caused increased immunoreactivity against iNOS, CAPE treatment reduced the immunoreactive hepatocytes and CAPE administration normalized the tissue the GSH level and the XO activity. CAPE treatment significantly reduced cisplatin-induced alanine transaminase (ALT) and aminotransferase (AST) in the serum. The results suggested that cisplatin causes oxidative and nitrosative damage to hepatocytes, where cisplatin-induced increases in XO and NO could contribute to oxidative stress in the hepatotoxicity. CAPE shows it protected against cisplatin-associated biochemical and histopathological alterations. Recently, it was reported that CAPE-mediated free radical scavenging activity decreased the total number of chromosomal aberrations and abnormal metaphases induced by cisplatin [248]. CAPE also prevented cisplatin-induced oxidative changes in the liver via boosting the antioxidant defense system and reducing ROS [249]. In addition, CAPE protects against cisplatin-cytotoxicity in rats via minimizing the disturbance in ADA, CAT, and the XO/xanthine dehydrogenase (XD) system [250]. Because XD and XO catalyze the same reaction at the end steps of the purine catabolic pathway, and XD enzyme may be converted to XO to produce more superoxide radicals in oxidizing microenvironment of tissues, the pathway is also involved in cisplatin-cytotoxicity.

Another methotrexate (MTX) is also a widely used antimetabolite in cancer malignancy chemotherapy, primarily in leukemia. The efficacy of MTX is often limited by severe side effects and toxic sequelae, as it induces hepatotoxicity and renal toxicity in animal and human patients [251]. The cytotoxic effects of MTX are not selective for cancer cells, but also affect normal tissues that have a high rate of proliferation, including the hematopoietic cells of the bone marrow and the actively dividing cells of the gut mucosa. CAPE protects against MTX-induced hepatic and renal injury in experimental rats [251], reducing proinflammatory cytokines TNF-α and IL-1β levels and lipid peroxidation. The report suggested that the GSH levels and Na^+K^+-ATPase activities in hepatic and renal tissues are restored during the repeated administration of CAPE, thereby protecting the tissue membranes. Moreover, it was indicated that CAPE protected the peroxidation and total oxidant and antioxidant systems in MTX-induced spinal cord injury models. It alleviated SOD, GPx, ADA activity, and NO levels that are involved in the pathogenesis of MTX-induced spinal cord toxicity [252]. CAPE is a potential protector of cerebellar-oxidative damage induced by MTX via its antioxidant properties [253]. CAPE protects against MTX-induced testicular toxicity and oxidative injury on the main elimination organs [254].

2.4. Anti-Viral Activity

Virus infection is one of the major threats to human health, thus there is a need to prevent and treat infectious viral diseases. In a routine life style, because virus infection has possibly occurred, effective and safe antiviral or virucidal agents are needed. Virus infection treatment by antiviral or virucidal agents using natural products is the specific concern of this section. Medicinal plants are increasingly of interest as a novel source of drugs for antiviral agents, because: 1) many virus continue to evolve to resistance to commonly or specifically used antiviral drugs; 2) because viral clearance of natural products may solve the viral infectious outcome; and 3) because people routinely consume food or drinks. The antiviral agents from plant source may have an advantage over synthetic drugs for safety concerns. Especially, some natural products, such as coffee or propolis, have antiviral and virucidal activities. To date, several hundred of plants have been reported to have potential antiviral activity and several phytochemicals have been identified with a relationship between the structure-antiviral activity [255, 256, 257, 258]. CA has multi-antiviral activities against HIV [259, 260], adenovirus [259, 261], HSV [262], influenza [263], and HCV [10, 264]. As CA derivatives, caffeoylquinic acid analogues are also active against HIV [265], HIV-1 integrase [266] and respiratory syncytial virus (RSV) [267, 268]. In this section, antiviral and virucidal activities and the mechanism of CA and derivatives are described.

2.4.1. Anti-Influenza Viral Activity

The issue of influenza A virus infection calls to the forefront, socio-economic demands. Influenza viruses are specifically susceptible to the environmental responses resulting in genetic mutation and eventually spreading with a possible avian-swine-human transinfection. The infection of influenza A viruses requires its interaction with the cell surface using the surface glycoproteins hemagglutinin (HA) and neuraminidase (NA). The virus envelope HA and NA are the major targets of the protective host antibody response. Influenza A is antigenically subtyped into 16 HA and 9 NA subtypes. By the interchangeable mutation susceptibility of the virus, there were three worldwide influenza pandemics with a specificity (H1H1, H2N2, and H3N2) [269]. Since the first outbreak of the pathogenic avian influenza A (H5N1) virus in 1997, the viral strains evolve to the recent emerging strains of the 2009 swine-origin H1N1 virus and the 2013 Chinese H7N9 virus. Especially, the highly pathogenic avian influenza A (H5N1) virus is a potential threat to humans because of the spread of H5N1 avian influenza virus in birds and the bird-to-human transmission [270].

For the infection of influenza A viruses, surface glycoproteins HA binds the *N*-acetylneuraminic acid or sialic acid-linked receptor and NA cleaves the sialic acid-linked receptor, where the human-type receptor has α2,6-linked sialic acid and the avian receptor carries α2,3-linked sialic acid, respectively [271]. The site for human virus infection is in the airway epithelialium of the upper respiratory tract [269]. For efficient cell attachment and viral entry, the human virus has to recognize the α2,6-linked sialic acids, and cleavage of the 2,6-linked sugars is required for viral release. Thus, HA and NA activity are essential for a viral life cycle and transmissibility. For cleavage of an *N*-acetylneuraminic acid or sialic acid residue in glycoproteins, influenza neuraminidase (NA) enzymes are divided into two groups: group-1 NAs containing N1, N4, N5, N8; and group-2 NAs containing N2, N3, N6, N7, N9. The influenza NA has been an important target to block the life cycle of influenza virus and for anti-influenza drugs design [272]. Until now, three NA inhibitors, zanamivir, oseltamivir, and peramivir are used for the treatment and prophylaxis of influenza. Among them, oseltamivir administered orally is the first line defense drug against influenza. However, more and more influenza virus strains are resistant to oseltamivir, such as the seasonal H1N1 viruses [273] and avian H5N1 strains [374]. In 2009, the U.S. Food and Drug Administration

(FDA) issued an emergency use authorization (EUA) for the use of peramivir to treat the novel swine-origin H1N1 (A) influenza virus that was also resistant to oseltamivir. At present, this drug is mainly administered intravenously to patients but with inconvenience.

Recently, many natural products, especially some plant-derived polyphenols, have been found to exert antiviral effects against influenza virus and show inhibitory activities on NA. In particular, relatively small plant-derived polyphenols exhibited good antiviral effects against the influenza virus and show moderate inhibitory activity against NA. From natural dietary resources, the antiviral and virucidal activities of various natural products have been reported [37]. Among them, CA and CA derivatives also presented anti-influenza virus activity and NA-inhibitory activity *in vitro* [275, 276]. In order to know that the CA structure is necessary for the NA inhibitory activity, the structure–activity relationship was examined. CA inhibits the multiplication of influenza A virus *in vitro* in host cells, whereas caffeine, quinic acid, and CGA do not. The multiplication in the infected cells of the influenza virus in the presence of CA was eclipsed, while the progeny virus yield was markedly decreased in the presence of CA [277]. CA decreases the progeny virus yield, virus-induced cytopathic damage, and apoptotic nuclear fragmentation. The potential pharmacological use of CA or its derivatives is admitted for an antiviral drug against the influenza A virus. By structural modifications of the CA fragment, it was proven that CA fragment played an important role in the NA inhibitory activity. CA acts at an early time point post-infection for extensive antiviral activity, hence indicating acting in the early stages of infection. Then, CA has been used as the basic unit to build a various structured compound library. Some modified compounds exhibited strong activities against the N1 and N2 of NA-group. The 3,4-dihydroxyphenyl group of CA was important for an NA-inhibitory activity [278]. More potent NA inhibitors from CA derivatives can possibly cope with the influenza virus. Several CA derivatives, which acted as non-competitive inhibitors, exhibited potent antiviral activities in cells. Different structured compounds such as flavonoids [279], xanthones [280, 281], and diarylheptanoids [282] are also suitable to be leads for the design of novel influenza NA inhibitors. Unlike the classical ones, the structures of these compounds are various and, interestingly, most of them act as noncompetitive NA inhibitors, indicating that they likely inhibit the NA activity by binding on some other sites of the enzyme, rather than the catalytic center. From the recent viral evolution, it is thought that the H5N1 influenza virus is a significant threat to human and many isolated strains of the virus have been resistant to oseltamivir. Because CA and CA derivatives have a variety of potential pharmacological effects, such as anti-inflammatory, anti-cancer, and antiviral activities as well as anti-influenza virus activities and anti-NA activities, they could be new candidates for developing new anti-influenza agents. Until now, multiple NA inhibitors, using CA, are being developed for anti-influenza infection, however, the rapid occurrence of drug-resistant viral variants needs further multifunctional anti-influenza drugs.

2.4.2. Anti-Herpes Simplex Virus

It was known that CA inhibited HSV and adenovirus replication after infection without preventing virus adsorption. CA has also been known to exhibit antiviral and virucidal activities against HSV type 1 (DNA virus) and the polio virus (RNA virus) [38]. CA inhibited the multiplication of HSV-1 *in vitro* without any direct virucidal effect. At an early time post-infection only inhibited the formation of progeny infectious virus in the infected cells [38]. Hence, CA seems to inhibit HSV-1 multiplication before the completion of viral DNA replication. CA inhibits the multiplication of HSV-1 *in vitro* at an early stage of HSV-1 multiplication in infected cells. Concerning the anti-viral pharmacological properties of CA, Utsunomiya et al. (2008) recently reported that coffee extracts inhibited the multiplication of HSV-1 and poliovirus *in vitro* [283]. The hot water extracts of coffee grinds and commercial instant coffee solutions exhibited antiviral and virucidal activities against HSV-1, an enveloped DNA virus [283]. As the coffee ingredients, caffeine [284] and *N*-methyl-pyridinium formate [285], inhibited the multiplication of HSV-1 in HEp-2 cells. However, little is known about the mechanism underlying these effects.

2.4.3. Anti-HBV and Anti- HCV

Hepatitis B virus (HBV) is the most common serious liver infection in the world and is the major epidemiological agent of acute and chronic liver disease, worldwide [286]. HBV carriers have a high risk of development of cirrhosis, causing necroinflammation, hepatic fibrosis, end-stage liver disease, and hepatocellular carcinoma. Two therapeutic agents of immunomodulators and nucleoside analogs are medicated for chronic HBV treatment, although there is a clinical limitation with side effects of interferon -α and the substantial resistance of the virus to early nucleosidic inhibitors [287]. The current anti-hepatitis B drugs are interferon-α and lamivudine with problems of a rebound phenomenon of HBV replication after withdrawal of treatment, drug resistance during the long-term use, and some side effects. Since there is no complete cure for chronic hepatitis B, natural CA analogues possess prominent antiviral activity, especially anti-HBV and anti-HIV effects. Thus, safe and effective anti-hepatitis B agents, as alternative drugs, are needed.

In comparing the world HBV distribution map (Center of Disease Control and Prevention, 2006) [288] with the world coffee markets (U.S. Dept. of Agriculture, 2007) [289], an interesting phenomenon interested these researchers; that is, some countries with a high consumption of coffee have relatively low incidence of HBV infection, such as Finland, Sweden, and Norway, although other factors including race, environment, temperature, culture, health care, and dietary habits should be taken into consideration. A growing body of evidence from epidemiological investigations suggests that coffee consumption is inversely associated with the risk of chronic liver disease, cancer, and cirrhosis [290, 291, 292, 293, 294]. Since CA is abundant in coffee beans, coffee extracts have been conjectured to have many anti-viral activities including anti-HBV activity. The crude water extract of coffee beans, which had a high content of CA, inhibited HBV-DNA replication as well as hepatitis B surface antigen (HBsAg) and hepatitis B envelope antigen (HBeAg) secretion *in vitro*, and its polyphenol component possibly contributed to this effect. Caffeine seemed irrelevant to the antiviral activity, as the extracts of decaffeinated coffee showed an even stronger effect than that of regular coffee such as Americano. CA is an abundant plant polyphenol that has diverse antiviral activity. The antiviral activity of CA against HBV is to suppress the synthesis of HBsAg, as previously reported [295, 296]. In HepG2.2.15 cells and duck HBV infection model, the anti-HBV activity of CA was observed. In the cell model, CA inhibited HBV-DNA replication as well as HBsAg production. CA also reduced the serum duck hepatitis B virus (DHBV) level in a DHBV-infected duckling model. Both the extracts of regular coffee and that of decaffeinated coffee showed an inhibitory effect on HBV replication [296]. To clarify whether CA affects HBV replication, the antiviral activity was assessed in a HepG2.2.15 cell line [297] and in the DHBV-infected duckling model. CA inhibited HBV replication *in vitro* and *in vivo*. The antiviral activity of CA in a HepG2.2.15 cell line showed that CA inhibited HBV-DNA replication *in vitro*. CA possessed a substantial inhibitory activity against HBV-DNA replication in HepG2.2.15 cells. After eight days of treatment, both the virions in the culture medium and the core-associated DNA in the cytoplasm were reduced, and the level of HBsAg secreted in the culture medium was also down-regulated. However, the amount of HBeAg in the culture medium was decreased only by CA, suggesting that CA might perform an antiviral activity. The levels of both HBsAg and HBeAg secreted in culture medium were effectively reduced by CA, suggesting it may interfere with the capsid envelopment or the viral egression. The precise modes of action of CA need to be further explored thoroughly, although CA showed efficacy in the DHBV-infected duckling model.

On the other hand, cichoric acid as a CA derivative from *Cichorium intybus*, showed an anti-hepatitis B viral property in the D-galactosamine (D-GalN)-induced normal human HL-7702 hepatocyte injury model, the DHBV-infected duck fetal hepatocytes, and the HBV-transfected cell line HepG2.2.15 cells. Cichoric acid attenuated the D-GalN-induced HL-7702 hepatocyte injury. Moreover, cichoric acid inhibited DHBV DNA replication in infected duck fetal hepatocytes. Also, cichoric acid reduced significantly the HBsAg and HBeAg levels in HepG2.2.15 cells. It inhibited HBV DNA replication [298]. Thus, an anti-hepatitis B effect of cichoric acid could be used to design antiviral agents. Cichoric acid and dicaffeoyltartaric acid as derivatives of both CA and tartaric acid are found from *Cichorium intybus*, *Echinacea purpurea*, dandelion leaves, basil, lemon balm, and in aquatic plants, including algae and sea grasses [299]. The antiviral effects of cichoric acid against HIV-1 and HSV-I are interesting [300, 301, 302, 303]. CA analogues also exhibit a variety of bioactivities, especially anti-HBV and anti-HIV effects [266, 304]. Cichoric acid, as a natural CA derivative, stimulates cell-mediated immunity, exhibits anti-vesicular stomatitis virus and antioxidant activities, inhibits hyaluronidase, an enzyme involved in infection and inflammation, and has a stimulant effect on phagocytes and an inhibitive effect on the HIV infection [302, 305, 306].

On the other hand, in the anti-HCV activity of CA, previous human studies on coffee consumption and HCV disease in patients suggested that a specific substance, such as polyphenols in coffee, can inhibit HCV infection or propagation [2, 4, 307]. Absorbed CA is important for the inhibition of HCV infection and HCV propagation, as coffee consumption is associated with a response to peginterferon and ribavirin therapy in patients with chronic hepatitis C [4]. CA has an anti-propagation activity of HCV in an *in vitro* naïve HCV particle-infection and production system within human hepatoma-derived Huh7.5.1-8 cells. When cells were treated with CA after an HCV infection, the HCV particles at 3 and 4 days post-infection were considerably decreased. HCV-infected cells treated with CA for 4 days exhibited the decreased HCV particles. CA inhibited the initial infection between virion entry and the translation of the RNA genome, regardless of the viral genotype. Thus, CA inhibited the HCV propagation and replication of hepatitis C virus [4, 264, 308]. Because viral intracellular processes are essential for HCV propagation, CA may target the related viral pathway such as polyprotein processing, RNA replication, virion assembly, transport, release, viral entry, and viral polyprotein translation [4, 264, 309, 310]. It is possible that CA inhibits in part RNA-replication in HCV-infected cells, since it has been reported that CA inhibits HCV-replication.

2.5. Anti-Diabetic, Anti-Obese and Anti-Fatty Liver Activity

In 2013, a report from the International Diabetes Federation, stated that 380 million people are considered to have diabetes ("Diabetes Atlas," http://www.idf.org/diabetesatlas); hence this disease should be regarded as a worldwide epidemic. Diabetes is a chronic inflammatory and systemic disease. Type 2 diabetes (noninsulin-dependent diabetes mellitus, NIDDM) or (T2D) tends to strongly be induced by over-nutrition and physical inactivity. In progressed patients, a defect in insulin secretion is detected with insulin resistance in liver and adipose tissues. In parallel, the defected insulin-secretion of the pancreatic β-cells cause the onset and progression of Type 2 diabetes. Thus, because novel anti-inflammatory treatments for diabetes need to be involved, the screening and development of anti-diabetic substances have long been of interested to clinical medical scientists focusing on mainly T2D treatment from natural materials. T2D is a dysregulated metabolic disease with hyperglycemia, insulin resistance, and dysfunction of the pancreatic islet cells [311]. A prolonged disease status of T2D is linked with obesity and inflammation that cause insulin resistance and damaged pancreatic islet cells [311]. Thus, polyphenolic compounds from plant sources have been considered as preventive and therapeutic agents with the triple activities of anti-oxidation, anti-inflammation, and glucose homeostasis [312]. Although there is currently no available anti-diabetic agent from plant sources, CA has been considered as a potential therapy for T2D [313, 314, 315].

CA is easily absorbed the intestine after administration [316]. In a recent study by Adisakwattana et al. (2008), cinnamic acid exhibited insulin secreting activity in perfused rat pancreas and pancreatic β-cells (INS-1) with an increase in $[Ca^{2+}]i$ *in vitro* [317]. The presence of hydroxy residues on cinnamic acid was important as an effective insulin-releasing agent, because cinnamic acid regulated the blood glucose level by stimulating insulin secretion from pancreatic β-cells. In addition, CA is involved in antihyperglycemic activity through stimulation of insulin secretion and alleviation of insulin resistance in db/db mice [49]. Huang et al. (2009) reported that CA has effects on the glucose uptake of the insulin-resistant cells [318]. The phenolic acid enhances insulin receptor tyrosyl phosphorylation, up-regulates the expression of insulin signal proteins such as insulin receptor, phosphatidylinositol-3 kinase, glycogen synthase, and glucose transporter-2, as well as increasing glucose uptake. Jung et al. (2006) reported that CA is anti-hyperglycemic with the stimulation of insulin secretion from pancreatic β-cells [49]. When FL83B cells, derived from mouse liver, were treated with TNF-α, the induced insulin resistance was greatly reduced by CA treatment, showing the antihyperglycemic effects of CA. CA was also effective in the reduction of glucose level in C57BL/KsJ mice and in the augmentation of glucose uptake in C2C12 cells. CA increased glucose uptake in normal FL83B cells via a pathway involving no increase in the expression and tyrosyl phosphorylation of insulin receptor (IR). CA alleviates insulin resistance by improving insulin signaling and enhancing glucose uptake in insulin-resistant cells. Alleviation in insulin resistance is hence involved in the antihyperglycemic effect of CA. In experimental animals, a dietary supplement of CA improved glycemic control, blood glucose level, and lipid metabolism in diabetic mice indicating the anti-diabetic effects of CA [319, 320]. CA could protect cardiac tissue against diabetes associated dyslipidemia.

CA has been shown to have an anti-diabetic activity for insulin resistance animals because of its strong antioxidant property [321], as it was revealed that CA could also provide anti-inflammatory protection by the down-regulation of TNF-α, IL-6, and IL-1β mRNA and protein expressions in the cardiac tissue of diabetic mice [24]. CA also possesses beneficial effects on insulin secretion and on the gene expression in β cell survival, and improving glucose sensitivity. CA exhibited a promising potential to improve both insulin secretion and insulin gene expression, together with restoring the regulation of genes associated with lipotoxicity in β cells. CA and CAPE also ameliorate glucose metabolism and uptake [322, 323]. CA induces glucokinase (GK) mRNA in rat liver cells *in vitro* [324], where GK has an antihyperglycemic effect [325]. CA increases insulin secretion under hyperglycemic conditions during acute and chronic exposure in rat insulinoma cell line INS-1E cells, and raised the expression of genes involved in β-cell survival and function under normoglycemic and glucotoxic conditions, suggesting a possible role in preventing hypoglycemia [326]. CA significantly increased the expression of GLUT-2, GK, insulin 1 gene (Ins1), transcription factors-β2, and pancreatic duodenal homeobox gene-1. The increase in insulin gene expression correlates well with the increase in the expression of the transcription factors responsible for its regulation. Several other transcription factors, such as Akt1 and Akt2, are responsible for insulin gene expression in the insulin-signaling pathway and β cell function [327]. Akt1 and Akt2 gene expression was upregulated by CA. In addition, downregulation of the pro-apoptotic genes Casp3 and BAX by CA suggests a cyto-protective effect of CA on β cells, as glucotoxicity is based on the aberrant gene expression of β cell specific genes for apoptosis [328]. The gene expression of the antiapoptotic Bcl-2 induced by CA indicates an induction of pro-survival genes.

Interestingly, the heat shock protein, Hsp70 gene expression was also induced by CA. The heat shock proteins function as molecular chaperones during cellular injury and stress, to protect the cell survival. Thus, the induction of Hsp70 is known to be helpful for human diseases such as neurodegeneration, ischemic heart disease, cancer, and diabetes. Inversely, in cancer cells, Hsp90 expression is suppressed [329]. Upon CA treatment, however, hsp90 mRNA expression was up-regulated by CA, indicating the protecting role of CA on β cell survival during glucotoxicity. The expression of acetyl-CoA carboxylase (ACC)-1, an enzyme catalyzing the first step of lipogenesis, was also increased by CA, thus indicating its importance in the lipotoxicity induced β cell dysfunction. In addition, although CA exhibited insulin mimetic effects on 3T3-L1 cells and insulin resistant mouse hepatocytes [330, 322], the elucidation of the process of CA in the exocytotic status of β cells is needed for future investigations.

Another aspect of diabetes-complicated diseases is obesity, a chronic metabolic disorder caused by an imbalance of energy because of the excess consumption of nutrients and inadequate physical activity [331, 332, 233]. One of the most common characteristics among people with obesity is the development of a fatty liver [334, 335]. High-fat feeding has commonly been used to induce visceral obesity in rodent animal models because its pathogenesis of obesity is similar to that in humans [336]. The prevalence of obesity has been increasing sharply and has become a global concern of healthcare systems [337]. Excess energy intake, accompanied with low energy expenditure, induces lipid accumulation in both liver and adipose tissue leading to the development of metabolic disturbances [338]. There are various strategies to control obesity or overweight, including dietary control, exercise, and medication. Antiobesity drugs, such as orlistat and sibutramine, have modest clinical efficacy and serious side effects [339]. Thus, some botanicals might serve as a safe and natural way to manage body weight and an effective alternative to synthetic drugs [339, 340]. There is a growing interest in exploiting the potential of plants and naturally occurring materials for treating obesity. Recent studies demonstrated that a natural phenolic compound like CA can treat obesity in an obese mouse model [341]. The protective effect of CA was reported on high-fat diet (HFD) induced obesity mice and as a possible role in ameliorating the development of fatty liver. In addition, the effects of CA on HFD-induced fat accumulation in the adipose tissue of C57BL/6 mice were carefully examined. Body weight gain, adipose tissue weight, cholesterol, and TG serum levels were lowered in the CA-treated group, compared to the HFD group. The quantifications of triglycerides (TG) and cholesterol revealed that the amount of lipids in the liver of CA-treated mice was lowered compared with that of the HFD group.

Finally, an important aspect of diabetes-complicated diseases is hepatic fatty liver disease. Hepatic esterification of free fatty acid (FFA) to TG leads to the formation of a fatty liver that is accelerated by an increased lipogenesis as a result of hyperinsulinemia and decreased FFA oxidation. In a recent study [342], CA lowered the fasting blood glucose and FFA, compared with HFD mice. Thus, the possible mechanism for the antidiabetic action of CA is suggested to be the enhanced transport of blood glucose to adipose tissue. An adipose GLUT4 overexpression is known to eliminate insulin resistance and pancreatic defects in db/db mice [343]. Again, CA enhanced the GLUT 4 protein expression in adipose tissue [344]. An activation of adenosine monophosphate-activated protein kinase (AMPK), a major kinase in glucose and lipid metabolism to control metabolic disorder such as diabetes, obesity, and cancer [345], was observed by CA treatment. In that study, CA recovered the phosphorylation of AMPK, which has been reduced in C57BL/6 mice that were fed the HFD. Furthermore, AMPK activation increases fatty acid oxidation by reducing the malonyl-CoA through the inhibition of ACC, and this process upregulated carnitine palmitoyltransferase (CPT)-1a expression [346]. The activity of 3-hydroxy-3-methylglutaryl (HMG)-CoA reductase of cholesterol synthesis and the sterol regulatory element-binding protein (SREBP)-2, an adipose tissue lipogenesis regulator for fatty acid biosynthetic fatty acid synthase (FAS) and acetyl CoA carboxylase (ACC) gene, were significantly lowered by CA, compared with obese mice. These results suggest that CA activates the AMPK and lowers lipid metabolism via inhibition of the cholesterol biosynthesis in HFD-induced mice. The SREBP-1c level was also decreased by CA. In addition, FAS expression was decreased in the CA group, compared to the HFD group, and CA raised HDL levels in high-fat animals (other phenolic compounds are known to elevate the HDL concentration) [347].

Liao et al. (2013) suggested that CA exhibits a significant potential as an antiobesity agent by the suppression of lipogenic enzymes and hepatic lipid accumulation. CA administration to mice with HFD-induced obesity reduced body weight, adipose weight, and serum cholesterol and TG, hence inhibiting lipogenesis. In addition, CA increased β-oxidation through AMPK activation and improved HFD induced obesity by β-oxidation and lipolysis in liver [341]. The effect of CA on obesity in mice, which were fed a HFD, is attributed to adipose and hepatic lipid lipogenesis. The C57BL/6 mice, administered with a HFD and CA for six weeks, were examined for hyperlipidemia, hyperglycemia, visceral fat accumulation, and related enzyme activities. The levels of serum glutamyl oxaloacetic transaminase (GOT) and glutamic-pyruvate transaminase (GPT), as clinical indicators for hepatic damage, were improved in the CA-treated HFD. Thus, CA improved the high fat liver status and is more effective than those of simvastatin and silymarin. CA lowered body weight, visceral fat mass, FAS activity, and free fatty acid, compared to the HFD group. CA also lowered TG and cholesterol levels in the plasma and liver. Furthermore, CA inhibited cholesterol biosynthesis, as evidenced by HMG-CoA reductase in the liver. CA suppressed the activity of lipogenesis via SREBP-1c and its target enzyme FAS. In addition, CA resulted in increased phosphorylation of AMPK and decreased ACC, a target of AMPK, which are related to fatty acid β-oxidation in the liver.

2.6. Protection of Hypertensive and Cardiovascular Diseases

The phenolic CA has been used for treatment of coronary artery disease, myocardial infraction, cerebrovascular disease, and chronic renal failure. CA was also reported to improve microcirculation in human body as well as prevent atherosclerosis via inhibition of both LDL oxidation and ROS production [348, 349] through NF-κB regulation. The oral supplementation of CAPE also attenuates the atherosclerotic process through reduction of NF-κB activity and expression of NF-κB-regulated genes. Thus, these biological activities for CA concern the prevention of cardiovascular disease [350]. CA in serum has been shown to be incorporated into LDL particles, and as a result, the oxidation-resistance of LDL was increased [351]. CA also provide plasma triglyceride (TG)-glucose lowering and anticoagulatory protection through an increase in the plasma antithrombin-III and protein C activities in the cardiac tissue of mice. The cardiac content of CA was increasingly maintained in the cardiac tissue and plasma. CA lowered cardiac levels of IL-1β, IL-6, TNF-α and monocyte chemoattractant protein (MCP)-1. The positive activities were accompanied with the enhanced up-regulation of cardiac mRNA expression of GPX, SOD, and CAT and down-regulation of IL-1β, IL-6, TNF-α, and MCP-1 mRNA expression in mice [352].

Bhulla et al. (2014) recently reported that CA and its derivatives are effective in the treatment of hypertension through target modulation of the renin–angiotensin–aldosterone system (RAAS) [52]. Aldosterone, a sodium retaining steroid, is a downstream member. Principally, plant secondary metabolites such as flavonoids (i.e., quercetin) [353] and phenolic acids exhibit antihypertensive activities [354]. CA also has an anti-hypertensive action in *in vitro* [355] and *in vivo* models [356]. A recent study [52] reported that CA, a potential multi-target RAAS blockade, regulated multi RAAS targets, including renin, angiotensin-converting enzyme (ACE), and aldosterone *in vitro*. To assess a multi-target RAAS inhibition, the inhibition of aldosterone production in adrenal glands of spontaneously hypertensive rats (SHR) was examined. The SHR are used for pharmacological studies on hypertension and removal of the adrenal gland from SHR results in a fall of blood pressure, indicating the involvement of adrenal hormones on the development of hypertension in SHR [357, 358]. The kidneys of SHR and adrenal cultures were then treated with CA and its derivatives (500 mM) in comparison with a drug, captopril. The results obtained with CA and derivatives had a significant effect on the aldosterone levels and they emerged as inhibitors of aldosterone activity *in vitro*, although captopril was strong in inhibiting aldosterone activity.

Additionally, the renin inhibitory activity is regarded as a key anti-hypertension drug. The structural modification of CA had a significant impact on the rennin inhibitory activity of its derivatives. Although Aliskiren, as a direct renin inhibitor, exhibited strong renin inhibition with IC_{50} value of 0.1 mM, whereas CA was a weak renin inhibitor. Addition of a phenyl group in the CAPE structure significantly improved the property of CA through better interaction of the CAPE with renin enzyme upon its esterification. CA inhibited renin and ACE activities and production of aldosterone. CA modulated aldosterone production without toxicity against human fibroblasts WI-38 cells. Previously, CA was reported as a potent anti-hypertensive agent [359] and an inhibitor of ACE. Hidalgo et al. (2012) [460] showed similar ACE inhibitory activities of CA and derivatives using a different system [361, 362, 363]. Quercetin and potent antihypertensive compounds, such as epicatechin–tetramer, epicatechin-hexamer and quercetin-3-O-glucuronic acid [364, 365], showed ACE inhibitory activity. CA skeleton as the parent phytochemical structure acted as a novel dual RAAS inhibitor. Further studies on human clinical models are required for the *in vivo* biochemical and pharmaceutical potential of the novel compound. Direct renin inhibition by polyphenols and other bioactives was also reported. Baicalin, a flavone member of the flavonoid class of polyphenols, exhibits a strong renin inhibition with IC_{50} value of 120 mM [366]. Catechin has a strong inhibitory action against renin *in vitro* [24]. Catechins (flavan-3-ols), including epigallocatechin gallate (IC_{50}= 45 mM), (-)-epicatechin gallate (IC_{50}=619 mM), and (-)-epigallocatechin (IC_{50}=2175 mM) have exhibited antihypertensive potential through renin inhibition *in vitro* [367]. Similarly, saponins (triterpenoids) showed the human renin inhibitory activity [368, 369]. These suggest that CA derivatives, together with CA, exerted strong anti-renin activity, comparable to other renin inhibitors. Moreover, the addition of large molecular weight moieties to the CA skeleton improved renin inhibition.

2.7. Neuroprotective Activity of CA

There is a growing interest about the neuroprotective effects of commonly used food-derived active phenolic substances. Dietary polyphenols present in a wide variety of plant derived foods and beverages have been investigated for their potential to provoke protection against neuronal injury [370, 371]. Although the neuroprotective actions of flavonoids are well reported, those of HCs and other phenolic acids are less established. For example, the consumption of red wine, containing flavonoids, HCs, and other polyphenols, has been associated with a reduction in the neurological disorders, such as AD [372], and cerebral aging [373]. This protection is mainly attributed to the anti-inflammatory properties and the controlled signaling cascades for NO production and neuronal survival [374]. Thus it is a reasonable hypothesis that the phenolic components contribute to neuroprotection.

The neuroprotective potentials of a natural polyphenolic CA have recently received much attention because of its pharmacological effect on cognitive dysfunction, AD, memory, and degenerative neurodysfunction. First, CA has been reported to protect against Aβ-induced neurotoxicity and peroxynitrite-induced neuronal damages *in vitro*, exhibiting potential neuroprotective effects [375]. It was shown that the hydroxycinnamates CA and *p*-coumaric acid and hydroxyphenethyl alcohol tyrosol protect neurons against injury induced by 5-S-cysteinyl-dopamine [376]. Tyrosol and CA improved the neurological dysfunction by focal cerebral ischemia, greater than those observed for the flavonoids, (+)-catechin, (-)-epicatechin and quercetin. This indicates that HCs, phenolic acids, and phenolic alcohol induce neuroprotection to a similar extent to that of flavonoids [377]. Huang et al. (2013) investigated the action mechanisms of CA and CAPE on acrolein-induced cell damage as well as AD-like pathologies in immortalized mouse hippocampal neurons and HT22 cells [378]. Interestingly, CA and CAPE exerted a bell-shaped dose- and time-dependent protection to the HT22 cells. The best neuroprotection doses were at CA 25 μM and CAPE 30 μM, respectively. Similarly, Scapagnini et al. (2002) also observed that CAPE increases HO-1 enzyme (a redox-sensitive protein against oxidation) expression [379]. Nevertheless, although there are many experimental approaches, it is not clear as to the exact mechanism for the action mode of CA on neuroprotection. It is hypothesized that CA and CAPE can prevent acrolein-induced neuronal cell death in animals and also restore the acrolein-induced AD pathologies in HT22 cells. Thus, it is possible that CA and CAPE can provide a promising approach for neurodegenerative diseases, as shown in experimental brain injury models.

Furthermore, behavioral and antioxidative examination *in vivo* was performed for CA against the D-Gal-induced neurotoxicity in mice [380]. CA significantly attenuated cytotoxicity and improved the behavioral performance of the mice in both the Morris water maze and step-down avoidance test. In an action mode, CA decreased the antioxidative parameters such as SOD, GPx, CAT, and MDA. Although CA partially inhibits the 5-LP, the precise processes are not explained. The same group also confirmed that CA derivatives protect neuronal disorders such as stroke, neurodegenerative diseases, hypoxic-ischemic brain or spinal injury, as these compounds prevented NO and glutamate-induced neuronal loss cerebellar granule neuron cells.

2.7.1. Dementia Protection

Brain ischemia-accompanied inflammation and oxidative stress lead to a neuronal loss and cognitive deficits. When CA was treated after ischemia at the different doses of 2, 20, and 60 mg/kg/day, i.p., CA improved the infarct and neurological deficit scores in the parameters of the working, spatial, and long-term aversive memory deficits, as well as synaptic formation and function [381]. Thus, CA exerts neuroprotective and antidementia effects through protection of the neural cells and synapses in ischemic brain injury.

Age-dependent neurodegenerative AD causes dementia by the deposition of Aβ peptide and intraneuronal neurofibrillary tangles, as well as loss of the cholinergic synapses. The accumulated Aβ peptide plaques cause AD pathology including neurotoxicity, oxidative damage, and inflammation. The accelerated Aβ accumulation induces the aggregation and tau hyperphosphorylation, resulting in neurotoxicity [382]. Until now, several medications with CA have been evaluated for their anti-AD effects [383, 394]. For example, CA protected the Aβ-induced neurotoxicity through inhibition of calcium influx and tau phosphorylation, which is upregulated in AD. The Aβ-induced tau hyperphosphorylation induces neurocytotoxicity to neurons [385]. Hence, the inhibition of tau phosphorylation is a therapeutic target for AD. Although antioxidant, estrogens, and nonsteroidal anti-inflammatory drugs, such as ibuprofen and COX-2 inhibitors, were suggested as therapeutic candidates, all these were noneffective. Hence, CA as a potential AD drug reduced the Aβ-induced neurotoxicity and CA against Aβ-induced toxicity is an example that could be of use. Recently, CA also protected the PC12 cells against Aβ-induced toxicity, and attenuated the intracellular calcium influx and tau phosphorylation by deactivation of glycogen synthase kinase-3β (GSK-3β).

The fact that CA decreases the oxidative stress, calcium influx, and tau hyperphosphorylation induced by Aβ, thereby suggests an antioxidant capacity against the Aβ-induced oxidative stress. The early depletion of antioxidant compounds, such as glutathione, develops AD pathology [386]. It was also reported that CA protects rat cortical neurons by decreasing the glutamate-induced increase in intracellular calcium [387]. Thus, inhibition of calcium influx is one of the mechanisms by which CA exerts a neuroprotective effect against Aβ in neuronal cells.

2.7.2. Protection from Neurodegenerative Diseases and the Cholinergic System

AD is age-related for a neurodegenerative disease and also is a recently growing health problem. At present, for the first line pharmacotherapy for AD, patients receive cholinergic drugs as a cholinesterase inhibitor including donepezil, galantamine, and rivastigmine [388]. AChE inhibitors, such as tacrine or donepezil, are the general medications for the AD because they suppress AD symptoms in the memory deficits [389]. The cognitive dysfunctions in AD patients are caused by deficits of the cholinergic neurotransmission in cortical regions of the brain [390]. The AChE degrades acetylcholine (ACh) and AChE is distributed throughout the neuronal and non-neuronal tissues. Among the two cholinesterases of AChE (E.C. 3.1.1.7) and BChE, AChE hydrolyzes acetylcholine after its binding to receptors [391]. AChE is ubiquitously present in the brain, erythrocytes, lymphocytes, and neuromuscular junction for the turnover of acetylcholine at the cholinergic synapses and the neuromuscular junction [392]. AChE plays a role in the cellular adhesion, neurite extension, and postsynaptic differentiation in the central nervous system (CNS) [393]. AChE contributes to the muscle functions for the contractile, fatigue and myasthenia gravis [394, 395]. Thus, AChE is a therapeutic target in neurodegenerative diseases [396, 397] and so AChE inhibitors can be used for treatment of the symptoms.

Aluminium-exposed experimental AD animals exhibit AD-like neurodegenerative disorders through oxidative inflammation in the brain, thereby increasing Aβ protein expression, tau protein hyperphosphorylation, cholinergic degeneration in the cortex and hippocampus [398, 399, 400]. In the aluminium chloride ($AlCl_3$)-induced dementia rats, CA protected them from the brain inflammation and neurodegeneration [401]. An *in vivo* effect of CA in aluminium toxicity was to protect the oxidative stress-mediated neurodegeneration, neuroinflammation, and cognition capacity. The neuroprotective effect of CA was also confirmed by behavioral and AChE activity in the brain cortex and hippocampus region. In the Morris water maze test, acquisition, consolidation, and recall of learned task, as memory assessment parameters, were specifically recovered by CA [401]. The anti-dementia and memory-enhancing effect of CA was dependent on the AChE [402] and 5-LP modulation [403]. An *in vivo* CA decreased the AChE activity in the cerebral cortex and striatum, thus showing improved memory and interfered cholinergic signaling. CA-promoted cholinergic signaling parameters in CNS are an alternative source for the treatment of cholinergic dysfunction. Thus, CA improved the memory and learning with AChE inhibition, and CA is considered to be potentially therapeutic in the cholinergic diseases [402]. Since acetylcholine-keeping AChE inhibitors are used in the memory and cognitive deficits, *in vivo* decreased AChE activity by CA in the cortex and striatum is meaningful. As the cholinergic system is involved in brain circulation [404], CA contributes to the cerebral circulation and improvement of cognitive performance. In another study [402], CA *in vitro* and *in vivo* increased the AChE activity and behavioral parameters in rats. An *in vivo* 50 and 100 mg/kg of CA increased the AChE activity in the cerebellum, hippocampus, hypothalamus, pons, lymphocytes, and muscles. The amount of 100 mg/kg of CA improved the step-down latencies in the inhibitory avoidance. Similarly, CGA also altered the AChE activity in CNS [405]. CA can inhibit *in vitro* the AChE activity from the muscle. CA may modify acetylcholine homeostasis, inducing the cholinergic deficits in the neuromuscular junction. In parallel, CA protects against hydrogen peroxide-induced oxidative damage in the brain [405], and cerebral ischemia [406], and prevents brain damage, as well as behavioral and biochemical changes caused by aluminum [407].

To date, there are currently two different licensed AChE inhibitors of galantamine and rivastigmine from natural products, which can alleviate the cognitive symptoms in dementia [408]. Considering biologically active drug candidates, such as paclitaxel or taxol, are identified from plants, there is a need for more promising AChE inhibitors from the plant kingdom.

2.7.3. Protection from Parkinson's Disease and Epileptogenesis

A recent study showed that CA has a dopaminergic neuroprotective and dopamine-restorative capacity for Parkinson's disease, where neuro-inflammation is a key mechanism for Parkinson's disease [409] and proinflammatory cytokines such as TNF-α, IL-1β, IL-6, and NO are increased in Parkinson's disease. Thus, CA is an effective preventive agent against the development of neurodegenerative diseases such as Parkinson's disease. Parkinson's disease represents a default of dopaminergic neurons in the substantia nigra associated with a massive astrogliosis and excessive microglial activation [410]. Glial fibrillary acidic protein (GFAP) and an ionized calcium binding adaptor molecule are associated with the inflammation [411]. Thus, these markers are related to inflammatory damage in Parkinson's disease.

On the other hand, CA functioned as a potent neuroprotective agent against the development of Parkinson's disease [153]. CA reduces chemical dopamine-induced neurotoxicity in mouse cortical neurons [412]. CA can penetrate the blood-brain barrier (BBB) and CA intake elevated the brain CA content in experimental animals, thereby recovering inflammatory damage and dopamine loss [153]. Inflammatory injury in brain enhances the expression of IL-1β, IL-6, TNF-α, IL-4, IL-10, iNOS, neuronal NOS, COX-2, GFAP, ionized calcium binding adaptor molecule, and PGE2, while CA intake declined the those expression in the brain striatum of an animal [412]. CA intake maintained tyrosine hydroxylase (TH) activity and dopamine synthesis. The brains consequently retained neurotransmitters such as dopamine and 3,4-dihydroxyphenylacetic acid (DOPAC) [453]. TH catalyzes for dopamine synthesis with dopaminergic functions where TH converts tyrosine to L-DOPA-dopamine [413]. Inflammation decreased the expression of TH, hence causing the inflammatory dopamine loss. CA counteracted neuro-toxic effects of inflammation and reserved the expression of brain neurotrophic factors and TH for dopamine synthesis in brain. In summary, CA could prevent Parkinson's disease progression through regulating neurotrophic factors, stabilizing TH, and protecting dopaminergic neurons.

In the study of Coelho et al. (2015) on the investigation to prevent epileptogenesis from natural compounds, CA inhibited principal inhibitory neurotransmitter γ-aminobutyric acid (GABA) transaminase activity *in vitro* [414]. Although epilepsy is characterized by spontaneous recurrent seizures and represents one of the most common neurological diseases, the cellular and neurocircuit bases of epilepsy are poorly understood and are subsequently without appropriate medications. The pentylenetetrazole (PTZ) kindling model of epilepsy is the prevalent model of primary generalized epilepsy, which entails chemical kindling in mice, as tonic seizures are induced by the intraperitoneal (i.p.) administration of PTZ. CA positively improved the parameters of free radicals, SOD, and genotoxic activity when the effect of CA was examined on the tonic seizures induced by PTZ in mice. The positive controls, such as diazepam or vigabatrin, were comparable with the CA outcome. CA showed neuroprotective action against oxidative DNA damage in the epilepsy model, however a direct antiepileptogenic effect *in vivo* was not observed. Although CA was unable to prevent the epileptogenesis induced by PTZ [414], CA decreased the percentage and increased the latency of seizures in an acute model of seizure induced by pilocarpine [415]. In addition, CA increased the latency of seizures in a different model of PTZ kindling [416]. Thus, CA presents minimizes an exacerbated increase of free radicals occurring in epileptogenesis. Although CA was not directly able to prevent the establishment of a kindling state, CA reduces free radicals and the genotoxic damage caused by the PTZ-kindling model.

2.8. Cell Differentiation, UV-Irradiation Protection and Anti-Skin Aging

Previously, Jin et al. (2014) reported the megakaryocytic differentiation of leukemic cells by CAPE [210]. CAPE induced the differentiation of human chronic myelogenous K562 cells into the megakaryocytic lineage by ganglioside GM3-mediated mechanism. CAPE-induced differentiation of K562 cells with the expression of megakaryocytic markers, requiring Erk1/2 activation and p38 MAPK inhibition. Together, lactosylceramide α-2,3-sialyltransferase (hST3Gal V) and ganglioside GM3 levels were in parallel with CAPE effect via activation of the cAMP response element-binding protein (CREB). Because leukemic cells acquire immortality, the cells are frequently used as suitable model systems to study the mechanisms of cell differentiation of unipotent and multipotent hematopoietic cells. For example, the K562 cell line, as a model of differentiation, is capable of differentiating into erythrocytic, monocytic, granulocytic, and megakaryocytic lineages [417]. The most well-studied inducer is PMA and it induces the megakaryocytic differentiation [418]. In the study of CA, CAPE enhanced ATRA-induced granulocytic differentiation in K562 cells, and also induced apoptosis in a human promyelocytic HL-60 cell line [419]. The CAPE-mediated megakaryocytic differentiation of the leukemia K562 cells is potentially applicable to leukemia treatment. Interestingly, ganglioside GM3 has a crucial role in the differentiation of induction by CAPE treatment. The ganglioside GM3 is synthesized by ST3Gal V enzyme and GM3 is associated with megakaryocytic differentiation, but not with an erythrocyte differentiation [417]. Also, hST3Gal V activity is remarkably elevated in a time-dependent manner during megakaryocytic differentiation [418]. CAPE suppresses the production of leptin during differentiation to adipocytes in 3T3-L1 cells [420]. As indirect evidence, rosmarinic acid, an ester of CA and 3,4-dihydroxyphenyllacticacid, potentiates all-trans retinoic acid-induced macrophage differentiation in acute promyelocytic leukemia cells [421]. Moreover, the cells had a typical expression of CD11b+ and CD14+ cells depended on ERK-NF-κB axis activation. Thus, it was suggested that CA-containing rosmarinic acid also plays an important role as a differentiation agent for the functional macrophage differentiation in acute promyelocytic leukemia cells. Additionally, the differentiated macrophages might have anormal life span and they could die. These data indicate that co-treatment with rosmarinic acid and retinoic acid has potential as an anti-leukemic therapy in acute promyelocytic leukemia cells.

Recently, Kim et al. (2014) screened novel peroxisome proliferator-activated receptors (PPAR)-α agonists from natural compounds and found that CA induces keratinocyte differentiation to restore skin barrier homeostasis [422]. CA is a potent stimulator of keratinocyte differentiation and epidermal permeability barrier. During epidermal differentiation, keratinocytes synthesize lipids for the epidermis [423]. Thus, epidermal homeostasis is a target for the promotion of skin barrier function. The potential of CA can be applicable as a skin protection cosmetic having epidermal permeability barrier function. CA activated the PPAR-α-responsive element (PPRE) and cornified envelope (CE) formation in keratinocytes. The CA-mediated PPAR-α transactivation promoted the expression of CE formation-related proteins such as involucrin and transglutaminase-1. CA is regarded as a skin therapeutic agent for xerosis and atopic dermatitis. CA showed similar activity for synthetic compounds such as clofibrate and WY14643, thus suggesting that CA can be a candidate for epidermal permeability barrier recovery. On PPRE transactivation activity, CA was an efficient PPAR-α agonist candidate. CA protects skin cells through anti-inflammatory and anticancer properties during exposure to ultraviolet radiation [154]. In cells, CA reduces DNA methylation in human cancer cells, resulting in retarded tumor growth and regulated epigenetics [424].

UV radiation causes inflammation, gene mutation and immunosuppression in the skin cells, inducing photocarcinogenesis and photoaging. Normal lymphocytes are highly sensitive to the damaging effect of UV-radiation. In the study of Prasad et al. (2009), the photoprotective effect of CA has been reported in the UVB with a wavelength range (280–320 nm) irradiated human blood lymphocytes [425]. Lymphocytes, pretreated with increasing concentrations of CA (l, 5, and 10 μg/mL) for 30 min, were irradiated and lipid peroxidation, antioxidant defense status, cell viability, and DNA damage were examined. UVB-irradiation causes increased levels of lipid peroxidation, DNA damage and decreased antioxidant status, cell viability in human lymphocytes. CA reduced the levels of lipid peroxidation markers, i.e., thiobarbituric acid reactive substance, lipid hydroperoxide, conjugated diene, and decreased DNA damage in UVB-irradiated lymphocytes. Furthermore, CA maintains antioxidant status and decreased UVB-induced cytotoxicity. The maximum dose of CA (l0 μg/mL) normalized the UVB induced cellular changes indicating the photoprotective effect of CA in irradiated lymphocytes. UV, in particular UVB affects human health [426] and the ongoing thinning of the ozone layer results in greater UVB transmission to earth. The effects of UVB irradiation on normal human cellular system include sunburn, inflammation, tanning, immunosuppression, photoaging, and photocarcinogenesis [271]. The pathologies of UVB are the formation of ROS, apart from direct excitation of DNA [428]. The imbalance between oxidants and antioxidants shifts the cellular redox state to the oxidized site and subsequently modulates redox-sensitive cellular signal transduction pathways and gene expression [429]. The use of antioxidants to inhibit photooxidative toxicity is important in preventing and treating photodamage [430]. Natural antioxidants appear to be promising in photoprotectors [431]. CA exhibits radioprotective effects in human lymphocytes [432].

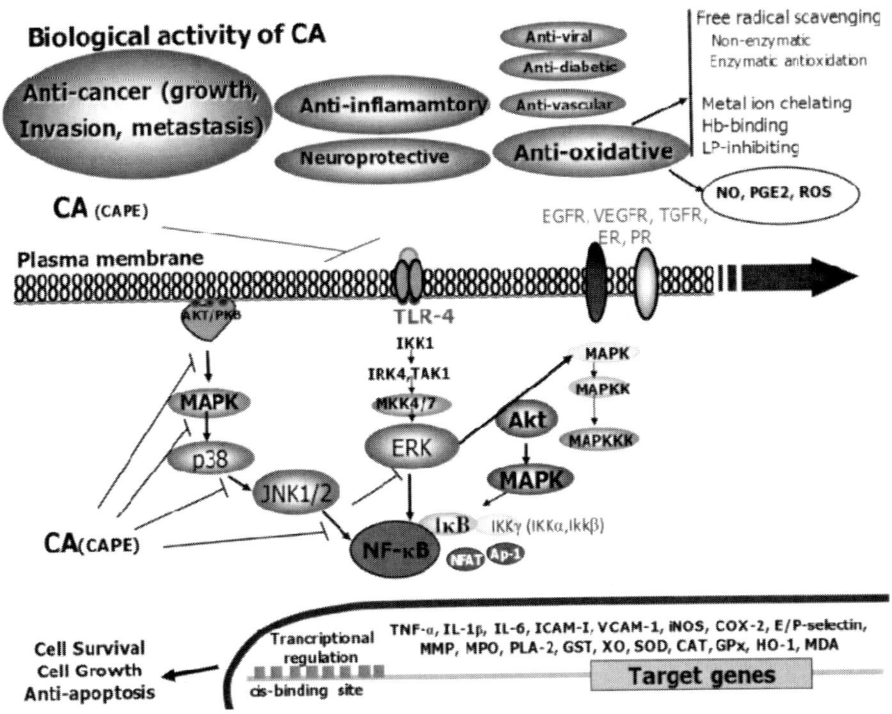

Figure 4. Biological activities of caffeic acid in cellular phenotype and human diseases.

In contemporary times, anti-skin ageing and anti-melanogenic activity have been a focus for the so-called beauty wellbeing industry and cosmetics business. The caffeoyl moiety is structurally similar to the natural substrate of tyrosinase and thus interacts with the enzyme [433]. Given the fact that hydroxamic acid moiety is a chelator of the dinuclear copper in the enzyme's active site, double- tyrosinase inhibition may be possible. Because it's a natural antioxidant, CA can be applicable for cosmetic and medicinal additives. Physicochemically CA is partially water-soluble, and water-soluble CA derivatives are useful in vitamin C-harboring cosmetics. Melanogenesis, by which melanins are produced, is a skin protective reaction during solar radiation. Specific melanin isotypes induce skin pigmentation with free radicals. HCAs, as inhibitors of melanin production, are used for hydroxyl cinnamoyl-amino acid hybrid with demelanizing activity [422, 434]. The CA-based inhibited tyrosinase activity, by which tyrosinase is involved in the first step of the melanin biosynthesis by catalyzing the two-step oxidation of L-tyrosine to L-dopaquinone. The compounds were safe and suitable for the cosmetic treatment of skin pigmentation. Thus, it is suggested that CA derivatives are applicable with potential application in food, beverage, and cosmetic industries. Recently, the antimelanogenic activity of CA and CAPE in α-melanocyte stimulating hormone (α-MSH)-stimulated B16-F10 melanoma cells was observed and mechanistically elucidated [61]. CAPE reduced α-MSH-stimulated melanin synthesis by suppressing expression of melanogenic enzymes such as tyrosinase, tyrosinase-related protein-1 (TRP-1), and tyrosinase-related protein-2 (TRP-2). On the other hand, the expression and nuclear translocation of microphthalmia-associated transcription factor (MITF) is a key transcription factor for tyrosinase expression. CAPE inhibited transcriptional activity of a tyrosinase promoter by suppressing the interaction of MITF protein with an M-box containing a CATGTG motif on the tyrosinase promoter. Most antimelanogenic agents act specifically to reduce the melanin synthesis by means of several processes: interfering with transcription or post-translational modification of melanogenic enzymes, inhibiting enzyme activity of the enzymes, reducing byproducts such as ROS, inhibiting transfer of melanins, and accelerating skin turnover [435, 436]. Among these mechanisms, the transactivation function of MITF was reported as an initial and crucial regulator on the transcription of melanogenic enzymes, including tyrosinase, TRP-1, and TRP-2 [437, 438]. Some substances such as all-trans-retinoic acid, C2-ceramide, and dioxinodehydroeckol are reported to be able to inhibit MITF expression through regulating ERK or Akt signaling pathways related to melanogenesis [439, 440, 441]. The suppression of MITF transactivation activity has been recognized as being applicable to tyrosinase-mediated melanogenesis as a more potent molecular target. CAPE inhibited the MITF-transactivation for target genes, including melanogenic tyrosinase, TRP-1, and TRP-2, suggesting the new therapeutic treatment of hyperpigmentation disorders.

Discussion and Conclusion

Health and beauty are significant parts of human concerns that continue to seek the best way to maintain the best status. The interest in human health and well-being, including beauty and longevity, have been continual subject and attraction to different cultures, which human beings have pursued from ancient eras to the present time. Research in biotechnology has changed dramatically in recent years by many different philosophical concepts. Several pharmaceutical and cosmetic industries followed a new research revolution: the return to nature from the artificial. The pharmaceutical and cosmetic production have been motivated with a new ideology for the next generations. The old bottle is now filled with new and sweet pharmacological contents.

Dietary natural polyphenols as biologically active constituents in plants, plant-derived foods, and food supplements have received considerable attention because of their beneficial effects for human health as antioxidants. Since polyphenols shows cell protection against environmental toxicants, the administration of antioxidants may thereby help to remove ROS and thus improve clinical outcomes. Dietary antioxidants can enhance cellular defense and help to prevent oxidation damage to cellular components. Consumption of polyphenols prevents chronic human diseases such as cancer, coronary heart disease, inflammatory disease, and osteoporosis. Polyphenols, mainly classified into phenolic acids and flavonoids, are readily oxidizable compounds in nature, and thus show potent antioxidant activities. Among them, HC or CA is a ubiquitous component of plants and propolis, and thus it belongs to the well-known phenylpropanoids. CA is esterified with quinic acid for the CGA. Antioxidant CA has exhibited biological and pharmacological properties such as immunomodulatory, anti-inflammatory, anti-cancer, anti-diabetic, anti-cardiovascular, neuroprotective, and intracellular signaling cascades-regulating activities, thereby, suggesting its use as a natural source for safe anti-human disease drugs. CA and its derivatives are ingredients of foods and drinks that are normally consumed and hence have an advantage over synthetic drugs. However, little is still known about the molecular and biological mechanisms of the above, as well as the full spectrum of their healthy effects and action. The supplement of CA might be helpful for the prevention or alleviation of diabetes, cardiomyopathy, cancer, viral and neurological disorders, however, further study is necessary to corroborate the efficacy and safety of CA before using it on humans. In addition, the fusion sciences using chemiobiological modification technology is needed for efficient drug development using the CA as a leader compound.

REFERENCES

[1] Touaibia, M., Jean-François, J. & Doiron, J. (2011). Caffeic Acid, a versatile pharmacophore: an overview. *Mini Rev Med Chem.*, *11*(8), 695-713.

[2] Modi, A. A., Feld, J. J., Park, Y., Kleiner, D. E., Everhart, J. E., Liang, T. J. & Hoofnagle, J. H. (2010). Increased caffeine consumption is associated with reduced hepatic fibrosis. *Hepatology*, *51*, 201-209.

[3] Bravi, F., Bosetti, C., Tavani, A., Gallus, S. & La Vecchia, C. (2013). Coffee reduces risk for hepatocellular carcinoma: an updated meta-analysis. *Clin Gastroenterol Hepatol.*, *11*, 1413-1421.

[4] Freedman, N. D., Everhart, J. E., Lindsay, K. L., Ghany, M. G., Curto, T. M., Shiffman, M. L., Lee, W. M., Lok, A. S., Di Bisceglie, A. M., Bonkovsky, H. L., Hoefs, J. C., Dienstag, J. L., Morishima, C., Abnet, C. C. & Sinha, R. (2009). Coffee intake is associated with lower rates of liver disease progression in chronic hepatitis C. *Hepatology*, *50*, 1360-1369.

[5] Higdon, J. V. & Frei, B. (2006). Coffee and health: a review of recent human research. *Critical Reviews in Food Science and Nutrition*, *46*(2), 101–123.

[6] Nardini, M., Cirillo, E., Natella, F. & Scaccini, C. (2002). Absorption of phenolic acids in humans after coffee consumption. *J Agric Food Chem.*, *50*, 5735-5741.

[7] Renouf, M., Guy, P., Marmet, C., Longet, K., Fraering, A. L., Moulin, J., Barron, D., Dionisi, F., Cavin, C., Steiling, H. & Williamson, G. (2010). Plasma appearance and correlation between coffee and green tea metabolites in human subjects. *Br J Nutr.*, *104*, 1635-1640.

[8] Manach, C., Scalbert, A., Morand, C., Remesy, C. & Jimenez, L. (2004). Polyphenols: food sources and bioavailability. *Am. J. Clin. Nutr.*, *79*, 727−747.

[9] Radtke, J., Linseisen, J. & Wolfram, G. (1998). Phenolic acid intake of adults in a Bavarian subgroup of the national food consumption survey. Z. *Ernaehrungswiss*, *37*, 190−197.

[10] Tanida, I., Shirasago, Y., Suzuki, R., Abe, R., Wakita, T., Hanada, K. & Fukasawa, M. (2015). Caffeic acid, a coffee-related organic acid, inhibits the propagation of hepatitis C virus. *Jpn J Infect Dis.*, In Press.

[11] Radoï, L., Paget-Bailly, S., Menvielle, G., Cyr, D., Schmaus, A. & Carton, M. (2013). Tea and coffee consumption and risk of oral cavity cancer: results of a large population-based case-control study, the ICARE study. *Cancer Epidemiology*, *37*(3), 284–289.

[12] Sang, L. X., Chang, B., Li, X. H. & Jiang, M. (2013). Consumption of coffee associated with reduced risk of liver cancer: a meta-analysis. *BMC Gastroenterology*, *13*, 34, 23433483.

[13] Tian, C., Wang, W., Hong, Z. & Zhang, X. (2013). Coffee consumption and risk of colorectal cancer: a dose–response analysis of observational studies. *Cancer Causes Control*, *24*(6), 1265–1268.
[14] Wilson, K. M., Bälter, K., Möller, E., Adami, H. O., Andrén, O. & Andersson, S. O. (2013). Coffee and risk of prostate cancer incidence and mortality in the cancer of the prostate in Sweden Study. *Cancer Causes & Control*, *24*(8), 1575–1581.
[15] Brown, J. C. & Jiang, X. (2013). Activities of muscadine grape skin and polyphenolic constituents against *Helicobacter pylori*. *Journal of Applied Microbiology*, *114*(4), 982–991.
[16] Rashed, K. N. & Butnariu, M. (2014). Isolation and antimicrobial and antioxidant evaluation of bio-active compounds from *Eriobotrya japonica* stems. *Advanced Pharmaceutical Bulletin*, *4*(1), 75–81.
[17] Darvesh, A. S. & Bishayee, A. (2013). Chemopreventive and therapeutic potential of tea polyphenols in hepatocellular cancer. *Nutrition and Cancer*, *65*(3), 329–344.
[18] Malerba, S., Galeone, C., Pelucchi, C., Turati, F., Hashibe, M. & La Vecchia, C. (2013). A meta-analysis of coffee and tea consumption and the risk of glioma in adults. *Cancer Causes & Control.*, *24*(2), 267–276.
[19] Sirota, R., Gorelik, S., Harris, R., Kohen, R. & Kanner, J. (2013). Coffee polyphenols protect human plasma from postprandial carbonyl modifications. *Molecular Nutrition & Food Research*, *57*(5), 916–919.
[20] Harborne, J. B. In Natural Products. Their chemistry and biological significance. Longman Scientific and Technical Publisher, *Co. England*, 1995.
[21] Kalonia, H., Kumar, P., Kumar, A. & Nehru, B. (2009). Effect of caffeic acid and rofecoxib and their combination against intrastriatal quinolinic acid induced oxidative damage, mitochondrial and histological alterations in rats. *Inflammopharmocology*, *17*, 211–219.
[22] Rezg, R., Mornagui, B., Santos, J. S., Dulin, F., El-Fazaa, S., El-Haj, N. B., Bureau, R. & Gharbi, N. (2015). Protective effects of caffeic acid against hypothalamic neuropeptides alterations induced by malathion in rat. *Environ Sci Pollut Res Int.*, *22*(8), 6198-6207.
[23] Fazili, N. A. & Naeem, A. (2015). Anti-fibrillation potency of caffeic acid against an antidepressant induced fibrillogenesis of human α-synuclein: Implications for Parkinson's disease. *Biochimie*, *108*, 178-185.
[24] Chao, P. C., Hsu, C. C. & Yin, M. C. (2009). Anti-inflammatory and anti-coagulatory activities of caffeic acid and ellagic acid in cardiac tissue of diabetic mice. *Nutr Metab (Lond)*, *6*, 33.
[25] Fernandez, M. A., Saenz, M. T. & Garcia, M. D. (1998). Antiinflammatory activity in rats and mice of phenolic acids isolated from Scrophularia frutescens. *J. Pharm. Pharmacol.*, *50*, 1183–1186.
[26] Maurya, D. K. & Devasagayam, T. P. (2010). Antioxidant and prooxidant nature of hydroxycinnamic acid derivatives ferulic and caffeic acids. *Food Chem Toxicol.*, *48*(12), 3369-3373.
[27] Chan, J. H. & Ho, C. T. (1997). Antioxidant activities of caffeic acid and its related hydroxy cinnamic acid compounds. *J Agric Food Chem*, *45*, 2374–2378.
[28] Rice-Evans, C. A., Miller, N. J. & Paganga, G. (1996). Structure-antioxidant activity relationships of flavonoids and phenolic acids. *Free Radic Biol Med*, *20*, 933-956.

[29] Mori, H. & Iwahash, H. (2009). Antioxidant activity of caffeic acid through a novel mechanism under uva irradiation. *J Clin Biochem Nutr*, *45*, 49-55.

[30] Gulcin, I. (2006). Antioxidant activity of caffeic acid (3,4-dihydroxycinnamic acid). *Toxicology*, *217*, 213-220.

[31] Nardini, M., Natella, F., Gentili, V. Di Felice, M. & Scaccini, C. (1999). Effect of caffeic acid dietary supplementation on the antioxidant defense system in rat: an in vivo study, Arch. *Biochem. Biophys.*, *342*, 157–160.

[32] Rice-Evans, C. A., Miller, N. J. & Paganga, G. (1997). Antioxidant properties of phenolic compounds. *Trends in Plant Science*, *2*, 152–159.

[33] Kerman, M., Kanter, M., Coşkun, K. K., Erboga, M. & Gurel, A. (2012). Neuroprotective effects of caffeic acid phenethyl ester on experimental traumatic brain injury in rats. *J Mol Histol.*, *43*(1), 49–57.

[34] Mehrotra, A., Shanbhag, R., Chamallamudi, M. R., Singh, V. P. & Mudgal, J. (2011). Ameliorative effect of caffeic acid against inflammatory pain in rodents. *Eur J Pharmacol.*, *666*(1-3), 80-86.

[35] Búfalo, M. C. & Sforcin, J. M. (2015). The modulatory effects of caffeic acid on human monocytes and its involvement in propolis action. *J Pharm Pharmacol*. In press.

[36] Búfalo, M. C., Ferreira, I., Costa, G., Francisco, V., Liberal, J., Cruz, M. T., Lopes, M. C., Batista, M. T. & Sforcin, J. M. (2013). Propolis and its constituent caffeic acid suppress LPS-stimulated pro-inflammatory response by blocking NF-κB and MAPK activation in macrophages. *J Ethnopharmacol.*, *149*(1), 84-92.

[37] Arakawa, T., Yamasaki, H., Ikeda, K., Ejima, D., Naito, T. & Koyama, A. H. (2009). Antiviral and virucidal activities of natural products. *Curr Med Chem*, *16*, 2485-2497.

[38] Ikeda, K., Tsujimoto, K., Uozaki, M., Nishide, M., Suzuki, Y., Koyama, A. H. & Yamasaki, H. (2011). Inhibition of multiplication of herpes simplex virus by caffeic acid. *Int J Mol Med*, *28*, 595-598.

[39] Johnson, A. A., Marchand, C. & Pommery, Y. (2004). HIV-1 integrase inhibitors a decade of research and two drugs in clinical trial, Curr. Top. *Med. Chem.*, *4* (10), 1059–1077.

[40] Kashiwada, Y., Nishizawa, M., Yamagishi, T., Tanaka, T., Nonaka, G., Cosentino, L. M., Snider, J. V. & Lee, K. (1995). Anti-AIDS agents, 18. Sodium and potassium salts of caffeic acid tetramers from Arnebia euchroma as anti-HIV agents. *J. Nat. Prod.*, *58*, 392–400

[41] Fesen, M. R., Kohn, K. W., Leteurtre, F. & Pommier, Y. (1993). Inhibitors of human immunodeficiency virus integrase. *Proc. Natl. Acad. Sci. USA*, *90*, 2399–2403

[42] Fesen, M. R., Pommier, Y., Leteurtre, E., Hiroguchi, S., Yung, J. & Kohn, K. W. (1994). Inhibition of HIV-1 integrase by flavones, caffeic acid phenyl ester (CAPE), related compound. *Biochem Pharmacol*, *3*, 595–608.

[43] Takeda, H., Tsuji, M., Inazu, M., Egashira, T. & Matsumiya, T. (2002). Rosmarinic acid and caffeic acid produce antidepressive-like effect in the forced swimming test in mice. *Eur J Pharmacol*, *449*, 261–267.

[44] Rajendra-Prasad, N., Karthikeyan, A., Karthikeyan, S. & Reddy, B. V. (2011). Inhibitory effect of caffeic acid on cancer cell proliferation by oxidative mechanism in human HT 1080 fibrosarcoma cell line. *Mol Cell Biochem*, *349*, 11-19.

[45] Hudson, E. A., Dinh, P. A., Kokubun, T., Simmonds, M. S. & Gescher, A. (2000). Characterization of potentially chemopreventive phenols in extracts of brown rice that

inhibit the growth of human breast and colon cancer cells. *Cancer Epidemiol Biomark Prev*, *9*, 1163–1170.

[46] Park, E. H. & Kahng, J. H. (1995). Suppressive effects propolis in rate adjuvant arthritis. *Arch Pharmacol Res*, *22*, 554–558.

[47] Challis, B. C. & Bartlett, C. D. (1995). Possible carcinogenic effects of coffee constituents. *Nature*, *254*, 532–533.

[48] Chao, C. Y., Mong, M. C., Chan, K. C. & Yin, M. C. (2010). Anti-glycative and anti-inflammatory effects of caffeic acid and ellagic acid in kidney of diabetic mice. *Mol Nutr Food Res*, *54*, 388–395.

[49] Jung, U. J., Lee, M. K., Park, Y. B., Jeon, S. M. & Choi, M. S. (2006). Antihyperglycemic and antioxidant properties of caffeic acid in db/db mice. *J Pharmacol Exp Ther*, *318*, 476–483.

[50] Stocker, R. (1999). Dietary and pharmacological antioxidants in atherosclerosis. *Curr. Opin. Lipidol.* 10, 589–597.

[51] Morton, L. W., Croft, K. D., Puddey, I. B. & Byrne, L. (2000). Phenolic acids protect low density lipoproteins from peroxynitrite-mediated modification in vitro. *Redox Rep*, *5*, 124–125.

[52] Bhullar, K. S., Lassalle-Claux, G., Touaibia, M. & Rupasinghe, H. P. (2014). Antihypertensive effect of caffeic acid and its analogs through dual renin-angiotensin-aldosterone system inhibition. *Eur J Pharmacol.*, *730*, 125-132.

[53] Chuang, S. T., Kuo, Y. H. & Su, M. J. (2015). KS370G, a caffeamide derivative, attenuates unilateral ureteral obstruction-induccd renal fibrosis by the reduction of inflammation and oxidative stress in mice. *Eur J Pharmacol.*, *750*, 1-7.

[54] Holtomo, O., Nsangou, M., Fifen, J. J. & Motapon, O. (2014). DFT study of the effect of solvent on the H-atom transfer involved in the scavenging of the free radicals (·)HO2 and (·)O2(-) by caffeic acid phenethyl ester and some of its derivatives. *J Mol Model.*, *20*(11), 2509.

[55] Psotova. J., Lasovsky, J. & Vicor, J. (2003). Metal chelating properties, electrochemical Scavenging and cytoprotective activities of six natural phenolics. *Biomed Pap*, *147*, 147–153.

[56] Simsek, M., Quezada-Calvillo, R., Ferruzzi, M. G., Nichols, B. L. & Hamaker, B. R. (2015). Dietary Phenolic Compounds Selectively Inhibit the Individual Subunits of Maltase-Glucoamylase and Sucrase-Isomaltase with the Potential of Modulating Glucose Release. *J Agric Food Chem.*, *63*(15), 3873-3879.

[57] Tseng, T. H., Shen, C. H., Huang, W. S., Chen, C. N., Liang, W. H., Lin, T. H & Kuo H. C. (2014). Activation of neutral-sphingomyelinase, MAPKs, and p75 NTR-mediating caffeic acid phenethyl ester-induced apoptosis in C6 glioma cells. *J Biomed Sci.*, *21*,61.

[58] Bassil, D., Makris, D. P. & Kefalas, P. (2005). Oxidation of caffeic acid in the presence of Lcysteine: isolation of 2-S-cysteinyl caffeic acid and evaluation of its antioxidant properties, *Food Res. Int.*, *38*, 395–402.

[59] Piazzon, A., Vrhovsek, U., Masuero, D., Mattivi, F., Mandoj, F. & Nardini, M. (2012). Antioxidant activity of phenolic acids and their metabolites: synthesis and antioxidant properties of the sulfate derivatives of ferulic and caffeic acids and of the acyl glucuronide of ferulic acid. *J Agric Food Chem*, *60*, 12312–12323.

[60] Hirose, M., Takesada, Y., Tanaka, H., Tamano, S., Kato, T. & Shirai, T. (1998). Carcinogenicity of antioxidants BHA, caffeic acid, sesamol, 4-methoxyphenol and catechol at low doses, either alone or in combination, and modulation of their effects in a rat medium-term multi-organ carcinogenesis model, *Carcinogenesis*, *19*, 207–212.

[61] Lee, J. Y., Choi, H. J., Chung, T. W., Kim, C. H., Jeong, H. S. & Ha, K. T. (2013). Caffeic acid phenethyl ester inhibits alpha-melanocyte stimulating hormone-induced melanin synthesis through suppressing transactivation activity of microphthalmia-associated transcription factor. *J Nat Prod.*, *76*(8), 1399-1405.

[62] Pickens, C. A., Sordillo, L. M., Comstock, S. S., Harris, W. S., Hortos, K., Kovan, B. & Fenton, J. I. (2015). Plasma phospholipids, non-esterified plasma polyunsaturated fatty acids and oxylipids are associated with BMI. *Prostaglandins Leukot Essent Fatty Acids*, *95*, 31-40.

[63] Reuter, S., Gupta, S. C., Chaturvedi, M. M. & Aggarwal, B. B. (2010). Oxidative stress, inflammation, and cancer: how are they linked? *Free Radic Biol Med.*, *49*(11), 1603-1616.

[64] Sies, H. (2010). Polyphenols and health: update and perspectives. *Arch Biochem Biophys.*, *501*(1), 2-5.

[65] Yamada, Y., Yasui, H. & Sakurai, H. (2006). Suppressive effect of caffeic acid and its derivatives on the generation of UVA-induced reactive oxygen species in the skin of hairless mice and pharmacokinetic analysis on organ distribution of caffeic acid in ddY mice. *Photochem Photobiol*, *82*, 1668-1676.

[66] Vanella, A. (2000). Bioflavonoids as antiradicals, antioxidants and DNA cleavage protectors. *Cell Biol Toxicol*, *16*, 91-98.

[67] Sroka, Z. & Cisowski, W. (2003). Hydrogen peroxide scavenging, antioxidant and anti-radical activity of some phenolic acids. *Food Chem Toxicol*, *41*, 753–758.

[68] Kristinova, V., Mozuraityte, R., Storro, I. & Rustad, T. (2009). Antioxidant activity of phenolic acids in lipid oxidation catalyzed by different prooxidants. *Journal of Agricultural and Food Chemistry.*, *57*, 10377-10385.

[69] Medina, I., Gallardo, J. M., Gonzalez, M. J., Lois, S. & Hedges, N. (2007). Effect of molecular structure of phenolic families as hydroxycinnamic acids and catechins on their antioxidant effectiveness in minced fish muscle. *Journal of Agricultural and Food Chemistry.*, *55*, 3889–3895.

[70] Smith, C., Halliwell, B. & Aruoma, O. I. (1992). Protection by albumin against the pro-oxidant actions of phenolic dietary components. *Food Chem Toxicol*, *30*, 483–489.

[71] Koshihara, Y., Neichi, T., Murota, S., Lao, A., Fujimoto, Y. & Tatsuno, T. (1984). Caffeic acid is a selective inhibitor for leukotriene biosynthesis. Biochim. Biophys. *Acta*, *792*, 92–97.

[72] Krzyzankova, M., Chovanova, S., Chlapek, P., Radsetoulal, M., Neradil, J., Zitterbart, K., Sterba, J. & Veselska, R. (2014). LOX/COX inhibitors enhance the antineoplastic effects of all-trans retinoic acid in osteosarcoma cell lines. *Tumour Biol*, *35*(8), 7617-7627.

[73] Kumaran, K. S. & Prince, P. S. (2010). Protective effect of caffeic acid on cardiac markers and lipid peroxide metabolism in cardiotoxic rats: an in vivo and in vitro study. *Metabolism*, *59*(8), 1172-1780.

[74] Masuda, T., Shingai, Y., Takahashi, C., Inai, M., Miura, Y., Honda, S. & Masuda, A. (2014). Identification of a potent xanthine oxidase inhibitor from oxidation of caffeic acid. *Free Radic Biol Med.*, *69*, 300-307.

[75] Yang, W. S., Jeong, D., Yi, Y. S., Park, J. G., Seo, H., Moh, S. H., Hong, S. & Cho, J. Y. (1995). IRAK1/4-targeted anti-inflammatory action of caffeic acid. *Mediators Inflamm.*, 2013, 518183.

[76] Chan, W. S., Wen, P. C. & Chiang, H. C. (1995). Structure activity relationship of caffeic acid analogues on xanthine oxidase inhibition. *Anticancer Res.*, *15*, 703–707.

[77] Schefferlie, J. G. & van Bladeren, P. J. (1993). In vitro and in vivo reversible and irreversible inhibition of rat glutathione S-transferase isoenzymes by caffeic acid and its 2-S-glutathionyl conjugate. *Food Chem. Toxicol.*, *31*, 475–482.

[78] D'Antuono, I., Garbetta, A., Linsalata, V., Minervini, F. & Cardinali, A. (2015). Polyphenols from artichoke heads (Cynara cardunculus (L.) subsp. scolymus Hayek): in vitro bio-accessibility, intestinal uptake and bioavailability. *Food Funct.*, *6*(4), 1268-1277.

[79] Chen, L., Yin, H., Lan, Z., Ma, S., Zhang, C., Yang, Z., Li, P. & Lin, B. (2011). Anti-hyperuricemic and nephroprotective effects of Smilax china L. *J Ethnopharmacol.*, *135*(2), 399-405.

[80] Teke, Z., Bostanci, E. B., Yenisey, C, Sacar, M., Simsek, N. G. & Akoglu, M. (2012). Caffeic acid phenethyl ester alleviates mesenteric ischemia/reperfusion injury. *J Invest Surg.*, *25*(6), 354-365.

[81] Park, S. Y., Undeland, I., Sannaveerappa, T. & Richards, M. P. (2013). Novel interactions of caffeic acid with different hemoglobins: effects on discoloration and lipid oxidation in different washed muscles. *Meat Sci.*, *95*(1), 110-117.

[82] Everse, J. & Hsia, N. (1997). The toxicities of native and modified hemoglobins. *Free Radic Biol*, *22*(6), 1075-1099.

[83] Ziouzenkova, O., Asatryan, L. & Sevanian, A. (1999). Oxidative stress resulting from hemolysis and formation of catalytically active hemoglobin: protective strategies.*Int J Clin Pharmacol Ther.*, *37*(3), 125-132.

[84] Martins, J., Madeira, V., Almeida, L. & Laranjinha, J. (2002). Photoactivation of phthalocyanine-loaded low density lipoproteins induces a local oxidative stress that propagates to human erythrocytes: protection by caffeic acid. *Free Radic Res.*, *36*(3), 319-328.

[85] Larsson, K. J. & Undeland, I. K. (2010). Effect of caffeic acid on haemoglobin-mediated lipid and protein oxidation in washed cod mince during ice and frozen storage. *J Sci Food Agric.*, *90*(14), 2531-2540.

[86] Tolba, M. F., Omar, H. A., Azab, S. S., Khalifa, A. E., Abdel-Naim, A. B. & Abdel-Rahman, S. Z. (2014). Caffeic acid phenethyl ester: A review of its antioxidant activity, protective effects against ischemia-reperfusion injury and drug adverse reactions. *Crit Rev Food Sci Nutr.* In Press.

[87] Tolba, M. F., Omar, H. A., Azab, S. S., Khalifa, A. E., Abdel-Naim A. B. & Abdel-Rahman, S. Z. (2014). Caffeic acid phenethyl ester: A review of its antioxidant activity, protective effects against ischemia-reperfusion injury and drug adverse reactions, *Critical Reviews in Food Science and Nutrition.*, In Press.

[88] Göçer, H. & Gülçin, I. (2011). Caffeic acid phenethyl ester (CAPE): correlation of structure and antioxidant properties. *Int J Food Sci Nutr.*, *62*(8), 821-825.

[89] Flora, S. J. (2009). Structural, chemical and biological aspects of antioxidants for strategies against metal and metalloid exposure. *Oxid Med Cell Longev.*, *2*(4), 191-206.

[90] Michaluart, P., Masferrer, J. L., Carothers, A. M., Subbaramaiah, K., Zweifel, B. S., Koboldt, C., Mestre, J. R., Grunberger, D., Sacks, P. G., Tanabe, T. & Dannenberg, A. J. (1999). Inhibitory effects of caffeic acid phenethyl ester on the activity and expression of cyclooxygenase- 2 in human oral epithelial cells and in a rat model ofinflammation. *Cancer Res.*, *59*, 2347–2352.

[91] Wang, T., Chen, L., Wu, W., Long, Y &, Wang, R. (2008). Potential cytoprotection: antioxidant defence by caffeic acid phenethyl ester against free radical-induced damage of lipids, DNA, and proteins. Can *J Physiol Pharmacol.*, *86*(5), 279-287.

[92] Okutan, H., Ozcelik, N., Yilmaz, H. R. & Uz, E. (2005). Effects of caffeic acid phenethyl ester on lipid peroxidation and antioxidant enzymes in diabetic rat heart. *Clin Biochem.*, *38*(2), 191-196.

[93] Mansour, H. H. & Tawfik, S. S. (2012). Early treatment of radiation-induced heart damage in rats by caffeic acid phenethyl ester. *Eur J Pharmacol.*, *692*(1-3), 46-51.

[94] Ohori, M., Kinoshita, T., Okubo, M., Sato, K., Yamazaki, A., Arakawa, H., Nishimura, S., Inamura, N., Nakajima, H., Neya, M., Miyake, H. & Fujii, T. (2005). Identification of a selective ERK inhibitor and structural determination of the inhibitor-ERK2 complex. *Biochem Biophys Res Commun*, *336*, 357–363.

[95] Kim, S. J., Chung, T. W., Choi, H. J., Jin, U. H., Ha, K. T., Lee, Y. C. & Kim, C. H. (2014). Monosialic ganglioside GM3 specifically suppresses the monocyte adhesion to endothelial cells for inflammation. *Int J Biochem Cell Biol.*, *46*, 32-38.

[96] Gupta, S. C., Kim, J. H., Prasad, S. & Aggarwal, B. B. (2010). Regulation of survival, proliferation, invasion, angiogenesis, and metastasis of tumor cells through modulation of inflammatory pathways by nutraceuticals," *Cancer and Metastasis Reviews*, *29* (3), 405–434.

[97] Zhang, M., Zhou, J., Wang, L., Li, B., Guo, J., Guan, X., Han, Q. & Zhang, H. (2014). Caffeic acid reduces cutaneous tumor necrosis factor alpha (TNF-α), IL-6 and IL-1β levels and ameliorates skin edema in acute and chronic model of cutaneous inflammation in mice. *Biol Pharm Bull.*, *37*(3), 347-354.

[98] Song, H. S., Park, T. W., Sohn, U. D., Shin, Y. K., Choi, B. C., Kim, C. J. & Sim, S. S. (2008). The effect of caffeic acid on wound healing in skin-incised mice. *Korean J. Physiol. Pharmacol.*, *12*, 343-347.

[99] Wu, C. H., Huang, H. W., Lin, J. A., Huang, S. M. & Yen, G. C. (2011). The proglycation effect of caffeic acid leads to the elevation of oxidative stress and inflammation in monocytes, macrophages and vascular endothelial cells," *J. Nutritional Biochem.*, *22*(6), 585–594.

[100] Uto-Kondo, H., Ayaori, M. & Ogura, M. (2010). Coffee consumption enhances high-density lipoprotein-mediated cholesterol efflux in macrophages. *Circulation Research*, *106*(4), 779–787.

[101] Moon, M. K., Lee, Y. J., Kim, J. S., Kang, D. J. & Lee, H. S. (2009). Effect of caffeic acid on tumor necrosis factor-alpha-induced vascular inflammation in human umbilical vein endothelial cells," *Biological and Pharmaceutical Bulletin*, *32* (8), 1371–1377.

[102] Liu, Z., Nickel, C. H. & Cantley, L. G. (2001). HGF promotes adhesion of ATP-depleted renal tubular epithelial cells in a MAPK-dependent manner. American Journal of Physiology. *Renal Physiology*, *281*(1), F62–F70.

[103] Platanias, L. C. (2003). Map kinase signaling pathways and hematologic malignancies. *Blood*, *101*, 4667–4679.

[104] Muriel, P. (2009). NF-kappaB in liver diseases: a target for drug therapy. *J Appl Toxicol.*, *29*(2), 91-100.

[105] Tang, E. D., Wang, C. Y., Xiong, Y. & Guan, K. L. (2003). A role for NF-kappaB essential modifier/Ikappa B kinase-gamma (NEMO/ IKKgamma) ubiquitination in the activation of the Ikappa B kinase complex by tumor necrosis factor-alpha. *J. Biol. Chem.*, *278*,37297-37305.

[106] Barkett, M. & Gilmore, T. D. (1999). Control of apoptosis by Rel/NF-κB transcription factors. *Oncogene*, *18*(49), 6910–6924.

[107] Dutta, J., Fan, Y., Gupta, N., Fan, G. & Gélinas, C. (2006). Current insights into the regulation of programmed cell death by NF-κB. *Oncogene*, *25*(51), 6800–6816.

[108] Singh, R., Akhtar, N. & Haqqi, T. M. (2010). Green tea polyphenol epigallocatechin-3-gallate: inflammation and arthritis. *Life Sci*, *86*, 907–918.

[109] Jung, W. K., Choi, I., Lee, D. Y., Yea, S. S., Choi, Y. H., Kim, M. M., Park, S. G., Seo, S. K., Lee, S. W., Lee, C. M., Park, Y. M. & Choi, I. W. (2008). Caffeic acid phenethyl ester protects mice from lethal endotoxin shock and inhibits lipopolysaccharide-induced cyclooxygenase-2 and inducible nitric oxide synthase expression in RAW 264.7 macrophages via the p38/ERK and NF-kappaB pathways. *Int J Biochem Cell Biol*, *40*, 2572–2582.

[110] Nagasaka, R., Chotimarkorn, C., Shafiqul, I. M., Hori, M., Ozaki, H. & Ushio, H. (2007). Anti-inflammatory effects of hydroxycinnamic acid derivatives. *Biochem Biophys Res Commun*, *358*, 615–619.

[111] Tucsek, Z., Radnai, B., Racz, B., Debreceni, B., Priber, J. K., Dolowschiak, T., Palkovics, T., Gallyas, F. Jr., Sumegi, B. & Veres, B. (2011). Suppressing LPS-induced early signal transduction in macrophages by a polyphenol degradation product: a critical role of MKP-1. *J Leukoc Biol*, *89*, 105-111.

[112] Lee, K. M., Kang, H. S., Yun, C. H. & Kwak, H. S. (2012). Potential in vitro protective effect of quercetin, catechin, caffeic acid and phytic acid against ethanol-induced oxidative stress in SK-Hep-1 cells," *Biomolecules and Therapeutics.*, *20*(5), 492–498.

[113] Cario, E., Rosenberg, I. M., Brandwein, S. L., Beck,P. L., Reinecker, H. C. & Podolsky, D. K. (2000). Lipopolysaccharide activates distinct signaling pathways in intestinal epithelial cell lines expressing Toll-like receptors. *Journal of Immunology*, *164*, 966–972.

[114] Shi, Z. H., Li, N. G., Shi, Q. P., Tang, H., Tang, Y. P., Li, W., Yin, L., Yang, J. P. & Duan, J. A. (2013). Synthesis and structure activity relationship analysis of caffeic acid amides as selective matrix metalloproteinase inhibitors," *Bioorganic & Medicinal Chemistry Letters.*, *23*(5), 1206–1211.

[115] Anwar, J., R. M. Spanevello, V. C. Pimentel, Gutierres J, Thomé G, Cardoso A, Zanini D, Martins, C., Palma, H. E., Bagatini, M. D., Baldissarelli, J., Schmatz, R., Leal, C. A., da Costa, P., Morsch, V. M. & Schetinger M. R. (2013). Caffeic acid treatment alters the extracellular adenine nucleotide hydrolysis in platelets and lymphocytes of adult rats," *Food and Chemical Toxicology.*, *56*, 459–466.

[116] Kang, N. J., Lee, K. W., Shin, B. J., Jung, S. K., Hwang, M. K., Bode, A. M., Heo, Y. S., Lee, H. J. & Dong, Z. (2009). Caffeic acid, a phenolic phytochemical in coffee,

directly inhibits Fyn kinase activity and UVB-induced COX-2 expression," *Carcinogenesis.*, *30*(2), 321–330.

[117] Boot, J. R., Kitchen, E. A., Walker, J. R., Harvey, J. & Dawson, W. (1985). Directional movement of cells in vivo and in vitro under the influence of 5-lipoxygenase inhibitors. *Br. J. Dermatol.*, *113*, 168–173.

[118] Reyes, M. T., Mourelle, M., Hong, E. & Muriel, P. (1995). Caffeic acid prevents liver damage and ameliorates liver fibrosis induced by CCl4 in the rat. *Drug Dev. Res.*, *36*, 125–128.

[119] Pérez-Álvarez, V., Bobadilla, R. A & Muriel, P. (2001). Structure–hepatoprotective activity relationship of 3,4-dihydroxycinnamic acid (caffeic acid) derivatives. *J. Appl. Toxicol.*, *21*, 527–531.

[120] Fernández-Martínez, E., Bobadilla, R. A., Morales-Ríos, M. S., Muriel, P. & Pérez-Álvarez, V. M. (2007). Trans-3-phenyl-2-propenoic acid (Cinnamic Acid) derivatives: structure-activity relationship as hepatoprotective agents. *Med. Chem.*, *3*, 475–479.

[121] Natarajan, K., Singh, S., Burke, T. R, Grunberger, D. & Aggarwal, B. B. (1996). Caffeic acid phenethyl ester is a potent and specific inhibitor of activation of nuclear transcription factor NF-κB. Proc. *Natl Acad. Sci.*, *USA*, *93*, 9090–9095.

[122] Watabe, M., Hishikawa, K., Takayanagi, A., Shimizu, N. & Nakaki, T. (2004). Caffeic acid phenethyl ester induces apoptosis by inhibition of NFκB and activation of Fas in human breast cancer MCF-7 cells. *J. Biol. Chem.*, *279*, 6017–6026.

[123] Puangpraphant, S., Berhow, M. A., Vermillion, K., Potts, G. & Mejia, E. G. (2011). Dicaffeoylquinic acids in Yerbamat e(Ilex paraguariensis St. Hilaire) inhibit NF-κB nucleus translocation in macrophages and induce apoptosis by activating caspases-8 and-3 in human colon cancer cells. *Molecular Nutrition and Food Research*, *55*, 1509–1522.

[124] Lee, C. S., Lee, S. A., Kim, Y. J., Seo, S. J. & Lee, M. W. (2011). 3,4,5-Tricaffeoylquinicacid inhibits tumor necrosis factor-α-stimulated production of inflammatory mediators in keratinocytes via suppression of Akt- andNF-κB-pathways. *International Immunopharmacology*, *11*, 1715–1723.

[125] Greenberg, S. & Grinstein, S. (2002). Phagocytosis and innate immunity. *Curr Opin Immunol*, *14*, 136–145.

[126] Kawai, T. & Akira, S. (2009). The roles of TLRs, RLRs and NLRs in pathogen recognition. Int Immunol, *21*, 317–337.

[127] Mocsai, A. (2013). Diverse novel functions of neutrophils in immunity, inflammation, and beyond. *J Exp Med*, *210*, 1283–1299.

[128] Shi, C. & Pamer, E. G. (2011). Monocyte recruitment during infection and inflammation. *Nat Rev Immunol*, *11*, 762–774.

[129] Rossi, A. & Lord, J. (2013). Adiponectin inhibits neutrophil phagocytosis of Escherichia coli by inhibition of PKB and ERK 1/2 MAPK signalling and Mac-1 activation. *PLoS One*, *8*, e69108.

[130] Chan, G. C., Cheung, K. W. & Sze, D. M. (2013). The immunomodulatory and anticancer properties of propolis. *Clin Rev Allergy Immunol*, *44*, 262–273.

[131] Wang, L. C., Lin, Y. L., Liang, Y. C., Yang, Y. H., Lee, J. H., Yu, H. H., Wu, W. M. & Chiang, B. L. (2009). The effect of caffeic acid phenethyl ester on the functions of human monocyte-derived dendritic cells. *BMC Immunol*, *10*, 39–51.

[132] Wang, L. C., Chu, K. H., Liang, Y. C., Lin, Y. L. & Chiang, B. L. (2010). Caffeic acid phenethyl ester inhibits nuclear factor-kappa B and protein kinase B signalling pathways and induces caspase-3 expression in primary human CD4+T cells. *Clin Exp Immunol*, *160*, 223–232.

[133] Kim, S. Y., Koo, J. E., Seo, Y. J., Tyagi, N., Jeong, E., Choi, J., Lim, K. M., Park, Z. Y. & Lee, J. Y. (2013). Suppression of Toll-like receptor 4 activation by caffeic acid phenethyl ester is mediated by interference of LPS binding to MD2. *Br J Pharmacol*, *168*, 1933–1945.

[134] Zhao, H. P., Feng, J., Sun, K., Liu, Y. Y., Wei, X. H., Fan, J. Y., Huang, P., Mao, X. W., Zhou, Z., Wang, C. S., Wan, X. & Han, J. Y. (2012). Caffeic acid inhibits acute hyperhomocysteinemia-induced leukocyte rolling and adhesion in mouse cerebral venules. *Microcirculation*, *19*(3), 233-244.

[135] Kanakoudi-Tsakalidou, F., Debonera, F., Drossou-Agakidou, V., Sarafidis, K., Tzimouli, V., Taparkou, A. & Kremenopoulos, G. (2001). Flow cytometric measurement of HLA-DR expression on circulating monocytes in healthy and sick neonates using monocyte negative selection. *Clin Exp Immunol*, *123*, 402–407.

[136] Conti, B. J., Búfalo, M. C., Golim, M. A., Bankova, V. & Sforcin, J. M. (2013). Cinnamic acid is partially involved in propolis immunomodulatory action on human monocytes. *Evid Based Complement Alternat Med*, 2013, 109864.

[137] Chen, M. L., Wu, S., Tsai, T. C., Wang, L. K., Chou, W. M. & Tsai, F. M. (2014). The caffeic acid in aqueous extract of Tournefortia sarmentosa enhances neutrophil phagocytosis of Escherichia coli. *Immunopharmacol Immunotoxicol*, *36*(6), 390–396.

[138] Chung ,T. W., Choi, H. J., Lee, Y. C. & Kim, C. H. (2005). Molecular mechanism for transcriptional activation of ganglioside GM3 synthase and its function in differentiation of HL-60 cells. *Glycobiology*, *15*(3), 233-244.

[139] Ha, K. T., Lee, Y. C., Cho, S. H., Kim, J. K. & Kim, C. H. (2004). Molecular characterization of membrane type and ganglioside-specific sialidase (Neu3) expressed in E. coli. *Mol Cells*, *17*(2), 267-273.

[140] Choi, H. J., Chung, T. W., Kang, N. Y., Kim, K. S., Lee, Y. C. & Kim, C. H. (2003). Transcriptional regulation of the human GM3 synthase (hST3Gal V) gene during monocytic differentiation of HL-60 cells. *FEBS Lett.*, *555*(2), 204-208.

[141] Luo, H. R. &Loison, F. (2008). Constitutive neutrophil apoptosis: mechanisms and regulation. *Am J Hematol*, *83*, 288–295.

[142] Hauert, A. B., Martinelli, S., Marone, C. & Niggli, V. (2002). Differentiated HL-60 cells are a valid model system for the analysis of human neutrophil migration and chemotaxis. *Int J Biochem Cell Biol*, *34*, 838–854.

[143] Melzig, M. F., Tran, G. D., Henke, K., Selassie, C. D. & Verma, R. P. (2005). Inhibition of neutrophil elastase and thrombin activity by caffeic acid esters. *Pharmazie*, *60*, 869-873.

[144] Gougerot-Pocidalo, M. A. (2012). The neutrophil and systemic inflammation. *Joint Bone Spine.*, *79*, 209–211.

[145] Muller, W. A. (2009). Mechanisms of transendothelial migration of leukocytes. *Circ. Res.*, *105*, 223-230.

[146] Pet. J. (2011). Inflammation and coagulation. An overview. *Thromb. Res.*, *127*, S34-S37.

[147] Lomakina, E. B. & Waugh, R. E. (2010). Signaling and dynamics of activation of LFA-1 and Mac-1 by immobilized IL-8. *Cell. Mol. Bioeng.*, *3*, 106–116.

[148] Lee, E. S., Park, S. H., Kim, M. S., Han, S. Y., Kim, H. S. & Kang, Y. H. (2012). Caffeic acid disturbs monocyte adhesion onto cultured endothelial cells stimulated by adipokine resistin. *J Agric Food Chem.*, *60*(10), 2730-2739.

[149] Schaeffler, A., Gross, P., Buettner, R., Bollheimer, C., Buechler, C., Neumeier, M., Kopp, A., Schoelmerich, J. & Falk, W. (2009). Fatty acid induced induction of Toll-like receptor-4/nuclear factor-κB pathway in adipocytes links nutritional signalling with innate immunity. *Immunology*, *126*, 233-245.

[150] Lee, D. H., Kim, H. H., Cho, H. J., Bae, J. S., Yu, Y. B. & Park, H. J. (2014). Antiplatelet effects of caffeic acid due to Ca(2+) mobilizationinhibition via cAMP-dependent inositol-1, *4*, 5-trisphosphate receptor phosphorylation. *J Atheroscler Thromb.*, *21*(1), 23-37.

[151] Kupper, T. S. & Fuhlbrigge, R. C. (2004). Immune surveillance in the skin: Mechanisms and clinical consequences. *Nat. Rev. Immunol.*, *4*, 211–222.

[152] Hossen, M. A., Inoue, T., Shinmei, Y., Minami, K., Fujii, Y. & Kamei, C. (2006). Caffeic acid inhibits compound 48/80-induced allergic symptoms in mice. *Biol. Pharm. Bull.*, *29*, 64–66.

[153] Tsai, S. J., Chao, C. Y. & Yin, M. C. (2011). Preventive and therapeutic effects of caffeic acid against inflammatory injury in striatum of MPTP-treated mice. *Eur. J. Pharmacol.*, *670*, 441–447.

[154] Staniforth, V., Chiu, L. T. & Yang, N. S. (2006). Caffeic acid suppresses UVB radiation- induced expression of interleukin-10 and activation of mitogenactivated protein kinases in mouse. *Carcinogenesis*, *27*, 1803–1811.

[155] Stanley, P. L., Steiner, S., Havens, M. & Tramposch, K. M. (1991). Mouse skin inflammation induced by multiple topical applications of 12-Otetradecanoylphorbol-13-acetate. *Skin Pharmacol.*, *4*, 262–271.

[156] Kalyan Kumar, G., Dhamotharan, R., Kulkarni, N. M., Mahat, M. Y., Gunasekaran, J. & Ashfaque, M. (2011). Embelin reduces cutaneous TNF-alpha level and ameliorates skin edema in acute and chronic model of skin inflammation in mice. *Eur. J. Pharmacol.*, *662*, 63–69.

[157] Khan, A. Q, Khan, R., Qamar, W., Lateef, A., Ali, F., Tahir, M., Muneeb-U-Rehman. & Sultana, S. (2012). Caffeic acid attenuates 12-O-tetradecanoylphorbol-13-acetate (TPA)-induced NF-kB and COX-2 expression in mouse skin: abrogation of oxidative stress, inflammatory responses and proinflammatory cytokine production. *Food Chem. Toxicol.*, *50*, 175–183.

[158] Brito, G. A. C., Saraiva, S. N. R., Falcao, J. L. A. A., Vale, M. L., Lima, A. A. M., Cunha, F. Q. & Ribeiro, R. A. (2001). Dual effect of cAMP on the writhing response in mice. *Eur. J. Pharmacol.*, *416*, 223–230.

[159] de Campos Buzzi, F., Franzoi, C. L., Antonini, G., Fracasso, M., Filho, V.C., Yunes, R. A. & Niero, R. (2009). Antinociceptive properties of caffeic acid derivatives in mice. *Eur. J. Med. Chem.*, *44*, 4596–4602.

[160] Yonathan, M., Asres, K., Assefa, A. & Bucar, F. (2006). In vivo anti-inflammatory and antinociceptive activities of Cheilanthes farinosa. *J. Ethnopharmacol.*, *108*, 462–470.

[161] Hunskaar, S. & Hole, K. (1987). The formalin test in mice: dissociation between inflammatory and non-inflammatory pain. *Pain*, *30*, 103–114

[162] Sud'ina, G. F., Mirzoeva, O. K., Pushkareva, M. A., Korshunova, G. A., Sumbatyan, N. V. & Varfolomeev, S.D. (1993). Caffeic acid phenethyl ester as a lipoxygenase inhibitor with antioxidant properties. *FEBS Lett.*, *329*, 21–24.

[163] Toyoda, T., Tsukamoto, T., Takasu, S., Shi, L., Hirano, N., Ban, H., Kumagai, T. & Tatematsu, M. (2009). Anti-inflammatory effects of caffeic acid phenethyl ester (CAPE), a nuclear factor-kappaB inhibitor, on Helicobacter pylori-induced gastritis in Mongolian gerbils. *Int. J. Cancer*, *125*, 1786-1795.

[164] Ghosh, S., May, M. J. & Kopp, E. B. (1998). NF-kappa B and Rel proteins: evolutionarily conserved mediators of immune responses. *Annu. Rev. Immunol.*, *16*, 225–260.

[165] Ma, W. & Bisby, M.A. (1998). Increased activation of nuclear factor kappa B in rat lumbar dorsal root ganglion neurons following partial sciatic nerve injuries. *Brain Res.*, *797*, 243–254.

[166] Kay, A. B. (2001). Allergy and allergic diseases. First of two parts. *N Engl J Med*, *344*, 30e7.

[167] Sy, L. B., Yang, L. K., Chiu, C. J. & Wu,W. M. (2011). The Immunoregulatory Effects of Caffeic Acid Phenethyl Ester on the Cytokine Secretion of Peripheral Blood Mononuclear Cells From Asthmatic Children. *Pediatrics and Neonatology*, *52*, 327e331

[168] Park, J. H., Lee, J. K., Kim, H. S., Chung, S. T., Eom, J. H., Kim, K. A., Chung, S. J., Paik, S. Y. & Oh, H. Y. (2004). Immunomodulatory effect of caffeic acid phenethyl ester in Balb/c mice. *Int Immunopharmacol*, *4*, 429e36.

[169] Ansorge, S., Reinhold, D. & Lenedeckel, U. (2003). Propolis and some of its constituents down-regulate DNA synthesis and inflammatory cytokine production but induce TGF-beta1 production of human immune cells. *Z Naturforsch*, *C*, *58*,580e9.

[170] Marquez, N., Sancho, R., Macho, A., Calzado, M. A., Fiebich, B. L. & Mun, E. (2004). Caffeic acid phenethyl ester inhibits T-cell activation by targeting both NFAT and NF-kappa B transcription factors. *J Pharmacol Exp Ther*, *308*, 993e1001.

[171] Gremy, O., Benderitter, M. & Linard, C. (2006). Caffeic acid phenethyl ester modifies the TH1/TH2 balance in ileal mucosa after gammairradiation in the rat by modulating the cytokine pattern. *World J Gastroenterol*, *12*, 4996e5004.

[172] Mosmann, T. R., Cherwinski, H., Bond, M. W., Giedlin, M. A. & Coffman, R. L. (1986). Two types of murine helper T-cell clone: I-definition according to profiles of lymphokine activities and secreted proteins. *J Immunol*, *136*, 2348e57.

[173] Kuo, M. L., Huang, J. L., Yeh, K. W., Li, P. S. & Hsieh, K. H. (2001). Evaluation of Th1/Th2 ratio and cytokine production profile during acute exacerbation and convalescence in asthmatic children. *Ann Allergy Asthma Immunol*, *86*, 272e6.

[174] Arab, L. (2010). Epidemiologic evidence on coffee and cancer. *Nutr Cancer*, *62*, 271–283.

[175] Holick, C. N., Smith, S. G., Giovannucci, E. & Michaud, D. S. (2010). Coffee, tea, caffeine intake, and risk of adult glioma in three prospective cohort studies. *Cancer Epidemiol Biomarkers Prev*, *19*, 39–47.

[176] Abel, E. L., Hendrix, S. O., McNeeley, S. G., Johnson, K. C., Rosenberg, C. A. & Mossavar-Rahmani, Y. (2007). Daily coffee consumption and prevalence of nonmelanoma skin cancer in Caucasian women. European journal of cancer prevention: *the official journal of the European Cancer Prevention Organisation*, *16*, 446–452.

[177] Kang, N. J., Lee, K. W., Kim, B. H., Bode, A. M., Lee, H. J., Heo, Y. S., Boardman, L., Limburg, P., Lee, H. J. & Dong, Z. (2011). Coffee phenolic phytochemicals suppress colon cancer metastasis by targeting MEK and TOPK. *Carcinogenesis*, *32*, 921–928.

[178] Yu, X., Bao, Z., Zou, J. & Dong, J. (2011). Coffee consumption and risk of cancers: a meta-analysis of cohort studies. *BMC Cancer*, *11*, 96.

[179] Turati, F., Galeone, C., La Vecchia, C., Garavello, W. & Tavani, A. (2011). Coffee and cancers of the upper digestive and respiratory tracts: meta-analyses of observational studies. *Ann Oncol*, *22*, 536–544.

[180] Berwick, M., Lachiewicz, A., Pestak, C. & Thomas, N. (2008). Solar UV exposure and mortality from skin tumors. *Adv Exp Med Biol*, *624*, 117–124.

[181] Khan, N., Sharma, S. & Sultana, S. (2004). Attenuation of potassium bromate induced nephrotoxicity by coumarin (1,2-benzopyrone) in Wistar rats: chemoprevention against free radical-mediated renal oxidative stress and tumor promotion response. *Redox Rep*, *9*(1), 19–28.

[182] Tanaka, T., Kojima, T., Kawamori, T., Wang, A., Suzui, M., Okamoto, K. & Mori, H. (1993). Inhibition of 4-nitroquinoline- 1-oxide-induced rat tongue carcinogenesis by the naturally occurring plant phenolics caffeic, ellagic, chlorogenic and ferulic acids. *Carcinogenesis*, *14*, 1321–1325.

[183] Frenkel, K., Wei, H., Bhimani, R., Ye, J., Zadunaisky, J. A., Huang, M. T., Ferraro, T., Conney, A. H. & Grunberger, D. (1993). Inhibition of tumor promoter-mediated processes in mouse skin and bovine lens by caffeic acid phenethyl ester. *Cancer Res.*, *53*, 1255–1261.

[184] Nardini, M., Leonardi, F., Scaccini, C. & Virgili, F. (2001), Modulation of ceramide-induced NF-kappaB binding activity and apoptotic response by caffeic acid in U937 cells: comparison with other antioxidants. *Free Radic. Biol. Med.*, *30*, 722–733.

[185] Yang, G., Fu, Y., Malakhova, M., Kurinov, I., Zhu, F., Yao, K., Li, H., Chen, H., Li, W., Lim, do Y., Sheng, Y., Bode, A. M., Dong, Z. & Dong, Z. (2014). Caffeic Acid Directly Targets ERK1/2 to Attenuate Solar UV-Induced Skin Carcinogenesis. *Cancer Prevention Research*, *7*(10), 1056-1066.

[186] Lee, Y. J., Kuo, H. C., Chu, C. Y., Wang, C. J., Lin, W. C. & Tseng, T. H. (2003). Involvement of tumor suppressor protein p53 and p38 MAPK in caffeic acid phenethyl ester-induced apoptosis of C6 glioma cells. *Biochem Pharmacol*, *66*, 2281–2289.

[187] Grunberger, D., Banerjee, R., Eisinger, K., Oltz, E. M., Efros, L., Caldwell, M., Estevez, V. & Nakanishi, K. (1988). Preferential cytotoxicity on tumor cells by caffeic acid phenethyl ester isolated from propolis. *Experientia*, *44*, 230–232.

[188] Chung, T. W., Moon, S. K., Chang, Y. C., Ko, J. H., Lee, Y. C., Cho, G., Kim, S. H., Kim, J. G. & Kim, C. H. (2004). Novel and therapeutic effect of caffeic acid and caffeic acid phenyl ester on hepatocarcinoma cells: complete regression of hepatoma growth and metastasis by dual mechanism. *FASEB J*, *18*, 1670–1681.

[189] He, Y. J., Liu, B. H., Xiang, D. B., Qiao, Z. Y., Fu, T. & He, Y. H. (2006). Inhibitory effect of caffeic acid phenethyl ester on the growth of SW480 colorectal tumor cells involves beta-catenin associated signaling pathway downregulation. *World J Gastroenterol*, *12*, 4981–4985.

[190] Wang, D., Xiang, D. B., He, Y. J., Li, Z. P., Wu, X. H., Mou, J. H., Xiao, H. L. & Zhang, Q. H. (2005). Effect of caffeic acid phenethyl ester on proliferation and apoptosis of colorectal cancer cells in vitro. *World J Gastroenterol*, *11*, 4008–4012.

[191] Hung, M. W., Shiao, M. S., Tsai, L. C. Chang, G. G. & Chang, T. C. (2003). Apoptotic effect of caffeic acid phenethyl ester and its ester and amide analogues in human cervical cancer ME180 cells. *Anticancer Res*, *23*, 4773–4780.

[192] Su, Z. Z., Lin, J., Prewett, M., Goldstein, N. I. & Fisher, P. B. (1995). Apoptosis mediates the selective toxicity of caffeic acid phenethyl ester (CAPE) toward oncogene-transformed rat embryo fibroblast cells. *Anticancer Res*, *15*, 1841–1848.

[193] Chiao, C., Carothers, A. M., Grunberger, D., Solomon, G., Preston, G. A. & Barrett, J. C. (1995). Apoptosis and altered redox state induced by caffeic acid phenethyl ester (CAPE) in transformed rat fibroblast cells. *Cancer Res*, *55*, 3576–3583.

[194] Omene, C., Kalac, M., Wu, J., Marchi, E., Frenkel, K. & O'Connor, O. A. (2013). Propolis and its active component, caffeic acid phenethyl ester (CAPE), modulate breast cancer therapeutic targets via an epigenetically mediated mechanism of action. *J Cancer Sci Ther*, *5*, 334-342.

[195] Wu, J., Omene, C., Karkoszka, J., Bosland, M., Eckard, J., Klein, C. B. & Frenkel, K. (2011). Caffeic acid phenethyl ester (CAPE), derived from a honeybee product propolis, exhibits a diversity of antitumor effects in pre-clinical models of human breast cancer. *Cancer Lett*, *308*, 43-53.

[196] Scholzen, T. & Gerdes, J. (2000). The Ki-67 protein: from the known and the unknown. *J Cell Physiol*, *182*, 311–322.

[197] McCubrey, J. A., Steelman, L. S., Chappell, W. H., Abrams, S. L., Wong, E. W., Chang, F., Lehmann, B., Terrian, D. M., Milella, M., Tafuri, A., Stivala, F., Libra, M., Basecke, J., Evangelisti, C., Martelli, A. M. & Franklin, R. A. (2007). Roles of the Ra26f/MEK/ERK pathway in cell growth, malignant transformation and drug resistance. *Biochim Biophys Acta*, *1773*, 1263–1284.

[198] Wang, T., Chen, L. X., Long, Y., Wu, W. M. & Wang, R. (2008). DNA damage induced by caffeic acid phenyl ester in the presence of Cu(II) ions: potential mechanism of its anticancer properties. *Cancer Lett*, *263*(1), 77–88.

[199] Metodiewa, D., Jaiswal, A. K., Cenas, N., Dickancaite, E. & Aguilar, J. S. (1999). Quercetin may act as a cytotoxic pro-oxidant after its metabolic activation to semiquinone and quinoidal product. *Free Rad Biol Med*, *26*, 107–116.

[200] Bhat, S. H., Azmi, A. S. & Hadi, S. M. (2007). Prooxidant DNA breakage induced by caffeic acid in human peripheral lymphocytes: involvement of endogenous copper and a putative mechanism for anticancer properties. *Toxicol Appl Pharmacol*, *218*(3), 249–255.

[201] Sergediene, E, Jonsson, K., Szymusiak, H., Tyrakowska, B., Rietjens, I. M. & Cenas, N. (1999). Prooxidant toxicity of polyphenolic antioxidants to HL-60 cells description of quantitative structure activity relationships. *FEBS Lett*, *462*, 392–396.

[202] Indap, M. A., Radhika, S., Motiwale, L. & Rao, K. V. K. (2006). Inhibitory effect of cinnamoyl compounds against human malignant cell line. *Indian J Exp Biol*, *44*, 216–220.

[203] Kampa, M., Alexaki, V. I., Notas, G., Nifli, A. P., Nistikaki, A., Hatzoglou, A., Bakogeorgou, E., Kouimtzoglou, E., Blekas, G., Boskou, D., Gravanis, A. & Castanas, E. (2004). Antiproliferative and apoptotic effects of selective phenolic acids on T47D human breast cancer cells: potential mechanisms of action. *Breast Cancer Res*, *6*, R63–R74.

[204] Sakihama, Y., Cohen, M. F., Grace, S. C. & Yamasaki, H. (2002). Plant phenolic antioxidant and prooxidant activities: phenolics-induced oxidative damage mediated by metals in plants. *Toxicology*, *177*, 67–80.

[205] McMillan, T, J. & Steel, G. G. (1997). DNA damage and cell killing. In: Steel GG (ed) Basic clinical radiobiology. *Arnold*, *London*, 58–69.

[206] Chou, C. C., Yang, J. S., Lu, H. F., Ip, S. W., Lo, C., Wu, C. C., Lin, J. P., Tang, N. Y., Chung, J. G., Chou, M. J., Teng, Y. H. & Chen, D. R. (2010). Quercetin mediated cell cycle arrest and apoptosis involving activation of a caspase cascade through the mitochondrial pathway in human breast cancer MCF-7 cells. *Arch Pharm Res*, *33*(8), 1181–1191.

[207] Galati, G., Sabzevari, O., Wilson, J. X. & O'Brien, P. J. (2002). Prooxidant activity and cellular effects of the phenoxyl radicals of dietary flavonoids and other polyphenolics. *Toxicology*, *177*, 91-104.

[208] Watabe, M., Hishikawa, K., Takayanagi, A., Shimizu, N. & Nakaki, T. (2004). Caffeic acid phenethyl ester induces apoptosis by inhibition of NF-B and activation of Fas in human breast cancer MCF-7 cells. *J Biol Chem*, *279*, 6017–6026.

[209] Jin, U. H., Song, K. H., Motomura, M., Suzuki, I., Gu, Y. H., Kang, Y. J., Moon, T. C. & Kim, C. H. (2008). Caffeic acid phenethyl ester induces mitochondria-mediated apoptosis in human myeloid leukemia U937 cells. *Mol Cell Biochem.*, *310*(1-2), 43-48.

[210] Jin. U. H., Chung, T. W., Song, K. H., Kwak, C. H., Choi, H. J., Ha, K. T., Chang, Y. C., Lee, Y. C. & Kim, C. H. (2014). Ganglioside GM3 is required for caffeic acid phenethyl ester-induced megakaryocytic differentiation of human chronic myelogenous leukemia K562 cells. *Biochem Cell Biol.*, *92(*4), 243-249.

[211] Motomura, M., Kwon, K. M., Suh, S. J., Lee, Y. C., Kim, Y. K., Lee, I. S., Kim, M. S., Kwon, D. Y., Suzuki, I. & Kim, C. H. (2008). Propolis induces cell cycle arrest and apoptosis in human leukemic U937 cells through Bcl-2/Bax regulation. *Environ Toxicol Pharmacol.*, *26*(1), 61-67.

[212] Zheng, L. F., Dai, F., Zhou, B., Yang, L. & Liu, Z. L. (2008). Prooxidant activity of hydroxycinnamic acids on DNA damage in the presence of Cu(II) ions: mechanism and structure-activity relationship Food. *Chem Toxicol*, *46*(1), 149–156.

[213] Rehman, M. U. & Sultana, S. (2011). Attenuation of oxidative stress, inflammation and early markers of tumor promotion by caffeic acid in Fe-NTA exposed kidneys of Wistar rats. *Mol Cell Biochem.*, *357*, 115–124.

[214] Han, S. G., Han, S. S., Toborek, M. & Hennig, B. (2012). EGCG protects endothelial cells against PCB 126-induced inflammation through inhibition of AhR and induction of Nrf2-regulated genes. *Toxicology and Applied Pharmacology*, *261*(2), 181–188.

[215] Hao, E., Lang, F., Chen, Y., Zhang, H., Cong, X. & Shen, X. (2013). Resveratrol alleviates endotoxin-induced myocardial toxicity via the Nrf2 transcription factor. *PLOS One*, *8*(7), e69452.

[216] Zhao, R., Yang, B., Wang, L., Xue, P., Deng, B. & Zhang, G. (2013). Curcumin protects human keratinocytes against inorganic arsenite-induced acute cytotoxicity through an NRF2-dependent mechanism. *Oxidative Medicine and Cellular Longevity.*, 2013,412576.

[217] Sirota, R., Gibson, D. & Kohen, R. (2015). The role of the catecholic and the electrophilic moieties of caffeic acid in Nrf2/Keap1 pathway activation in ovarian carcinoma cell lines. *Redox Biol.*, *4*, 48-59.

[218] Zhang, D. D. (2006). Mechanistic studies of the Nrf2–Keap1 signaling pathway. *Drug Metabolism Reviews.*, *38*(4), 769–789.

[219] Zhang, M., An, C., Gao, Y., Leak, R. K., Chen, J. & Zhang, F. (2013). Emerging roles of Nrf2 and phase II antioxidant enzymes in neuroprotection. *Progress in Neurobiology*, *100*, 30-47.

[220] Bryan, H. K., Olayanju, A., Goldring, C.E. & Park, B. K. (2013). The Nrf2 cell defence pathway: Keap1-dependent and -independent mechanisms of regulation. *Biochemical Pharmacology.*, *85*(6), 705–717.

[221] Acharya, A., Das, I., Chandhok, D. & Saha, T. (2010). Redox regulation in cancer: a double-edged sword with therapeutic potential. *Oxidative Medicine and Cellular Longevity*, *3*(1), 23–34.

[222] Wang, X. J., Sun, Z., Villeneuve, N. F., Zhang, S., Zhao, F. & Li, Y. (2008). Nrf2 enhances resistance of cancer cells to chemotherapeutic drugs, the dark side of Nrf2. *Carcinogenesis*, *29*(6), 1235–1243.

[223] Pasello, M., Michelacci, F., Scionti, I., Hattinger, C. M., Zuntini, M. & Caccuri, A. M. (2008). Overcoming glutathione S-transferase P1-related cisplatin resistance in osteosarcoma. *Cancer Research*, *68*, (16), 6661–6668.

[224] Dourado, D. F., Fernandes, P. A. & Ramos, M. J. (2010). Glutathione transferase classes alpha, pi, and mu: GSH activation mechanism. *Journal of Physical Chemistry B.*, *114*(40), 12972–12980.

[225] Lu, D., Shi, H. C., Wang, Z. X., Gu, X. W. & Zeng, Y. J. (2011). Multidrug resistance-associated biomarkers PGP, GST-pi, Topo-II and LRP as prognostic factors in primary ovarian carcinoma. *British Journal of Biomedical Science.*, *68*(2), 69–74.

[226] Schumaker, L., Nikitakis, N., Goloubeva, O., Tan, M., Taylor, R. & Cullen, K. J. (2008). Elevated expression of glutathione S-transferase pi and p53 confers poor prognosis in head and neck cancer patients treated with chemoradiotherapy but not radiotherapy alone. *Clinical Cancer Research.*, *14*(18), 5877–5883.

[227] Weng, C. J. & Yen, G. C. (2012). Chemopreventive effects of dietary phytochemicals against cancer invasion and metastasis: phenolic acids, monophenol, polyphenol, and their derivatives. *Cancer Treat Rev.*, *38*(1), 76-87.

[228] MacDonald, N. J. & Steeg, P. S. (1993). Molecular basis of tumor metastasis. In: Lemoine NR, Wright NA, editors. The molecular pathology of cancer. Plainview NY: *Cold Spring Harbor Laboratory*, 175–199.

[229] Woodhouse, E. C, Chuaqui, R. F. & Liotta, L. A. (1997). General mechanisms of metastasis. *Cancer 80(Suppl.)*, 529–537.

[230] Chung, T. W., Kim, S. J., Choi, H. J., Kwak, C. H., Song, K. H, Suh, S. J., Kim, K. J., Ha, K. T., Park, Y. G., Chang, Y. C., Chang, H. W., Lee, Y. C. & Kim, C. H. (2013). CAPE suppresses VEGFR-2 activation, and tumor neovascularization and growth. *J Mol Med.*, *91*(2), 271-282.

[231] Liao, H. F., Chen, Y. Y., Liu, J. J., Hsu, M. L., Shieh, H. J., Liao, H. J., Shieh, C. J., Shiao, M. S. & Chen, Y. J. (2003). Inhibitory effect of caffeic acid phenethyl ester on angiogenesis, tumor invasion, and metastasis. *J Agric Food Chem*, *51*, 7907–7912.

[232] Izuta, H., Shimazawa, M., Tsuruma, K., Araki, Y., Mishima, S. & Hara, H. (2009). Bee products prevent VEGF-induced angiogenesis in human umbilical vein endothelial cells. *BMC Complement Altern Med*, *9*, 45.

[233] Chung, T. W., Kim, S. J., Choi, H. J., Kim, K. J., Kim, M. J., Kim, S. H., Lee, H. J., Ko, J. H., Lee, Y. C., Suzuki, A. & Kim, C. H. (2009). Ganglioside GM3 inhibits VEGF/VEGFR-2-mediated angiogenesis: direct interaction of GM3 with VEGFR-2. *Glycobiology*, *19*, 229–239.

[234] Kudugunti, S. K., Vad, N. M., Ekogbo, E. & Moridani, M. Y. (2011). Efficacy of caffeic acid phenethyl ester (CAPE) in skin B16-F0 melanoma tumor bearing C57BL/6 mice. *Invest New Drugs*, *29*, 52–62.

[235] Chung, T. W., Moon, S. K., Lee, Y. C., Kim, J. G., Ko, J. H. & Kim, C. H. (2002). Enhanced expression of matrix metalloproteinase-9 by hepatitis B virus infection in liver cells. *Arch Biochem Biophys.*, *408*(2), 147-154.

[236] Park, W. H., Kim, S. H. & Kim, C. H. (2005). A new matrix metalloproteinase-9 inhibitor 3,4-dihydroxycinnamic acid (caffeic acid) from methanol extract of Euonymus alatus: isolation and structure determination. *Toxicology*, *207*(3), 383-390.

[237] Lee, S. O., Jeong, Y. J., Kim, M., Kim, C. H. & Lee, I. S. (2008). Suppression of PMA-induced tumor cell invasion by capillarisin via the inhibition of NF-kappaB-dependent MMP-9 expression. *Biochem Biophys Res Commun.*, *366*(4), 1019-1024.

[238] Jin, U. H., Chung, T. W., Kang, S. K., Suh, S. J., Kim, J. K., Chung, K. H., Gu, Y. H., Suzuki, I. & Kim, C. H. (2005). Caffeic acid phenyl ester in propolis is a strong inhibitor of matrix metalloproteinase-9 and invasion inhibitor: isolation and identification. *Clin Chim Acta.*, *362*(1-2), 57-64.

[239] Huang, M. T., Smart, R. C., Wong, C. Q. & Conney, A. H. (1988). Inhibitory effect of curcumin, chlorogenic acid, caffeic acid, and ferulic acid on tumor promotion in mouse skin by 12-O-tetradecanoylphorbol-13-acetate. *Cancer Res*, *48*, 5941–5946.

[240] Bakır, S., Özbay, M., Gün, R., Yorgancılar, E., Kınış, V., Keleş, A., Abakay, A., Gökalp, O. & Topçu, İ. (2013). The protective role of caffeic acid phenethyl ester against streptomycin ototoxicity. *Am J Otolaryngol.*, *34*(1), 16-21.

[241] Ocak, S., Gorur, S., Hakverdi, S., Celik, S. & Erdogan, S. (2007). Protective effects of caffeic acid phenethyl ester, vitamin C, vitamin E and N-acetylcysteine on vancomycin-induced nephrotoxicity in rats. *Basic Clin Pharmacol Toxicol.*, *100*(5), 328-333.

[242] Sahin, A., Kaya, S., Türkcü, G., Cingü, A. K., Yüksel, H., Türkcü, F. M., Yüksel, H., Cına,r Y., Arı, S., Sahin, M., Caça, I. & Gökalp, O. (2013). The effects of caffeic acid phenethyl ester in acute methanol toxicity on rat retina and optic nerve. *Cutan Ocul Toxicol.*, *32(*4), 263-267.

[243] Sulaiman, G. M. (2012). Role of caffeic acid phenethyl ester on mitomycin C induced clastogenesis: analysis of chromosome aberrations, micronucleus, mitotic index and adenosine deaminase activity in vivo. *J Appl Genet.*, *53*(2), 213-219.

[244] Gokcimen, A., Cim, A., Tola, H. T., Bayram, D., Kocak, A., Ozgüner, F. & Ayata, A. (2007). Protective effect of N-acetylcysteine, caffeic acid and vitamin E on doxorubicin hepatotoxicity. *Hum Exp Toxicol.*, *26*(6), 519-525.

[245] Yagmurca, M., Erdogan, H., Iraz, M., Songur, A., Ucar, M. & Fadillioglu, E. (2004). Caffeic acid phenethyl ester as a protective agent against doxorubicin nephrotoxicity in rats. *Clin Chim Acta*, *348*(1-2), 27-34.

[246] Kart, A., Cigremis, Y., Karaman, M. & Ozen, H. (2010). Caffeic acid phenethyl ester (CAPE) ameliorates cisplatin-induced hepatotoxicity in rabbit. *Exp Toxicol Pathol.*, *62*(1), 45-52.

[247] Ozen, S., Akyol, O., Iraz, M, Söğüt, S., Ozuğurlu, F., Ozyurt, H., Odaci, E. & Yildirim, Z. (2004). Role of caffeic acid phenethyl ester, an active component of propolis, against cisplatin-induced nephrotoxicity in rats. *J Appl Toxicol.*, *24*(1), 27-35.

[248] Yilmaz, H. R., Uz, E., Altunbasak, A., Sakalli, E. & Ozçelik, N. (2010). Anticlastogenic effect of caffeic acid phenethyl ester on cisplatin-induced chromosome aberrations in rat bone marrow cells. *Toxicol Ind Health.*, *26*(1), 33-37.

[249] Iraz, M., Ozerol, E., Gulec, M., Tasdemir, S., Idiz, N., Fadillioglu, E., Naziroglu, M. & Akyol, O. (2006). Protective effect of caffeic acid phenethyl ester (CAPE) administration on cisplatin-induced oxidative damage to liver in rat. *Cell Biochem Funct.*, *24*(4), 357-361.

[250] Yilmaz, H. R., Sogut, S., Ozyurt, B., Ozugurlu, F., Sahin, S., Isik, B., Uz, E. & Ozyurt, H. (2005). The activities of liver adenosine deaminase, xanthine oxidase, catalase, superoxide dismutase enzymes and the levels of malondialdehyde and nitric oxide after cisplatin toxicity in rats: protective effect of caffeic acid phenethyl ester. *Toxicol Ind Health.*, *21*(3-4), 67-73.

[251] Çakır, T., Özkan, E., Dulundu, E., Topaloğlu, Ü., Şehirli, A. Ö., Ercan, F., Şener, E. & Şener, G. (2011). Caffeic acid phenethyl ester (CAPE) prevents methotrexate-induced hepatorenal oxidative injury in rats. *J Pharm Pharmacol.*, *63*(12), 1566-1571.

[252] Gocmez, C., Celik, F., Kamasak, K., Kaplan, M., Uzar, E., Arıkanoglu, A. & Evliyaoglu, O. (2015). Effects of intrathecal caffeic acid phenethyl ester and methylprednisolone on oxidant/antioxidant status in traumatic spinal cord injuries. *J Neurol Surg A Cent Eur Neurosurg.*, *76*(1), 20-24.

[253] Uzar, E., Sahin, O., Koyuncuoglu, H. R., Uz, E., Bas, O., Kilbas, S., Yilmaz, H. R., Yurekli, V. A., Kucuker, H. & Songur, A. (2006). The activity of adenosine deaminase and the level of nitric oxide in spinal cord of methotrexate administered rats: protective effect of caffeic acid phenethyl ester. *Toxicology*, *218*(2-3), 125-133

[254] Akyol, S., Ugurcu, V., Altuntas, A., Hasgul, R., Cakmak, O. & Akyol, O. (2014). Caffeic acid phenethyl ester as a protective agent against nephrotoxicity and/or oxidative kidney damage: a detailed systematic review. *Scientific World Journal.*, 2014, 561971.

[255] Chattopadhyay, D. & Naik, T. N. (2007). Antivirals of ethnomedicinal origin: structure–activity relationship and scope. *Mini Rev. Med. Chem.*, *7*, 275–301.

[256] Jassim, S. A. & Naji, M. A. (2003). Novel antiviral agents: a medicinal plant perspective. *J. Appl. Microbiol.*, *95*, 412–427.

[257] Naithani, R., Huma, L. C., Holland, L. E., Shukla, D., McCormick, D. L., Mehta, R. G. & Moriarty, R.M. (2008). Antiviral activity of phytochemicals: a comprehensive review. *Mini Rev. Med. Chem.*, *8*, 1106–1133.

[258] Chiang, L. C., Chiang, W., Chang, M. Y., Ng, L. T. & Lin, C. C. (2002). Antiviral activity of Plantago major extracts and related compounds in vitro. *Antiviral Res.*, *55*, 53–62.

[259] Cos, P., Maes, L., Vanden Berghe, D., Hermans, N., Pieters, L. & Vlietinck, A. (2004). Plant substances as anti-HIV agents selected according to their putative mechanism of action. *J. Nat. Prod.*, *67*, 284–293.

[260] Cos, P., Maes, L., Vlietinck, A. & Pieters, L. (2008). Plant-derived leading compounds for chemotherapy of human immunodeficiency virus (HIV) infection-an update (1998–2007). *Planta Med.*, *74*, 1323–1337.

[261] Lefkovits, I., Su, Z., Fisher, P. & Grunberger, D. (1997). Caffeic acid phenethyl ester profoundly modifies protein synthesis profile in type 5 adenovirus-transformed cloned rat embryo fibroblast cells. *Int J Oncol.*, *11*(1), 59-67.

[262] Khan, M. T., Ather, A., Thompson, K. D. & Gambari, R. (2005). Extracts and molecules from medicinal plants against herpes simplex viruses. *Antiviral Res.*, *67*, 107–119.

[263] Xie, Y., Gong, J., Li, M., Fang, H. & Xu, W. (2011). The medicinal potential of influenza virus surface proteins: hemagglutinin and neuraminidase. *Curr. Med. Chem.*, *18*, 1050-1066.

[264] Inoue, M., Kurahashi, N., Iwasaki, M, Shimazu, T., Tanaka, Y., Mizokami, M., Tsugane, S. & Japan Public Health Center-Based Prospective Study Group. (1009). Effect of coffee and green tea consumption on the risk of liver cancer: cohort analysis by hepatitis virus infection status. *Cancer Epidemiol Biomarkers Prev.*, *18*, 1746-1753.

[265] Tamura, H., Akioka, T., Ueno, K., Chujyo, T., Okazaki, K., King, P. J. & Robinson Jr., W. E. (2006) Anti-human immunodeficiency virus activity of 3,4,5-tricaffeoylquinic acid in cultured cells of lettuce leaves. *Mol. Nutr. Food Res.*, *50*, 396–400.

[266] McDougall, B., King, P. J., Wu, B. W., Hostomsky, Z., Reinecke, M. G. & Robinson Jr., W. E. (1998). Dicaffeoylquinic and dicaffeoyltartaric acids are selective inhibitors of human immunodeficiency virus type 1 integrase. Antimicrob. *Agents Chemother.*, *42*, 140–146.

[267] Ojwang, J. O., Wang, Y. H., Wyde, P. R., Fischer, N. H., Schuehly, W., Appleman, J. R., Hinds, S. & Shimasaki, C. D. (2005). A novel inhibitor of respiratory syncytial virus isolated from ethnobotanicals. *Antiviral Res.*, *68*(3), 163-172.

[268] Li, Y., But, P. P. & Ooi, V. E. (2005). Antiviral activity and mode of action of caffeoylquinic acids from Schefflera heptaphylla (L.) Frodin. *Antiviral Res.*, *68*, 1–9.

[269] Peiris, J. S., Hui, K. P. & Yen, H. L. (2010). Host response to influenza virus: protection versus immunopathology. *Curr Opin Immunol.*, *22*(4), 475-481.

[270] Song, L. (2014).It is Unlikely That Influenza Viruses Will Cause a Pandemic Again Like What Happened in 1918 and 1919. Front Public Health. May 7, 2, 39. doi: 10.3389/fpubh.2014.00039. eCollection 2014.

[271] Shinya, K., Ebina, M., Yamada, S., Ono, M., Kasai, N. & Kawaoka, Y. (2006). Avian flu: influenza virus receptors in the human airway. *Nature*, *440*, 435–436.

[272] Walther, T., Karamanska, R., Chan, R. W., Chan, M. C., Jia, N., Air, G., Hopton, C., Wong, M. P., Dell, A., Malik Peiris, J. S., Haslam, S. M. & Nicholls, J. M. (2013). Glycomic analysis of human respiratory tract tissues and correlation with influenza virus infection. *PLoS Pathog.*, *9*(3), e1003223.

[273] Wu, W. L., Lau, S. Y., Chen, Y., Wang, G., Mok, B. W., Wen, X., Wang, P., Song, W., Lin, T., Chan, K. H., Yuen, K. Y. & Chen, H. (2012). The 2008-2009 H1N1 influenza virus exhibits reduced susceptibility to antibody inhibition: Implications for the prevalence of oseltamivir resistant variant viruses. *Antiviral Res.*, *93*, 144-153.

[274] Takano, R., Kiso, M., Igarashi, M., Le, Q. M., Sekijima, M., Ito, K., Takada, A. & Kawaoka, Y. (2013). Molecular mechanisms underlying oseltamivir resistance mediated by an I117V substitution in the neuraminidase of subtype H5N1 avian influenza A viruses. *J. Infect. Dis.*, *207*, 89-97.

[275] Pollikoff, R., Liberman, M., Cochran, K. W. & Pascale, A. M. (1965). Effect of caffeic acid on mouse and ferret lung infected with influenza A virus. *Antimicrob Agents Chemother (Bethesda).*, *5*, 561-566.

[276] Xie, Y., Huang, B., Yu, K., Shi, F., Liu, T. & Xu, W. (2013). Caffeic acid derivatives: a new type of influenza neuraminidase inhibitors. *Bioorg Med Chem Lett.*, *23*(12), 3556-3560.

[277] Utsunomiya, H., Ichinose, M., Ikeda, K., Uozaki, M., Morishita, J., Kuwahara, T., Koyama, H. & Yamasaki, H. (2014). Inhibition by caffeic acid of the influenza A virus multiplication in vitro. *Int J Molecular Medicine*, *34*, 1020-1024.

[278] Xie, Y., Huang, B., Yu, K. & Xu, W. (2013). Furrther discovery of caffeic acid derivatives as novel influenza neuraminidase inhibitors. *Bioorg Med Chem.*, *21*(24), 7715-7723.

[279] Liu, A. L., Wang, H. D., Lee, S. M., Wang, Y. T. & Du, G. H. (2008). Structure-activity relationship of flavonoids as influenza virus neuraminidase inhibitors and their in vitro anti-viral activities. *Bioorg. Med. Chem.*, *16*, 7141-71417.

[280] Ryu, Y. B., Curtis-Long, M. J., Lee, J. W., Kim, J. H., Kim, J. Y., Kang, K. Y., Lee, W. S. & Park, K. H. (2009). Characteristic of neuraminidase inhibitory xanthones from Cudrania tricuspidata. *Bioorg. Med. Chem.*, *17*, 2744-2750.

[281] Dao, T. T., Dang, T. T., Nguyen, P. H., Kim, E., Thuong, P. T. & Oh, W. K. (2012). anthones from Polygala karensium inhibit neuraminidases from influenza A viruses. Bioorg. *Med. Chem. Lett.*, *22*, 3688-3692.

[282] Grienke, U., Schmidtke, M., Kirchmair, J., Pfarr, K., Wutzler, P., Durrwald, R., Wolber, G., Liedl, K. R., Stuppner, H. & Rollinger, J. M. (2010). Antiviral potential and molecular insight into neuraminidase inhibiting diarylheptanoids from *Alpinia katsumadai*. *J. Med. Chem.*, *53*, 778-786.

[283] Utsunomiya, H., Ichinose, M., Uozaki, M., Tsujimoto, K., Yamasaki, H. & Koyama, A. H. (2008). Antiviral activities of coffee extracts in vitro. *Food Chem Toxicol*, *46*, 1919-1924.

[284] Murayama, M., Tujimoto, K., Uozaki, M., Katsuyama, Y., Yamasaki, H., Utsunomiya, H. & Koyama, A. H. (2008). Effect of caffeine on the multiplication of DNA and RNA viruses. *Mol Med Rep*, *1*, 251-255.

[285] Tsujimoto, K., Sakuma, C., Uozaki, M., Yamasaki, H., Utsunomiya, H., Oka, K. & Koyama, A. H. (2010). Antiviral effect of pyridinium formate, a novel component of coffee extracts. *Int J Mol Med*, *25*, 459-463.

[286] Carey, W. D. (2009). The prevalence and natural history of hepatitis B in the 21st century. Cleve. *Clin. J. Med.*, *76* (Suppl. 3), S2–S5.

[287] Buti, M. & Esteban, R. (2005). *Drugs in development for hepatitis B. Drugs*, *65*(11), 1451-1460

[288] Center of Disease Control and Prevention, 2006. Prevalence of chronic infection with hepatitis B virus, by country. Available at http://wwwn.cdc.gov/travel/yellowBookCh4-HepB.aspx.

[289] United States Department of Agriculture, 2007. Tropital products:worldmarkets and trade. Available at http://www.fas.usda.gov/htp/tropical/2007/June%202007/June%20Tropical.pdf.

[290] Bravi, F., Bosetti, C., Tavani, A., Bagnardi, V., Gallus, S., Negri, E., Franceschi, S. & La Vecchia, C. (2007). Coffee drinking and hepatocellular carcinoma risk: a metaanalysis. *Hepatology*, *46*, 430–435.

[291] Corrao, G., Zambon, A., Bagnardi, V., D'Amicis, A. & Klatsky, A. (2001). Coffee, caffeine, and the risk of liver cirrhosis. *Ann. Epidemiol.*, *11*, 458–465.

[292] Nkondjock, A. (2009). Coffee consumption and the risk of cancer: an overview. *Cancer Lett.*, *277*, 121–125.

[293] Ruhl, C. E. & Everhart, J. E. (2005). Coffee and tea consumption are associated with a lower incidence of chronic liver disease in the United States. *Gastroenterology*, *129*, 1928–1936.

[294] Shimizu, M., Yoshimi, N., Yamada, Y., Matsunaga, K., Kawabata, K., Hara, A., Moriwaki, H. & Mori, H. (1999). Suppressive effects of chlorogenic acid on N-methyl-Nnitrosourea-induced glandular stomach carcinogenesis in male F344 rats. *J. Toxicol. Sci.*, *24*, 433–439.

[295] Namba, T. & Matsuse, T. (2002). A historical study of coffee in Japanese and Asian countries: focusing the medicinal uses in Asian traditional medicines. *Yakushigaku Zasshi*, *37*, 65–75.

[296] Wang, G. F., Shia, L. P., Rena, U. D., Liub, Q. F., Liu, H. F., Zhang, R. J., Li, Z., Zhu, F. H., He, P. L., Tang, W., Tao, P. Z., Chuan, L., Wei-Min Zhao, W. M. & Zuo, J. P. (2009). Anti-hepatitis B virus activity of chlorogenic acid, quinic acid and caffeic acid in vivo and in vitro. *Antiviral Research*, *83*, 186–190.

[297] Sells, M. A., Chen, M. L., Acs, G., 1987. Production of hepatitis B virus particles in HepG2 cells transfected with cloned hepatitis B virusDNA. Proc.Natl. *Acad. Sci.U.S.A.*, *84*, 1005–1009.

[298] Zhang, H. L., Dai, L. H., Wu, Y. H., Yu, X. P., Zhang, Y. Y., Guan, R. F., Liu, T. & Zhao, J. (2014). Evaluation of hepatocyteprotective and anti-hepatitis B virus properties of Cichoric acid from Cichorium intybus leaves in cell culture. *Biol Pharm Bull.*, *37*(7), 1214-1220.

[299] Lee, J. (2010). Caffeic acid derivatives in dried Lamiaceae and Echinacea purpurea products. *J. Funct. Foods*, 2, 158–162 .

[300] King, P. J. & Robinson, W. E. Jr. (1998). Resistance to the anti-human immunodeficiency virus type 1 compound L-cichoric acid results from a single mutation at amino acid 140 of integrase. *J. Virol.*, *72*, 8420–8424 .

[301] Robinson, W. E. Jr. (1998). L-Cichoric acid, an inhibitor of human immunodeficiency virus type 1 (HIV-1) integrase, improves on the in vitro anti-HIV-1 effect of zidovudine plus a protease inhibitor (AG1350). *Antiviral Res.*, *39*, 101–111.

[302] Reinke, R. A., Lee, D. J., McDougall, B. R., King, P. J., Victoria, J., Mao, Y., Lei, X., Reinecke, M. G. & Robinson, W. E. Jr. (2004). L-Cichoric acid inhibits human immunodeficiency virus type 1 integration in vivo and is a noncompetitive but reversible inhibitor of HIV-1 integrase in vitro. *Virology.*, *326*, 203–219.

[303] Crosby, D. C., Lei, X., Gibbs, C. G., McDougall, B. R., Robinson, W. E. Jr. & Reinecke, M. G. (2010). Design, synthesis, and biological evaluation of novel hybrid dicaffeoyltartaric/diketo acid and tetrazole-substituted L-cichoric acid analogue inhibitors of human immunodeficiency virus type 1 integrase. *J. Med. Chem.*, *53*, 8161–8175.

[304] Wu, Y. H., Hao, B. J., Cao, H. C., Xu, W., Li, Y. J. & Li, L. J. (2012). Anti-hepatitis B virus effect and possible mechanism of action of 3,4-O-dicaffeoylquinic acid in vitro and in vivo. Evid. Based Complement. *Alternat. Med.*, 2012, 356806.

[305] Cheminat, A., Zawatzky, R., Becker, H. & Brouillard, R. (1988). Caffeoyl conjugates from Echinacea species: structures and biological activity. *Phytochemistry*, *27*, 2787–2794.

[306] Facino, R. M., Carini, M., Aldini, G., Saibene, L., Pietta, P. & Mauri, P. (1995). Echinacoside and caffeoyl conjugates protect collagen from free radical-induced degradation: a potential use of Echinacea extracts in the prevention of skin photodamage. *Planta Med.*, *61*, 510–514.

[307] Costentin, C. E., Roudot-Thoraval, F., Zafrani, E. S., Medkour, F., Pawlotsky, J. M., Mallat, A & Hézode, C. (2010). Association of caffeine intake and histological features of chronic hepatitis C. *J Hepatol.*, *54*, 1123-1129.

[308] Shen, H., Yamashita, A., Nakakoshi, M., Yokoe, H., Sudo, M., Kasai, H., Tanaka, T., Fujimoto, Y., Ikeda, M., Kato, N., Sakamoto, N., Shindo, H., Maekawa, S., Enomoto, N., Tsubuki, M. & Moriishi, K. (2013). Inhibitory effects of caffeic acid phenethyl ester derivatives on replication of hepatitis C virus. *PLoS One*, *8*(12), e82299.

[309] Carreño, V. (2014). Review article: management of chronic hepatitis C in patients with contraindications to anti-viral therapy. *Aliment Pharmacol Ther.*, *39*(2), 148-162.

[310] Ohfuji, S., Fukushima, W., Tanaka, T., Habu, D., Tamori, A., Sakaguchi, H., Takeda, T., Kawada, N., Seki, S., Nishiguchi, S., Shiomi, S. & Hirota, Y. (2006). Coffee consumption and reduced risk of hepatocellular carcinoma among patients with chronic type C liver disease: A case-control study. *Hepatol Res.*, *36*(3), 201-208.

[311] Akash, M. S. H., Rehman, K. & Chen, S. (2013). Role of inflammatory mechanisms in pathogenesis of type 2 diabetes mellitus. *J. Cell. Biochem.*, *114*, 525–531.

[312] Donath, M. Y. (2014). Targeting inflammation in the treatment of type 2 diabetes: time to start. *Nat. Rev. Drug Discov.*, *13*, 465–476.

[313] Oboh, G., Agunloye, O.M., Adefegha, S.A., Akinyemi, A.J. & Ademiluyi, A. O. (2014). Caffeic and chlorogenic acids inhibit key enzymes linked to type 2 diabetes (in vitro): A comparative study. *J. Basic Clin. Physiol. Pharmacol.*, *26*, 165–170.

[314] Akash, M. S. H., Rehman, K. & Chen, S. (2014). Effects of coffee on type 2 diabetes mellitus. *Nutrition*, *30*, 755–763.

[315] Akash, M. S. H., Rehman, K. & Chen, S. (2014). Spice plant Allium cepa: Dietary supplement for treatment of type 2 diabetes mellitus. *Nutrition*, *30*, 1128–1137.

[316] Azuma, K., Ippoushi, K., Nakayama, M., Ito, H., Higashio, H. & Terao, J. (2000). Absorption of chlorogenic acid and caffeic acid in rats after oral administration. J. Agric. *Food Chem.*, *48*, 5496−500.

[317] Adisakwattana, S., Moonsan, P. & Yibchok-Anun, S. (2008). Insulin releasing properties of a series of cinnamic acid derivatives in vitro and in vivo. J. Agric. *Food Chem.*, *56*, 7838−44.

[318] Huang, D. W., Shen, S. C. & Wu, J. S. (2009). Effects of caffeic acid and cinnamic acid on glucose uptake in insulin-resistant mouse hepatocytes. *J. Agric. Food Chem.*, *57*, 7687−7692.

[319] Cheng, J. T., Liu, I. M., Tzeng, T. F., Chen, W. C., Hayakawa, S. & Yamamoto, T. (2003). Release of beta-endorphin by caffeic acid to lower plasma glucose in streptozotocin-induced diabetic rats. *Horm Metab Res*, *35*, 251-258.

[320] Jung, U. J., Lee, M. K., Park, Y. B., Jeon, S. M. & Choi, M. S. (2006). Antihyperglycemic and antioxidant properties of caffeic acid in db/db mice. *J Pharmacol Exp Ther*, *318*, 476-483.

[321] He, X. & Liu, R. H. (2008). Phytochemicals of apple peels: isolation, structure elucidation, and their antiproliferative and antioxidant activities. *J Agric Food Chem*, *56*, 9905–9910.

[322] Huang, D. W., Shen, S. C. & Wu, J. S. (2009). Effects of caffeic acid and cinnamic acid on glucose uptake in insulin-resistant mouse hepatocytes. *J Agric Food Chem*, *57*, 7687–7692.

[323] Tsuda, S., Egawa, T., Ma, X., Oshima, R., Kurogi, E. & Hayashi, T. (2012). Coffee polyphenol caffeic acid but not chlorogenic acid increases 5′ AMP-activated protein kinase and insulin-independent glucose transport in rat skeletal muscle. *J Nutrit Biochem*, *23*, 1403–149.

[324] Valentova, K., Truong, N., Moncion, A., De Waziers, I. & Ulrichova, J. (2007). Induction of glucokinase mRNA by dietary phenolic compounds in rat liver cells in vitro. *J Agric Food Chem*, *55*, 7726–7731.

[325] Fyfe, M. C. T., White, J. R., Taylor, A., Chatfield, R., Wargent, E., Printz, R. L., Sulpice, T., McCormack, J. G., Procter, M. J., Reynet, C., Widdowson, P. S., Wong-Kai-In, P. (2007). Glucokinase activator PSN-GK1 displays enhanced antihyperglycaemic and insulinotropic actions. *Diabetologia*, *50*, 1277–1287.

[326] Bhattacharya, S., Oksbjerg, N., Young, J. F. & Jeppesen, P. B. (2014). Caffeic acid, naringenin and quercetin enhance glucose-stimulated insulin secretion and glucose sensitivity in INS-1E cells. Diabetes, *Obesity and Metabolism*, *16*, 602–612.

[327] Habener, J. F., Kemp, D. M. & Thomas, M. K. (2005). Transcriptional regulation in pancreatic development. *Endocrinology*, *146*, 1025–1034.

[328] Robertson, R. P., Harmon, J., Tran, P. O. T. & Poitout, V. (2004). Beta-cell glucose toxicity, lipotoxicity, and chronic oxidative stress in type 2 diabetes. *Diabetes*, *53*, S119–124.

[329] Soti, C., Nagy, E., Giricz, Z., Vigh, L., Csermely, P. & Ferdinandy, P. (2005). Heat shock proteins as emerging therapeutic targets. *Br J Pharmacol*, *146*, 769–780.

[330] Bai, N. S., He, K., Roller, M., Zheng, B., Chen, X., Shao, Z., Peng, T. & Zheng, Q. (2008). Active compounds from Lagerstroemia speciosa, insulin-like glucose uptake-stimulatory/inhibitory and adipocyte differentiation-inhibitory activities in 3T3-L1 cells. *J Agric Food Chem*, *56*, 11668–11674.

[331] Isganaitis, E. & Lustig, R. H. (2005). Fast food, central nervous system insulin resistance, and obesity. Arterioscler., Thromb., *Vasc. Biol.*, 25, 2451−2462.

[332] Devlin, M. J., Yanovski, S. Z. & Wilson, G. T. (2000). Obesity: what mental health professionals need to know. *Am. J. Psychiatry*, *157*, 854−866.

[333] Fujioka, K. (2002). Management of obesity as a chronic disease: nonpharmacologic, pharmacologic, and surgical options. *Obes. Res.*, *10* (Suppl. 2), 116S−123S.

[334] Schaffer, J. E. (2003). Lipotoxicity: when tissues overeat. *Curr. Opin. Lipidol.*, *14*, 281−287.

[335] Wang, Y. X., Lee, C. H., Tiep, S., Yu, R. T., Ham, J., Kang, H. & Evans, R. M. (2003). Peroxisome-proliferator-activated receptor delta activates fat metabolism to prevent obesity. *Cell*, *113*, 159−170.

[336] Hansen, P. A., Han, D. H., Nolte, L. A., Chen, M. & Holloszy, J. O. (1997). DHEA protects against visceral obesity and muscle insulin resistance in rats fed a high-fat diet. *Am. J. Physiol.*, *273*, R1704–1708.

[337] Mokdad, A. H., Bowman, B. A., Ford, E. S., Vinicor, F., Marks, J. S. & Koplan, J. P. (2001). The continuing epidemics of obesity and diabetes in the United States. JAMA, *J. Am. Med. Assoc.*, *286*, 1195–200.

[338] Hossain, P., Kawar, B. e. & Nahas, M. (2007). Obesity and diabetes in the developing world–a growing challenge. *N. Engl. J. Med.*, *356*, 213–215.

[339] Hanif, M. W. & Kumar, S. (2002). Pharmacological management of obesity. Expert Opin. *Pharmacother.*, *3*, 1711–1718.

[340] Moreno, D. A., Ilic, N., Poulev, A., Brasaemle, D. L., Fried, S. K. & Raskin, I. (2003). Inhibitory effects of grape seed extract on lipases. *Nutrition*, *19*, 876–9.

[341] Liao, C. C., Ou, T. T., Wu, C. H. & Wang, C. J. (2013). Prevention of diet-induced hyperlipidemia and obesity by caffeic acid in C57BL/6 mice through regulation of hepatic lipogenesis gene expression. *J Agric Food Chem.*, *61*(46), 11082-11088.

[342] Park, S. H. & Min, T. S. (2006). Caffeic acid phenethyl ester ameliorate changes in IGFs secretion and gene expression in streptozotocininduced diabetic rats. *Life Sci.*, *78*, 1741–1747.

[343] Gibbs, E. M., Stock, J. L., McCoid, S. C., Stukenbrok, H. A., Pessin, J. E., Stevenson, R. W., Milici, A. J. & McNeish, J. D. (1995). Glycemic improvement in diabetic db/db mice by overexpression of the human insulin-regulatable glucose transporter (GLUT4). *J. Clin. Invest.*, *95*, 1512–1518.

[344] Pinent, M., Blay, M., Blade, M. C., Salvado, M. J., Arola, L. & Ardevol, A. (2004). Grape seed-derived procyanidins have an antihyperglycemic effect in streptozotocin-induced diabetic rats and insulinomimetic activity in insulin-sensitive cell lines. *Endocrinology*, *145*, 4985–4990.

[345] Carling, D. (2004). The AMP-activated protein kinase cascade–a unifying system for energy control. *Trends Biochem. Sci.*, *29*, 18–24.

[346] Merrill, G. F., Kurth, E. J., Hardie, D. G. & Winder, W. W. (1997). AICA riboside increases AMP-activated protein kinase, fatty acid oxidation, and glucose uptake in rat muscle. *Am. J. Physiol.*, *273*, E1107–1112.

[347] Daniel, R. S., Devi, K. S., Augusti, K. T. & Sudhakaran Nair, C. R. (2003). Mechanism of action of antiatherogenic and related effects of Ficus bengalensis Linn. flavonoids in experimental animals. *Indian J. Exp. Biol.*, *41*, 296–303.

[348] Wei, H. A., Lian, T. W., Tu, Y. C., Hong, J. T., Kou, M. C. & Wu, M. J. (2007). Inhibition of low-density lipoprotein oxidation and oxidative burst in polymorphonuclear neutrophils by caffeic acid and hispidin derivatives isolated from sword brake fern (Pteris ensiformis Burm.). *J. Agric. Food Chem.*, *55*, 10579–10584.

[349] Hishikawa, K., Nakaki, T. & Fujita, T. (2005). Oral flavonoid supplementation attenuates atherosclerosis development in apolipoprotein E-deficient mice. Arterioscler., Thromb., *Vasc.* Biol., *25*, 442–6.

[350] Landete, J. M. (2012). Updated knowledge about polyphenols: functions, bioavailability, metabolism, and health. *Crit Rev Food Sci Nutr*, *52*, 936-948.

[351] Natella, F., Nardini, M., Belelli, F. & Scaccini, C. (2007). Coffee drinking induces incorporation of phenolic acids into LDL and increases the resistance of LDL to ex vivo oxidation in humans, *Am. J. Clin. Nutr.*, *86*, 604–609.

[352] Chao, P. C., Hsu, C. C. & Yin, M. C. (2009). Anti-inflammatory and anti-coagulatory activities of caffeic acid and ellagic acid in cardiac tissue of diabetic mice. *Nutrition & Metabolism*, *6*, 3330.
[353] Perez-Vizcaino, F., Duarte, J., Jimenez, R., Santos-Buelga, C. & Osuna, A. (2009). Antihypertensive effects of the flavonoid quercetin. *Pharmacol. Rep.*, *61*, 67–75.
[354] Zhao, Y., Wang, J., Ballevre, O., Luo, H. & Zhang, W. (2011). Antihypertensive effects and mechanisms of chlorogenic acids. *Hypertens. Res.*, *4*, 370–374.
[355] Actis-Goretta, L., Ottavian, J. I., Fraga, C. G. (2006). Inhibition of angiotensin converting enzyme activityby flavanol-rich foods. *J. Agric. Food Chem.*, *54*, 229-234.
[356] Li, P. G., Xu, J. W., Ikeda, K., Kobayakawa, A., Kayano, Y., Mitani, T., Ikami, T. & Yamori, Y. (2005). Caffeic acid inhibits vascular smooth muscle cell proliferation induced by angiotensin II in stroke-prone spontaneously hypertensive rats. *Hypertens. Res.*, *4*, 369–377.
[357] Hashimoto, K., Makino, S., Hirasawa, R. Takao, T., Sugawara, M., Murakami, K., Ono, K. & Ota, Z. (1989). Abnormalities in the hypothalamo-pituitary-adrenal axis in spontaneously hypertensive rats during development of hypertension. *Endocrinology*, *125*, 1161–1167.
[358] Ruch W., Baumann J. B., Hausler A., Otten U. H., Siegl H. & Girard J. (1984). Importance of the adrenal cortex for development and maintenance of hypertension in spontaneously hypertensive rats. *Acta Endocrinol.*, *105*, 417–424.
[359] Li, P. G., Xu, J. W., Ikeda, K., Kobayakawa, A., Kayano, Y., Mitani, T., Ikami, T. & Yamori, Y. (2005). Caffeic acid inhibits vascular smooth muscle cell proliferation induced by angiotensin II in stroke-prone spontaneously hypertensive rats. *Hypertens. Res.*, *4*, 369–377.
[360] Hidalgo, M., Martin-Santamaria, S., Recio, I., Sanchez-Moreno, C., dePascual-Teresa, B., Rimbach, G. & dePascual-Teresa, S. (2012). Potentialanti-inflammatory, anti-adhesive, anti-estrogenic, and angiotensin-converting enzyme inhibitory activities of anthocyanins and their gut metabolites. *Genes Nutr.*, *7*, 295–306.
[361] Kumar, S., Prahalathan, P. & Raja, B. (2011). Antihypertensive and antioxidant potential of vanillic acid, a phenolic compound in L-NAME-induced hypertensive rats: a dose-dependence study. *Redox Rep.*, *16*, 208–215.
[362] Ohsaki, Y., Shirakawa, H., Koseki, T. & Komai, M. (2008). Novel effects of a single administration of ferulic acid on the regulation of blood pressure and the hepatic lipid metabolic profile in stroke-prone spontaneously hypertensive rats. J. Agric. *Food Chem.*, *56*, 2825–2830.
[363] Jiang, R. W., Lau, K. M., Hon, P. M., Mak, T. C., Woo, K. S., Fung, K. P. (2005). Chemistry and biological activities of caffeic acid derivatives from Salvia miltiorrhiza. *Curr. Med. Chem.*, *12*, 237–246.
[364] Balasuriya, B. W. N. & Rupasinghe, H. P. V. (2011). Plant flavonoids as angiotensin converting enzyme inhibitors in regulation of hypertension. Funct. Food Health *Dis.*, *5*, 172–188.
[365] Balasuriya, N. & Rupasinghe, H. P. (2012). Antihypertensive properties of flavonoid-rich apple peel extract. *Food Chem.*,*135*, 2320–2325.
[366] Deng, Y. F., Aluko, R. E., Jin, Q., Zhang, Y. & Yuan, L. J. (2012). Inhibitory activities of baicalin against renin and angiotensin-converting enzyme. *Pharm. Biol.*, *50*, 401–406.

[367] Li, F., Takahashi, Y. & Yamaki, K. (2013). Inhibitory effect of catechin-related compounds on renin activity. *Biomed.* Res., *34*,167–171.
[368] Takahashi, S., Hori, K., Hokari, M., Gotoh, T. & Sugiyama, T. (2010). Inhibitionofhuman renin activity by saponins. *Biomed. Res.*, *31*, 155–159.
[369] Takahashi, S., Hori, K., Shinbo, M., Hiwatashi, K., Gotoh, T. & Yamada, S. (2008). Isolation of human renin inhibitor from soybean: soyasaponin I is the novel human renin inhibitor in soybean. Biosci. *Biotechnol. Biochem*, *72*, 3232–3236.
[370] Esposito, E., Rotilio, D., Di Matteo, V., Di Giulio, C., Cacchio, M. & Algeri, S. (2002). A review of specific dietary antioxidants and the effects on biochemical mechanisms related to neurodegenerative processes. *Neurobiol. Aging*, *23*, 719–735.
[371] Joseph, J. A., Denisova, N. A., Arendash, G., Gordon, M., Diamond, D., Shukitt-Hale, B. & Morgan, D. (2003). Blueberry supplementation enhances signaling and prevents behavioral deficits in an Alzheimer disease model. *Nutr. Neurosci.*, *6*, 153–162.
[372] Commenges, D., Scotet, V., Renaud, S., Jacqmin-Gadda, H., Barberger-Gateau, P. & Dartigues, J. F. (2000). Intake of flavonoids and risk of dementia. *Eur. J. Epidemiol.*, *16*, 357–363
[373] Orgogozo, J. M., Dartigues, J. F., Lafont, S., Letenneur, L., Commenges, D., Salamon, R., Renaud, S. & Breteler, M. B. (1997). Wine consumption and dementia in the elderly: A prospective community study in the Bordeaux area. *Breteler Revue de Neurologie*, *153*, 185–192
[374] Youdim, K. A., Spencer, J. P., Schroeter, H. & Rice-Evans, C. (2002). Dietary flavonoids as potential neuroprotectants. *Biol. Chem.*, *383*, 503–519
[375] Vauzour, D., Vafeiadou, K., Corona, G., Pollard, S. E., Tzounis, X. & Spencer, J. P. (2007). Champagne wine polyphenols protect primary cortical neurons against peroxynitrite-induced injury. J. Agric. *Food Chem.*, *55*, 2854–2860
[376] Vauzour, D., Corona, G. & Spencer, J. P. (2011). Caffeic acid, tyrosol and p-coumaric acid are potent inhibitors of 5-S-cysteinyl-dopamine induced neurotoxicity. *Arch Biochem Biophys.*, *501*(1), 106-111.
[377] Bu, Y., Rho, S., Kim, J., Kim, M. Y., Lee, D. H., Kim, S. Y., Choi, H. & Kim, H. (2007). Neuroprotective effect of tyrosol on transient focal cerebral ischemia in rats. *Neurosci. Lett.*, *414*, 218–221.
[378] Huang, Y., Jin, M., Pi, R., Zhang, J., Chen, M., Ouyang, Y., Liu, A., Chao, X., Liu, P., Liu, J., Ramassamy, C. & Qin, J. (2013). Protective effects of caffeic acid and caffeic acid phenethyl ester against acrolein-induced neurotoxicity in HT22 mouse hippocampal cells. *Neurosci Lett. 535*, 146-151.
[379] Scapagnini, G., Foresti, R., Calabrese, V., Giuffrida Stella, A. M., Green, C. J. & Motterlini, R. (2002). Caffeic acid phenethyl ester and curcumin: a novel class of heme oxygenase-1 inducers Molecular Pharmacology, *61*, 554–561.
[380] Fu, W., Du, G., Liu, D. & Ruan, J. L. (2013). Neuroprotective effect of a caffeic acid derivative from Abacopteris penangiana. *Pharm Biol.*, *51*(3), 376-382
[381] Pinheiro Fernandes, F. D., Fontenele Menezes, A. P., de Sousa Neves, J. C., Fonteles, A. A., da Silva, A. T., de Araújo Rodrigues, P., Santos do Carmo, M. R., de Souza, C. M. & de Andrade, G. M. (2014). Caffeic acid protects mice from memory deficits induced by focal cerebral ischemia. *Behav Pharmacol.*, *25*(7), 637-647.

[382] Kawahara, M. & Kuroda, Y. (2000). Molecular mechanism of neurodegeneration induced by Alzheimer's beta-amyloid protein: channel formation and disruption of calcium homeostasis. *Brain Research Bulletin*, *53*, 389–397.

[383] Eom, T. K., Ryu, B., Lee, J. K., Byun, H. G., Park, S. J. & Kim, S. K. (2013). Beta-secretase inhibitory activity of phenolic acid conjugated chitooligosaccharides. Journal of Enzyme Inhibition and Medicinal Chemistry, *28*, 214–217.

[384] Sul, D., Kim, H. S., Lee, D., Joo, S. S., Hwang, K. W., Park, S. Y. (2009). Protective effect of caffeic acid against beta-amyloid-induced neurotoxicity by the inhibition of calcium influx and tau phosphorylation. *Life Sci*, *84*, 257–262.

[385] Leschik, J., Welzel, A., Weissmann, C., Eckert, A. & Brandt, R. (2007). Inverse and distinct modulation of tau-dependent neurodegeneration by presenilin 1 and amyloid-beta in cultured cortical neurons: evidence that tau phosphorylation is the limiting factor in amyloid-beta-induced cell death. *Journal of Neurochemistry*, *101*, 1303–1315.

[386] Muller, W. E., Romero, F. J., Perovic, S., Pergande, G. & Pialoglou, P. (1997). Protection of flupirtine on beta-amyloid-induced apoptosis in neuronal cells in vitro: prevention of amyloid-induced glutathione depletion. *Journal of Neurochemistry*, *68*, 2371–2377.

[387] Koo, K. A., Kim, S. H., Oh, T. H. & Kim, Y. C. (2006). Acteoside and its aglycones protect primary cultures of rat cortical cells from glutamate-induced excitotoxicity. *Life Sciences*, *79*, 709–716.

[388] J. Birks. Cholinesterase inhibitors for Alzheimer's disease. The Cochrane Database of Syst. *Rev.*, *25* (1), CD005593.

[389] Winstein, C. J., Bentzen, K. R., Boyd, L. & Schneider, L. S. (2007). Does the cholinesterase inhibitor, donepezil, benefit both declarative and non-declarative processes in mild to moderate Alzheimer's disease? *Current Alzheimer Research*, *4*, 273–276.

[390] Schuster, D., Spetea, M., Music, M., Rief, S., Fink, M., Kirchmair, J., Schütz, J., Wolber, J., Wolber, G., Langer, T., Stuppner, H., Schmidhammer, H. & Rollinger, J. M. (2010). Morphinans and isoquinolines: acetylcholinesterase inhibition, pharmacophore modeling, and interaction with opioid receptors. Bioorg. *Med. Chem.*, *18*, 5071–5080.

[391] Gaspersic, R., Horitnik, B., Crne-Finderle, N. & Sketelj, J. (1999). Acetylcholinesterase in the neuromuscular junction. *Chem Biol Interact 119*,–120, 301–308.

[392] Anglister, L., Etlin, A., Finkel, E., Durrant, A., Lev-Tov, A. (2008). Cholinesterases in development and disease. *Chem Biol Interact*, *175*, 92-100.

[393] Silman, I. & Sussman, J. (2005). Acetylcholinesterase: "classical" and "non-classical" functions and pharmacology. *Curr Opin Pharmacol*, *5*, 293–302.

[394] Mouisel, E., Blondet, B., Escourrou, P., Chatonnet, A., Molgó, J. & Ferry, A. (2006). Outcome of acetylcho linesterase deficiency for neuromuscular functioning. *Neurosci Res*, *55*, 389–396.

[395] Vignaud, A., Fougerousse, F., Mouisel, E., Bertrand, C., Bonafos, B., Molgo, J., Ferry, A. & Chatonnet, M. A. (2008). Genetic ablation of acetylcholinesterase alters muscle function in mice. *Chem Biol Interact*, *175*, 129–130.

[396] Rakonczay, Z. (2003). Potencies and selectivities of inhibitors of acetylcholinesterase and its molecular forms in normal and Alzheimer's disease brain. *Acta Biol Hung*, *54*, 183–189.

[397] Brenner, T., Hamra-Amitay, Y., Evron, T., Boneva, N., Seidaman, S. & Soreq, H. (2003). The role of readthrough acetylcholinesterase in the pathophysiology of myasthenia gravis. *FASEB J*, *17*, 214–222.

[398] Kawahara, M. (2005).Effects of aluminum on the nervous system and its possible link with neurodegenerative diseases. *J Alzheimers Dis*, *2*, 171–182.

[399] Walton, J. R. (2007). An aluminum-based rat model for Alzheimer's disease exhibits oxidative damage, inhibition of PP2A activity, hyperphosphorylated tau, and granulovacuolar degeneration. *J Inorg Biochem*, *9*, 1275–1284.

[400] Platt, B., Fiddler, G., Riedel, G. & Henderson, Z. (2001). Aluminium toxicity in the rat brain: histochemical and immunocytochemical evidence. *Brain Res Bull*, *2*, 257–267.

[401] Khan, K. A., Kumar, N., Nayak, P. G., Nampoothiri, M., Shenoy, R. R., Krishnadas, N., Rao, C. M. & Mudgal, J. (2013). Impact of caffeic acid on aluminium chloride-induced dementia in rats of Pharmacy and Pharmacology, *65*, 1745–1752.

[402] Anwar, J., Spanevello, R. M., Thomé, G., Stefanello, N., Schmatz, R., Gutierres, J., Vieira, J., Baldissarelli, J., Carvalho, F. B., da Rosa, M. M., Rubin, M. A., Fiorenza, A., Morsch, V. M. & Schetinger, M. R. (2012). Effects of caffeic acid on behavioral parameters and on the activity of acetylcholinesterase in different tissues from adult rats. *Pharmacol Biochem Behav*, *2*, 386–394.

[403] Yang, J. Q., Zhou, Q. X., Liu, B. Z. & He, B. C. (2008). Protection of mouse brain from aluminum-induced damage by caffeic acid. *CNS Neurosci Ther*, *1*, 10–16.

[404] Moser, K. V., Schimidt-Kastner, R., Hinterhuber, H. & Humpel, C. (2003). Brain capillaries and cholinergic neurons persist in organotypic brain slices in the absence of blood flow. *Eur J Neurosci*, *18*, 85–94.

[405] Pereira, P, Oliveira, P., Ardenghi, P., Rotta, L., Henriques, J. & Piccada, J. (2006). Neuropharmacological analysis of caffeic acid in rats. *Basic Clin Pharmacol Toxicol*, *99*, 374–378.

[406] Zhou, Y., Fang, G., Ye, Y., Chu, L., Zhang, W.,Wang, M. & Wei, E.Q. (2006). Caffeic acid ameliorates early and delayed brain injuries after focal cerebral ischemia in rats. *Acta Pharmacol Sin*, *27*, 1103–1110.

[407] Yang, Q., Zho, B., Liu, B. & He, B. (2008). Protection of mouse brain from aluminum-induced damage by caffeic acid. *CNS Neuros Ther*, *14*, 10–16.

[408] Williams, P., Sorribas, A. & Howes, M. J. R. (2011). Natural products as a source of Alzheimer's drug leads. *Nat. Prod. Rep.*, *28*, 48–77.

[409] Hirsch, E. C. & Hunot, S. (2009). Neuroinflammation in Parkinson's disease: a target for neuroprotection? *Lancet Neurol.*, *8*, 382–397.

[410] Hirsch, E. C., Breidert, T., Rousselet, E., Hunot, S., Hartmann, A. & Michel, P. P. (2003). The role of glial reaction and inflammation in Parkinson's disease. Ann. N. Y. *Acad. Sci.*, *991*, 214–228.

[411] Sugama, S., Takenouchi, T., Cho, B. P., Joh, T. H., Hashimoto, M. & Kitani, H. (2009). Possible roles of microglial cells for neurotoxicity in clinical neurodegenerative diseases and experimental animal models. Inflamm. *Allergy Drug Targets*, *8*, 277–284.

[412] Vauzour, D., Ravaioli, G., Vafeiadou, K., Rodriguez-Mateos, A., Angeloni, C. & Spencer, J. P. (2008). Peroxynitrite induced formation of the neurotoxins 5-S-cysteinyl-dopamine and DHBT-1: implications for Parkinson's disease and protection by polyphenols. Arch. Biochem. *Biophys.*, *476*, 145–151.

[413] Nakashima, A., Hayashi, N., Kaneko, Y.S., Mori, K., Sabban, E.L., Nagatsu, T. & Ota, A. (2009). Role of N-terminus of tyrosine hydroxylase in the biosynthesis of catecholamines. *J. Neural. Transm.*, *116*, 1355–1362.

[414] Coelho, V. R., Vieira, C. G., de Souza, L. P., Moysés, F., Basso, C., Papke, D. K., Pires, T. R., Siqueira, I. R., Picada, J. N. & Pereira, P. (2015). Antiepileptogenic, antioxidant and genotoxic evaluation of rosmarinic acid and its metabolite caffeic acid in mice. *Life Sci.*, *122*, 65-71.

[415] Dos Santos Sales, I. M., Do Nascimento, K. G., Feitosa, C. M., Saldanha, G. B., Feng, D. & Freitas, R. M. (2011). Caffeic acid effects on oxidative stress in rat hippocampus after pilocarpine-induced seizures, *Neurol. Sci.*, *32*, 375–380.

[416] Nugroho, A., Kim, M., Choi, J., Choi, J. S., Jung, W. T., Lee, K. & Park, H. (2012). Phytochemical studies of the phenolic substances in *Aster glehni* extract and its sedative and anticonvulsant activity, Arch. *Pharm. Res.*, *35*, 423–430.

[417] Nakamura, M., Kirito, K., Yamanoi, J., Wainai, T., Nojiri, H. & Saito, M. (1991). Ganglioside GM3 Can Induce Megakaryocytoid Differentiation of Human Leukemia Cell Line K562 Cells. *Cancer Res*, *51*, 1940-1945.

[418] Choi, H. J., Chung, T. W., Kang, N. Y., Kim, K. S., Lee, Y. C. & Kim, C. H. (2004). Involvement of CREB in the transcriptional regulation of the human GM3 synthase (hST3Gal V) gene during megakaryocytoid differentiation of human leukemia K562 cells. *Biochem Biophys Res Commun*, *313*, 142-147.

[419] Kuo, H. C., Kuo, W. H., Lee, Y. J., Wang, C. J. & Tseng, T. H. (2006). Enhancement of caffeic acid phenethyl ester on all-trans retinoic acid-induced differentiation in human leukemia HL-60 cells. *Toxicol Appl Pharmacol*, *216*, 80-88.

[420] Juman, S., Yasui, N., Okuda, H., Ueda, A., Negishi, H., Miki, T. & Ikeda, K. (2011). Caffeic acid phenethyl ester suppresses the production of adipocytokines, leptin, tumor necrosis factor-α and resistin, during differentiation to adipocytes in 3T3-L1 cells. *Biol. Pharm. Bull.*, *34*, 490−494.

[421] Heo, S. K., Noh, E. K., Yoon, D. J., Jo, J. C., Koh, S., Baek, J. H., Park, J. H., Min, Y. J. & Kim, H. (2015). Rosmarinic acid potentiates ATRA-induced macrophage differentiation in acute promyelocytic leukemia NB4 cells. *Eur J Pharmacol.*, *747*, 36-44.

[422] Kim, B., Kim, J. E. & Kim, H. S. (2014). Caffeic acid induces keratinocyte differentiation by activation of PPAR-α. *J Pharm Pharmacol.*, *66*(1), 84-92.

[423] Lim, S. W., Hong, S. P., Jeong, S. W., Kim, B., Bak, H., Ryoo, H. C., Lee, S. H. & Ahn, S. K. (2007). Simultaneous effect of ursolic acid and oleanolic acid on epidermal permeability barrier function and epidermal keratinocyte differentiation via peroxisome proliferator activated receptor-α. *J Dermatol*, *34*, 625–634.

[424] Lee, W. J. & Zhu, B. T. (2006). Inhibition of DNA methylation by caffeic acid and chlorogenic acid, two common catechol-containing coffee polyphenols. *Carcinogenesis* *27*, 269−277.

[425] Prasad, N. R., Jeyanthimala, K. & Ramachandran, S. (2009). Caffeic acid modulates ultraviolet radiation-B induced oxidative damage in human blood lymphocytes. J *Photochem Photobiol B.*, 2009, *95*(3), 196-203.

In: Caffeic Acid
Editor: Leanna Vaughn
ISBN: 978-1-63483-121-5

Chapter 6

THERAPEUTIC POTENTIAL OF CAFFEIC ACID PHENETHYL ESTER IN NEURODEGENERATIVE DISEASES

Mushfiquddin Khan, Mauhammad Baarine and Inderjit Singh*

Department of Pediatrics, Medical University of South Carolina, Charleston, SC

ABSTRACT

The lack of treatment options for neurodegenerative diseases challenges neuroscientists to investigate not only new mechanisms but also to explore the properties of promising non-toxic compounds for a targeted therapy. The plant-originated, naturally occurring flavonoid caffeic acid phenethyl ester (CAPE) has attracted considerable interest due its substantiated efficacy, low toxicity and multi-mechanistic activity in a number of animal models of neurodegenerative diseases. CAPE is commonly stored as the active component of propolis of honeybee hives and has been used traditionally in folklore medicine. It shows redox-balancing activity through its antioxidant property and confers neuroprotective effects through its transcription factor-modulating activity. Major mediators of CAPE activity are stimulation of Nrf2 and inhibition of both NF-κB and STAT3. The lipophilic nature of CAPE renders it suitable to cross the blood-brain barrier and to induce a neurorestorative process in neurodegenerative diseases, including stroke, traumatic brain injury, spinal cord injury, adrenoleukodystrophy, multiple sclerosis, amyotrophic lateral sclerosis, and Alzheimer's disease.

Keywords: CAPE, antioxidant, neurodegenerative disease, demyelinating diseases, mitochondrial function

* Corresponding Author: Inderjit Singh, Ph.D., Distinguished University Professor, Medical University of South Carolina, Department of Pediatrics, 516 Children's Research Institute, Charleston, SC 29425, Tel: (843) 792-7991, Fax: (843) 792-3653, Email: singhi@musc.edu

INTRODUCTION

Investigating the therapeutic amelioration of neurodegenerative diseases and brain trauma is challenging because of complications invoked by secondary injury from oxidative stress. In settings of neurodegenerative diseases, reactive oxygen species (ROS) are highly damaging to the neurons and contribute to both activation of the endothelium and blood brain barrier leakage. They also facilitate deleterious inflammation and tissue degeneration. Both oxidative stress and inflammation are major culprits responsible for the secondary injury in neurodegenerative diseases. Unlike a primary irreversible causative injury, the secondary injury is treatable. Therefore, a therapy directed towards interventions based on amelioration of oxidative stress and inflammation is the focus of research on several neurodegenerative diseases, including stroke, traumatic brain injury (TBI), spinal cord injury (SCI), Alzheimer's and demyelinating diseases. Caffeic acid phenethyl ester (CAPE) is one of the most potent lipophilic antioxidants [1, 2], and it is well-tolerated in humans. For centuries, it has been used as a folklore medicine. CAPE is a plant-originated, naturally occurring compound commonly stored as the healing component in propolis of honeybee hives (Figure 1). Several natural flavonoids and polyphenols found in fruits, vegetables and tea are documented to be potent antioxidants. Among these flavonoids, the most effective antioxidant is CAPE [3]. CAPE can be readily synthesized in a laboratory, and the activity of CAPE can also be up/down regulated either by an alteration in the chain length of phenethyl alcohol or by oxidation/substitution in the catechol ring [4, 5]. CAPE is rapidly absorbed and metabolized by plasmatic esterases to caffeic acid [6], which is also an effective antioxidant and inhibits the activity of 5-lipoxygenase (5-LOX) [7]. Caffeic acid protects against cerebral IR injury in a rat model of experimental stroke; however, its efficacy is low compared with CAPE [8, 9]. CAPE's documented potent antioxidant properties are realized through the following mechanisms: robust inhibition of NF-κB; signal transduction and activation of transcription 1/3 (STAT1/3), inducible nitric oxide synthase (iNOS), protein tyrosine kinase; and lipid peroxidation [3, 10-14]. In fact, among the flavonoids in propolis, CAPE is the most effective anti-inflammatory, anticancer and immunomodulatory component [3]. CAPE has also been identified as scavenger of deleterious and potent oxidant peroxynitrite, an instantaneous reaction product of superoxide and nitric oxide (NO) in biological systems [15]. Levels of acrolein, the most reactive lipid peroxidation product, and acrolein-induced neurotoxicity are also attenuated by CAPE [16]. These observations support the use of CAPE as a drug for anti-inflammatory and anti-oxidant activities. The broad range of beneficial activities associated with CAPE have been recently reviewed [4, 17], but the major focus of this review is to highlight the efficacy of CAPE as a neuroprotective drug for neurodegenerative diseases.

In general, CAPE confers neuroprotection in animal models of cerebral ischemia and reperfusion [9] and protects against neonatal brain hypoxic-ischemic injury [18]. It also stabilizes hemodynamics and the area-at-risk in a cardiac IR injury model [19]. However, the molecular mechanisms underlying CAPE-mediated reductions in brain injury and improvements of neurological functions are not clear. Recent studies from our [8, 9] and other laboratories [3, 20-22] suggest that CAPE invokes its protective effects mainly by maintaining mitochondrial integrity, likely via the activation of AMPK and down regulation of inflammation via the inhibition of NF-κB. Recently, Tsai et al. showed down regulation of inflammation in microglial cells by CAPE via the inhibition of iNOS and COX-2, activation of AMPK, and the induction of HO-1 [23]. These findings support the notion that CAPE may be used as a potent anti-inflammatory and antioxidant drug in acute as well as chronic neurodegenerative diseases.

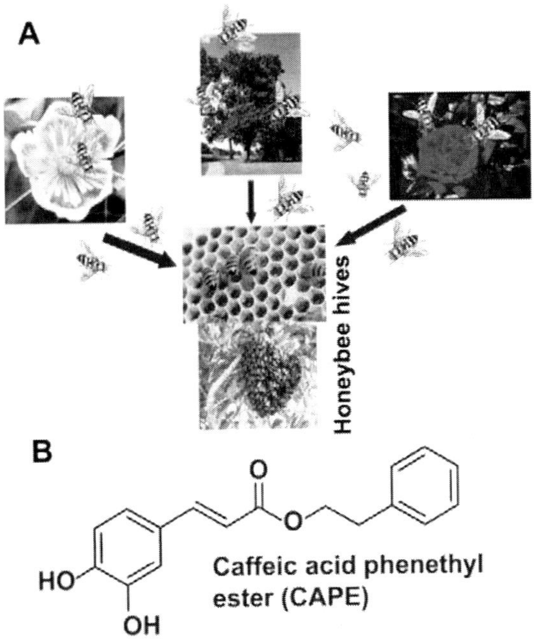

Figure 1. Schematic showing sources and structure of CAPE. The natural product propolis is collected by honeybees from botanical sources, including flowers, and stored in honeybee hives (A). CAPE (B), an ester of caffeic acid and phenethyl alcohol, is the one of the main constituents of propolis.

CAPE IN ACUTE BRAIN INJURY

Stroke: Stroke is the leading cause of long-term physical and cognitive disabilities, especially in elderly patients. It is caused by an obstruction in blood supply to the brain (ischemia) due to blood clot or rupture, or stenosis of vessels. Stroke deprives specific areas of the brain of oxygen and nutrients. Oxygen is essential for survival and the normal function of brain cells, including neurons. In an environment of low oxygen, neurons die within minutes, resulting in morbidity, disability and death. Consequently, following stroke, restoration of oxygen and nutrients to the brain is a must. This can be achieved by removing the obstruction using therapeutic agents or mechanical means. Currently, recombinant tissue plasminogen activator (tPA) is the only approved therapy for ischemic stroke. It degrades blood clots and restores blood flow. However, to be effective, it must be administered within a 3-hour time window. This therapeutic limitation stems mainly from oxidative stress and inflammation. We observed that CAPE not only restored the cerebral blood flow like tPA but also, unlike this current therapy, provided neuroprotection and extended the therapeutic window of treatment up to 12-18 hours (depending on severity of the injury), because it mitigated stroke's secondary injury from oxidative stress and inflammation [9]. A pretreatment with CAPE of IR animals increases NO levels in the plasma and in the brain, reducing brain infarction, structural changes and cerebral vasospasm [24-26]. We observed that CAPE treatment after IR inhibits NF-κB, iNOS, ICAM-1 and cytokines, blocks caspase-3 expression and macrophage infiltration, increases Bcl-xL levels, and inhibits apoptotic cell death in IR animals [9]. In a rabbit model of IR, CAPE significantly reduced serum levels of S-100B, an index of neuroinflammation and brain damage [27]. Supporting anti-inflammatory effects of CAPE, we observed inhibition of iNOS as well as NF-κB by CAPE in microglial BV2 cell lines (Figure 2).

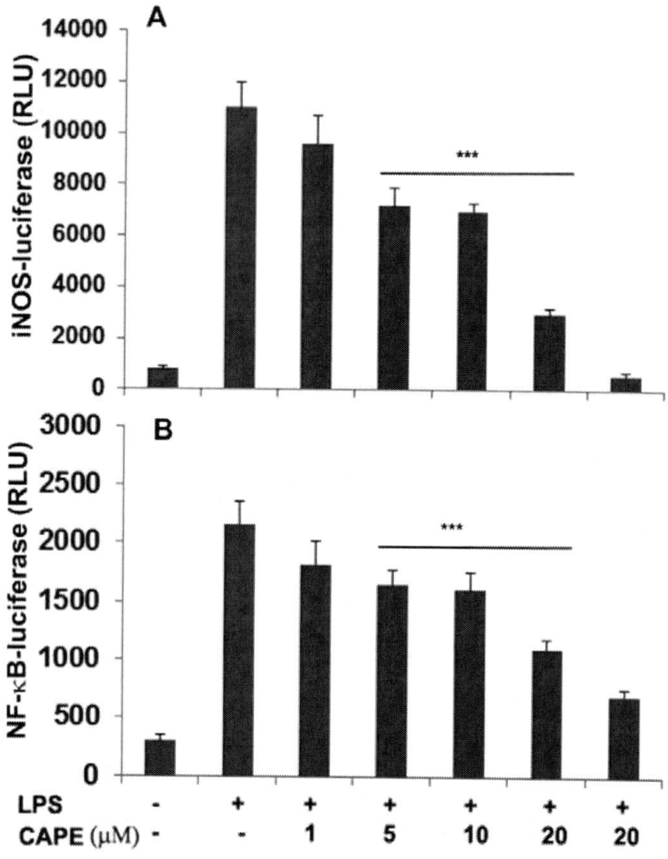

Figure 2. CAPE inhibits LPS-mediated iNOS gene expression and NF-κB reporter activity in microglial cell line BV2. (A): BV2 cells were transiently transfected with 1.5 μg of iNOS –luciferase with lipofectamine plus, according to the manufacturer's instructions, followed by stimulation for 6 h with indicated treatment with CAPE and LPS (1 μg/ml). (B): BV2 cells were transiently transfected with 1.5 μg of p(NF-κB)3LdLuc with lipofectamine plus, according to the manufacturer's instructions, followed by stimulation for 4 h with indicated treatment with CAPE and LPS (1 μg/ml). Data are mean±SD from three different experiments.

Furthermore, CAPE treatment of IR animals also increases levels of reduced glutathione and decreases the levels of oxidized glutathione as well as peroxynitrite [9]. However, many antioxidant drugs have failed in human trials of stroke therapy due to their inability to cross BBB and/or associated toxicity [28]. CAPE, administered > 30 mg/kg orally or ip, does not show toxicity in animal models [18, 29]. As a natural lipophilic small phenolic compound, it can penetrate cellular membranes and cross the BBB [1]. Resveratrol, structurally similar to CAPE, has been shown to cross the BBB not only in an animal model of cerebral ischemic injury [30] but also in normal animals [31]. The lipophilic property, safety and relatively higher antioxidant activity of CAPE [1], as well as its wide therapeutic time and dose windows, result in optimal neuroprotection, leading to functional recovery.

Traumatic brain injury (TBI): Among young adults driving on the highways and soldiers returning from the battleground, neuroinflammatory traumatic brain injury (TBI) is the leading cause of disability in the United States [32]. Furthermore, TBI is associated with significant morbidity/mortality and a cost of ~$60 billion annually. Following TBI, initial trauma causes necrotic neuronal death; however, even greater apoptotic neuronal loss occurs hours and days later, caused by the secondary injury of neuroinflammation and oxidative stress. Reactive oxygen/nitrogen species (ROS/RNS) alter the functions of multiple enzymes and worsen neurovascular injury following brain trauma [33, 34].

Presently, effective therapy for TBI patients is not available, mainly because of limited understanding of the crosstalk between functional recovery and oxidative/nitrosative stress and cell death [33]. In TBI, oxidative stress and inflammation are implicated in blood brain barrier (BBB) disruption, inflammation, neuronal cell death, and functional deficits [35-39]. A post-TBI administration of CAPE shows blood-brain barrier (BBB) protection and reduced cortical tissue loss [40]. In another TBI study, CAPE decreased oxidative stress, reduced neuronal loss and protected tissue structure. In this study, the efficacy of CAPE was shown to be associated with the protection of mitochondrial function [41]. That mitochondrial integrity is well protected by CAPE, likely via its antioxidant activity, has also been documented in ischemia reperfusion injury [42]. CAPE also protects against microglial inflammation by its HDAC inhibitory activity [43], indicating that CAPE is a promising drug for neuroinflammatory TBI therapy. Supporting this notion and the effectiveness of CAPE in an animal model of experimental stroke [9], we tested and observed that CAPE treatment of TBI rats 2 h after injury, induced by controlled cortical impact (CCI), improved motor and neuromuscular function when evaluated 14 days after the injury (Figure 3).

Spinal cord injury (SCI): In the US, approximately 11,000 new cases of spinal cord injury (SCI) occur each year. The cost of care for these patients is estimated to exceed billions of dollars per year [44, 45]. SCI results in the primary loss of cells (neurons and oligodendrocytes) and also sets into motion oxidative stress and inflammatory responses. Patients with SCI often have permanent and devastating neurological deficits [46-52]. A comprehensive review on SCI summarizes the progress in the field of SCI therapy [53]. Presently, treatment of SCI involves mainly minimizing damage at the cellular level (methylprednisolone). However, this drug has a short therapeutic window, and it is also associated with secondary damage from oxidative stress. Moreover, methylprednisolone is not a potent reducer of oxidative stress compared with CAPE as shown in a spinal cord ischemia reperfusion model [54]. In this study, CAPE reduced ischemia/reperfusion damage and provided better neurologic outcomes. A comparative study of CAPE and methylprednisolone showed that both decrease oxidant levels; however, the decrease in oxidant levels was more pronounced in CAPE group [55]. Methotrexate-induced neurotoxicity in spinal cord damage is blunted by CAPE through its antioxidant activity [56]. In a recent report, CAPE was shown to enhance the recovery of locomotor function and to reduce the lesion size in a rat model of SCI by suppressing inflammatory mediators including iNOS and COX-2 [57], indicating the premose of CAPE therapy in SCI.

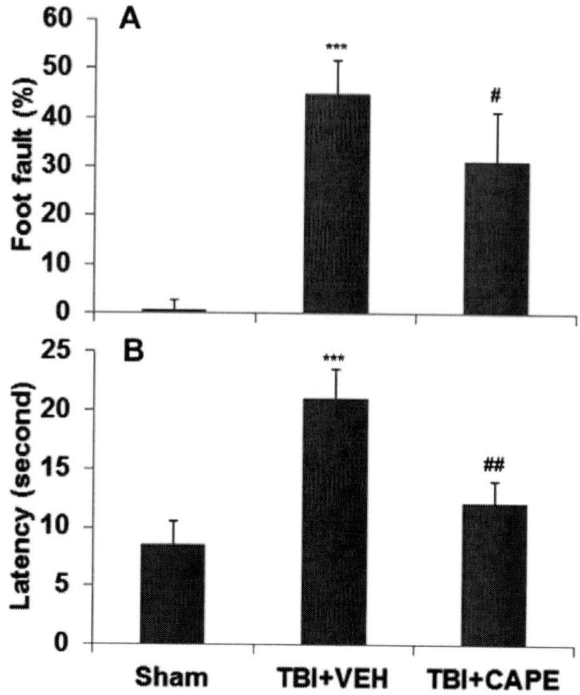

Figure 3. CAPE treatment of TBI mice improved foot fault and beam walking time. TBI was induced by controlled cortical impact, and CAPE (5 mg/kg, ip) was administered 2 h after TBI. The treatment with CAPE (every 24 h) was continued for 14 days. Foot fault and walking latency was determined using beam task. Data are presented as mean±SD from 5 different animals in each group.***$p<0.001$ vs. Sham, #$p<0.05$ vs. TBI+VEH, ##$p<0.01$ vs. TBI+VEH. VEH; vehicle.

CAPE in Chronic Neurodegenerative and Demyelinating Diseases

X-linked adrenoleukodystrophy (X-ALD): X-ALD is a white matter demyelinating disease with prominent inflammatory components. X-ALD is considered to be a neurometabolic disease with progressive demyelination in the central nervous system (CNS), axonopathy in the spinal cord, and adrenal insufficiency [58]. It is the most common peroxisomal disorder, occurring in approximately in 1/17,000 live births [58, 59]. X-ALD is caused by mutations/deletions in the ATP-binding cassette transporter (ABCD1) gene encoding peroxisomal ABC transporter adrenoleukodystrophy protein (ALDP) [60]. Metabolic dysfunction in X-ALD is characterized by an excessive accumulation of very long chain fatty acids, especially hexacosanoic acid (C26:0) in brain, tissues and plasma of X-ALD patients. Studies in different models of X-ALD showing oxidative and nitrosative stress imply that ROS-mediated and reactive nitrogen species (RNS)-mediated mechanisms participate in ALD pathology [61-65]. The relationship between ABCD1-loss-induced VLCFA accumulation and cellular ROS/RNS was further established by the observed ROS in Abcd1/Abcd2-silenced astrocytes [66], oligodendrocytes [65, 67-69], and C6 cells [68, 70]. Antioxidants, including N-acetylcysteine, are shown to reduce not only nitroxidative stress but also the levels of VLCFA [70]. These observations indicate that VLCFA-induced oxidative stress may play a role in the pathobiology of X-ALD. Increased activity of 5-LOX and COX-2 in the human X-ALD brain further indicates lipotoxicity in X-ALD [71]. The human X-ALD brain and a mouse model of X-ALD also show profound loss of plasmalogens, an endogenous antioxidant lipid, supporting redox alterations in X-ALD [72]. Therefore, we hypothesized and investigated CAPE as an antioxidant for its therapeutic potential in X-ALD. CAPE treatment reduced not only the levels of VLCFA but also the levels of ROS in X-ALD cell culture models [13]. The treatment reduced the accumulation of both saturated VLCFA (C26:0) and mono-unsaturated VLCFA (C26:1) and normalized the peroxisomal β-oxidation in cultured human skin fibroblasts from X-ALD patients [13]. The expression of ELOVL1, the single elongase catalyzing the synthesis of both saturated VLCFA (C26:0) and mono-unsaturated VLCFA (C26:1), was also reduced by CAPE treatment. Importantly, CAPE upregulated Abcd2 (redundant functional gene of Abcd1) expression and peroxisomal β-oxidation activity, and decreased the VLCFA levels in Abcd1-deficient U87 astrocytes and B12 oligodendrocytes. In addition, in Abcd1/Abcd2-silenced mouse primary astrocytes, CAPE reduced the VLCFA-induced inflammatory response, including that of 5-LOX. CAPE treatment also decreased the inflammatory response, as indicated by the reduced expression of inducible nitric oxide synthase (iNOS), inflammatory cytokine, and the activation of NF-κB. The observations indicate that CAPE corrects both lipoxidative metabolic and neuroinflammatory diseases.

Because X-ALD disease is specifically associated with mitochondrial dysfunction as reported from our laboratory [65], we investigated whether CAPE reverses mitochondrial dysfunctions using cell culture models of astrocytes and oligodendrocytes. Cells were silenced for the Abcd1 gene to mimic X-ALD disease, as previously reported [13], and then treated with CAPE (2.5 μM) for 72 h. We investigated mitochondrial homeostasis, the mitochondrial mass (Figure 4A-C), and the levels of mitochondrial superoxide (Figure 4D). CAPE induced mitochondrial biogenesis, as evidenced by increased mitochondria mass (Figure 4A) and the expression of PGC1α (Figure 4C), the master regulator of mitochondria biogenesis. CAPE also enhanced the expression of ND1, a mitochondrial genome mediator (Figure 4B). Furthermore, CAPE decreased the levels of mitochondrial superoxide (Figure 4D). Collectively, these data indicate CAPE is a drug candidate to reduce mitochondrial dysfunction not only in X-ALD disease but also in other demyelinating and neurodegenerative diseases.

Other neurodegenerative diseases: As an oxidant, CAPE reduces protein carbonyl and lipid peroxidation levels during glutathione depletion conditions [73] such as Alzheimer's disease, Parkinson's disease, multiple sclerosis (MS), and amyotrophic lateral sclerosis (ALS). In experimental autoimmune encephalomyelitis, an animal model of MS, CAPE significantly inhibited reactive oxygen species (ROS) production induced by EAE and ameliorated clinical symptoms in rats [74]. CAPE treatment of a cell culture model of Parkinson disease reduced mitochondrial dysfunction, blocked neurotoxicity and provided neuroprotection [75, 76], indicating potential efficacy in Parkinson's disease. Supporting the efficacy of CAPE, Fontanilla et al found that CAPE treatment of MPTP (1-methyl-4-phenyl-1,2,3,6-tetrahydropyridine) in a mouse model of Parkinson's disease attenuated dopaminergic neuronal death and dopamine loss [77]. CAPE treatment also reduced the expression of iNOS and decreased apoptotic cell death [77]. A screening of 2000 small molecules identified the 9 most potent antioxidant molecules which protected motor neurons in an *in vitro* model of ALS. Among them, CAPE was the most effective antioxidant as determined by NF-κB inhibition activity and reduction of ROS levels [78]. In an ALS mouse model, CAPE treatment increased total lifespan, helped motor neurons to survive and decreased glial cell activation in lumbar spinal cord [79]. CAPE, acting as an electrophile, was also able to activate the Nrf2–ARE pathway (Figure 5), a pathway we have previously shown to be down-regulated in motor neuron cells expressing mutant SOD1 isolated from cases of familial SOD1-associated ALS [78]. The structure of CAPE (Figure 1B), including a hydrophobic/aromatic moiety with a hydroxyl group, is consistent with known Nrf2 activators as an electrophile, which reacts with Keap1 (Kelch ECH-associating protein 1; the cytoplasmic Nrf2 regulator), thereby releasing Nrf2, which translocates to the nucleus as depicted in Figure 5. Genes up-regulated by Nrf2 include the antioxidants glutathione peroxidase, glutathione reductase, and heme oxygenase 1; enzymes involved in glutathione synthesis; and NADPH-regenerating enzymes [80].

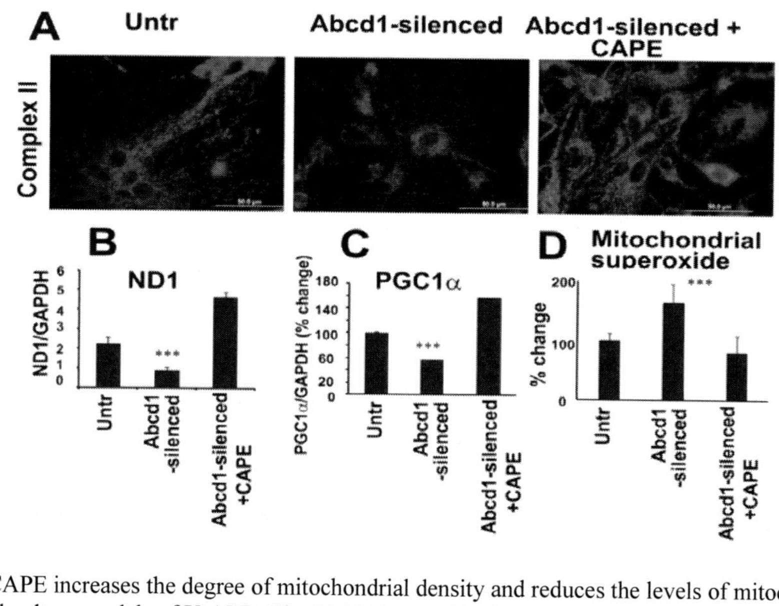

Figure 4. CAPE increases the degree of mitochondrial density and reduces the levels of mitochondrial ROS in cell culture models of X-ALD. The X-ALD gene Abcd1 was silenced using a set of three human (SK-009605-00-10) and rat (SK-098142-00-10) specific SMART vector 2.0 lentiviral shRNA particles (108 TU/ml) for Abcd1. SMARTvector 2.0 non-targeting shRNA control particles (NT) (108 TU/ml, Thermo Fisher Scientific), designed so that no known genes targeted in human and mouse or rat, were used as negative controls (A-D). Astrocytes (U87, A-C) and oligodendrocytes (B12, D) were treated with CAPE (2.5 μM) once daily for 72 h before the measurements. (A): Mitochondrial density was determined by immunostaining of mitochondria using Anti-MTCO2 antibody. (B): The mRNA expression of ND, a mitochondrial gene, was measured as a marker of mitochondrial genome activity. (C): The mRNA expression of PGC1α was quantitated as a marker of mitochondrial biogenesis. (D): The levels of mitochondrial superoxide were measured using MitoSOX staining (15min, 37°C) and quantitated by spectrofluorometry. Data are presented as mean ± SD from three different experiments. ***$p<0.001$ vs. Untr and Abcd1-silenced+CAPE. Magnification 40X. , Untr; untreated, ND1; NADH dehydrogenase subunit 1, PGC1α; peroxisome proliferator-activated receptor gamma co-activator-alpha.

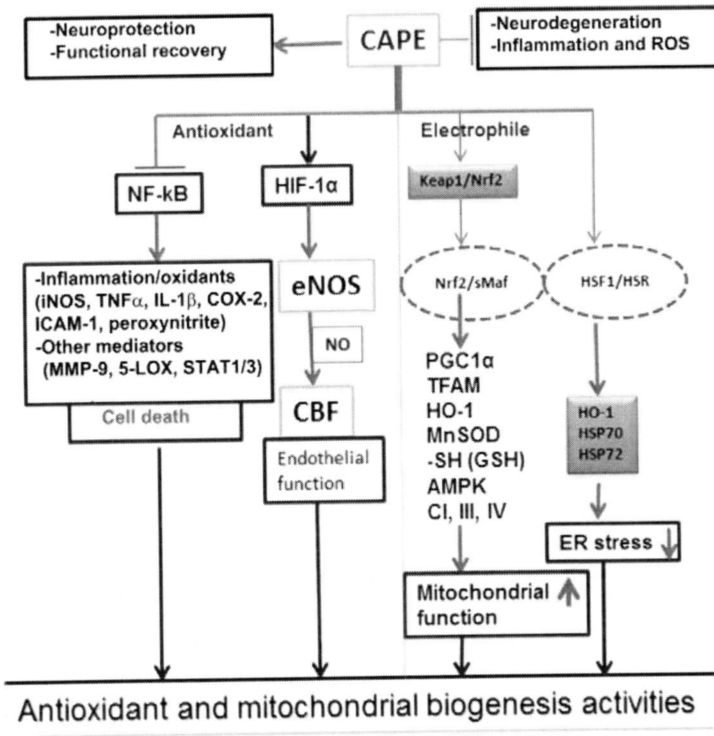

Figure 5. Schematic showing the effects of CAPE and the mechanisms involved therein, ameliorating neurodegenerative diseases. Based on reported efficacy and mechanisms of CAPE in neurodegenerative diseases, the pathways are illustrated. CAPE modulates two distinct antioxidant pathways leading to mitochondrial integrity, neuroprotection and functional recovery. While antiinflammatory pathway is mediated mainly by the inhibition NF-κB and its downstream signaling mediators, mitochondrial integrity pathway is mediated by CAPE as an electrophile through Nrf2-dependent downstream signaling molecules. CBF; cerebral blood flow, CI; complex I, CIII; complex III, CIV; complex IV, COX-2; cyclooxygenase-2, eNOS; endothelial nitric oxide synthase, HSR; hand skill relative, HSF1; heat shock transcription factor 1, HIF-1α; hypoxia-inducible factor-1alpha, iNOS; inducible nitric oxide synthase, ICAM-1; intercellular adhesion molecule, IL-1β; interleukin-1 beta, Keap1; kelch-like ECH-associated protein 1, sMaf; v-maf musculoaponeurotic fibrosarcoma oncogene family, MMP-9; matrix metalloproteinase-9, NO; nitric oxide, Nrf2; Nuclear factor (erythroid-derived 2)-like 2, NF-κB; nuclear factor kappa B, PGC1α; peroxisome proliferator activated receptor gamma coactivator 1 alpha, STAT; signal transducer and activator of transcription, TFAM; transcription factor A mitochondrial, TNFα; tumor necrosis factor alpha.

Collectively, the nature of oxidative/nitrosative stress and inflammation in neurodegenerative diseases is multifaceted, originating from multiple sources; therefore, the drug candidate must possess multimechanistic effects targeting affected components in all affected cell types. CAPE seems to possess such properties as depicted in Figure 5. Its lipophilicity, structural composition, and non-toxic presence in the human diet make CAPE suitable for a thorough investigation in neurodegenerative diseases.

Conclusion: Available data support that intervention by CAPE confers neuroprotection irrespective of the nature and origin of neurodegenerative diseases. However, the mechanisms of CAPE-mediated amelioration of a specific neurodegenerative disease processes are not clear. More mechanism-based studies are required to exploit the potential of CAPE as a therapy for neurodegenerative diseases. In clinical settings, CAPE deserves to be investigated alone to reduce the severity of a disease and as an adjunct to accelerate and enhance the efficacy of an approved therapy for human CNS trauma and neurodegenerative diseases.

ACKNOWLEDGMENTS

This work was supported by grants NS-22576, NS-72511 and NS-37766 from the NINDS of the NIH. We would like to thank Dr. Avtar K Singh for her intellectual inputs. We are also grateful to Dr. Thomas G. Smith from the MUSC Writing Center for his valuable editing and correction of the manuscript.

REFERENCES

[1] Son S, Lewis BA: Free radical scavenging and antioxidative activity of caffeic acid amide and ester analogues: structure-activity relationship. *J. Agric. Food Chem.* 2002, 50:468-472.

[2] Zheng Z, Yenari MA: Post-ischemic inflammation: molecular mechanisms and therapeutic implications. *Neurol. Res.* 2004, 26:884-892.

[3] Natarajan K, Singh S, Burke TR, Jr., Grunberger D, Aggarwal BB: Caffeic acid phenethyl ester is a potent and specific inhibitor of activation of nuclear transcription factor NF-kappa B. *Proc. Natl. Acad. Sci. USA* 1996, 93:9090-9095.

[4] Zhang P, Tang Y, Li NG, Zhu Y, Duan JA: Bioactivity and chemical synthesis of caffeic acid phenethyl ester and its derivatives. *Molecules* 2014, 19:16458-16476.

[5] Kim H, Kim W, Yum S, Hong S, Oh JE, Lee JW, Kwak MK, Park EJ, Na DH, Jung Y: Caffeic acid phenethyl ester activation of Nrf2 pathway is enhanced under oxidative state: structural analysis and potential as a pathologically targeted therapeutic agent in treatment of colonic inflammation. *Free Radic. Biol. Med.* 2013, 65:552-562.

[6] Celli N, Mariani B, Dragani LK, Murzilli S, Rossi C, Rotilio D: Development and validation of a liquid chromatographic-tandem mass spectrometric method for the determination of caffeic acid phenethyl ester in rat plasma and urine. *J. Chromatogr. B. Analyt. Technol. Biomed. Life Sci.* 2004, 810:129-136.

[7] Teixeira J, Soares P, Benfeito S, Murphy MP, Oliveira PJ, Borges F: Bridging the gap between nature and antioxidant setbacks: delivering caffeic acid to mitochondria. *Methods Mol. Biol.* 2015, 1265:73-83.

[8] Jatana M, Giri S, Ansari MA, Elango C, Singh AK, Singh I, Khan M: Inhibition of NF-kB activation by 5-lipoxygenase inhibitors protects brain against injury in a rat model of focal cerebral ischemia. *J. Neuroinflammation* 2006, 3:12.

[9] Khan M, Elango C, Ansari MA, Singh AK, Singh I: Caffeic acid phenethyl ester reduces neurovascular inflammation and protects rat brain following transient focal cerebral ischemia. *J. Neurochem.* 2007, 102:365-377.

[10] Sud'ina GF, Mirzoeva OK, Pushkareva MA, Korshunova GA, Sumbatyan NV, Varfolomeev SD: Caffeic acid phenethyl ester as a lipoxygenase inhibitor with antioxidant properties. *FEBS Lett.* 1993, 329:21-24.
[11] Kimura Y, Okuda H, Okuda T, Hatano T, Agata I, Arichi S: Studies on the activities of tannins and related compounds from medicinal plants and drugs. VII. Effects of extracts of leaves of Artemisia species, and caffeic acid and chlorogenic acid on lipid metabolic injury in rats fed peroxidized oil. *Chem. Pharm. Bull (Tokyo)* 1985, 33:2028-2034.
[12] Laranjinha J, Vieira O, Madeira V, Almeida L: Two related phenolic antioxidants with opposite effects on vitamin E content in low density lipoproteins oxidized by ferrylmyoglobin: consumption vs regeneration. *Arch. Biochem. Biophys.* 1995, 323:373-381.
[13] Singh J, Khan M, Singh I: Caffeic acid phenethyl ester induces adrenoleukodystrophy (Abcd2) gene in human X-ALD fibroblasts and inhibits the proinflammatory response in Abcd1/2 silenced mouse primary astrocytes. *Biochim. Biophys. Acta* 2013, 1831:747-758.
[14] Choi EY, Choe SH, Hyeon JY, Choi JI, Choi IS, Kim SJ: Effect of caffeic acid phenethyl ester on Prevotella intermedia lipopolysaccharide-induced production of proinflammatory mediators in murine macrophages. *J. Periodontal. Res.* 2015.
[15] Kassim M, Mansor M, Kamalden TA, Shariffuddin, II, Hasan MS, Ong G, Sekaran SD, Suhaimi A, Al-Abd N, Yusoff KM: Caffeic acid phenethyl ester (CAPE): scavenger of peroxynitrite in vitro and in sepsis models. *Shock* 2014, 42:154-160.
[16] Huang Y, Jin M, Pi R, Zhang J, Chen M, Ouyang Y, Liu A, Chao X, Liu P, Liu J, et al: Protective effects of caffeic acid and caffeic acid phenethyl ester against acrolein-induced neurotoxicity in HT22 mouse hippocampal cells. *Neurosci. Lett.* 2013, 535:146-151.
[17] Tolba MF, Azab SS, Khalifa AE, Abdel-Rahman SZ, Abdel-Naim AB: Caffeic acid phenethyl ester, a promising component of propolis with a plethora of biological activities: a review on its anti-inflammatory, neuroprotective, hepatoprotective, and cardioprotective effects. *IUBMB Life* 2013, 65:699-709.
[18] Wei X, Zhao L, Ma Z, Holtzman DM, Yan C, Dodel RC, Hampel H, Oertel W, Farlow MR, Du Y: Caffeic acid phenethyl ester prevents neonatal hypoxic-ischaemic brain injury. *Brain* 2004, 127:2629-2635.
[19] Ozer MK, Parlakpinar H, Acet A: Reduction of ischemia--reperfusion induced myocardial infarct size in rats by caffeic acid phenethyl ester (CAPE). *Clin. Biochem.* 2004, 37:702-705.
[20] Lee ES, Uhm KO, Lee YM, Han M, Lee M, Park JM, Suh PG, Park SH, Kim HS: CAPE (caffeic acid phenethyl ester) stimulates glucose uptake through AMPK (AMP-activated protein kinase) activation in skeletal muscle cells. *Biochem. Biophys. Res. Commun.* 2007, 361:854-858.
[21] Liao CC, Ou TT, Huang HP, Wang CJ: The inhibition of oleic acid induced hepatic lipogenesis and the promotion of lipolysis by caffeic acid via up-regulation of AMP-activated kinase. *J. Sci. Food Agric.* 2013.
[22] Tsuda S, Egawa T, Ma X, Oshima R, Kurogi E, Hayashi T: Coffee polyphenol caffeic acid but not chlorogenic acid increases 5'AMP-activated protein kinase and insulin-independent glucose transport in rat skeletal muscle. *J. Nutr. Biochem.* 2012, 23:1403-1409.

[23] Tsai CF, Kuo YH, Yeh WL, Wu CY, Lin HY, Lai SW, Liu YS, Wu LH, Lu JK, Lu DY: Regulatory effects of caffeic Acid phenethyl ester on neuroinflammation in microglial cells. *Int. J. Mol. Sci.* 2015, 16:5572-5589.

[24] Aladag MA, Turkoz Y, Ozcan C, Sahna E, Parlakpinar H, Akpolat N, Cigremis Y: Caffeic acid phenethyl ester (CAPE) attenuates cerebral vasospasm after experimental subarachnoidal haemorrhage by increasing brain nitric oxide levels. *Int. J. Dev. Neurosci.* 2006, 24:9-14.

[25] Tsai SK, Lin MJ, Liao PH, Yang CY, Lin SM, Liu SM, Lin RH, Chih CL, Huang SS: Caffeic acid phenethyl ester ameliorates cerebral infarction in rats subjected to focal cerebral ischemia. *Life Sci.* 2006, 78:2758-2762.

[26] Cengiz N, Colakoglu N, Kavakli A, Sahna E, Parlakpinar H, Acet A: Effects of caffeic acid phenethyl ester on cerebral cortex: structural changes resulting from middle cerebral artery ischemia reperfusion. *Clin. Neuropathol.* 2007, 26:80-84.

[27] Serarslan Y, Bal R, Altug ME, Kontas T, Melek IM: Caffeic acid phenethyl ester decreases the level of S-100B protein after middle cerebral artery [correction for after] occlusion in rabbits. *Pak. J. Pharm. Sci.* 2009, 22:313-316.

[28] Gilgun-Sherki Y, Rosenbaum Z, Melamed E, Offen D: Antioxidant therapy in acute central nervous system injury: current state. *Pharmacol. Rev.* 2002, 54:271-284.

[29] Hishikawa K, Nakaki T, Fujita T: Oral flavonoid supplementation attenuates atherosclerosis development in apolipoprotein E-deficient mice. *Arterioscler Thromb. Vasc. Biol.* 2005, 25:442-446.

[30] Wang Q, Xu J, Rottinghaus GE, Simonyi A, Lubahn D, Sun GY, Sun AY: Resveratrol protects against global cerebral ischemic injury in gerbils. *Brain Res.* 2002, 958:439-447.

[31] Asensi M, Medina I, Ortega A, Carretero J, Bano MC, Obrador E, Estrela JM: Inhibition of cancer growth by resveratrol is related to its low bioavailability. *Free Radic. Biol. Med.* 2002, 33:387-398.

[32] Langlois JA, Rutland-Brown W, Wald MM: The epidemiology and impact of traumatic brain injury: a brief overview. *J. Head Trauma Rehabil* 2006, 21:375-378.

[33] Jain KK: Neuroprotection in traumatic brain injury. *Drug Discov. Today* 2008, 13:1082-1089.

[34] Wu A, Ying Z, Gomez-Pinilla F: Dietary curcumin counteracts the outcome of traumatic brain injury on oxidative stress, synaptic plasticity, and cognition. *Exp. Neurol.* 2006, 197:309-317.

[35] Avila MA, Sell SL, Kadoi Y, Prough DS, Hellmich HL, Velasco M, Dewitt DS: L-Arginine decreases fluid-percussion injury-induced neuronal nitrotyrosine immunoreactivity in rats. *J. Cereb Blood Flow Metab.* 2008, 28:1733-1741.

[36] Deng-Bryant Y, Singh IN, Carrico KM, Hall ED: Neuroprotective effects of tempol, a catalytic scavenger of peroxynitrite-derived free radicals, in a mouse traumatic brain injury model. *J. Cereb. Blood Flow Metab* 2008, 28:1114-1126.

[37] Hall ED, Kupina NC, Althaus JS: Peroxynitrite scavengers for the acute treatment of traumatic brain injury. *Ann. N Y Acad. Sci.* 1999, 890:462-468.

[38] Singh IN, Sullivan PG, Hall ED: Peroxynitrite-mediated oxidative damage to brain mitochondria: Protective effects of peroxynitrite scavengers. *J. Neurosci. Res.* 2007, 85:2216-2223.

[39] Besson VC: Drug targets for traumatic brain injury from poly(ADP-ribose)polymerase pathway modulation. *Br. J. Pharmacol.* 2009, 157:695-704.
[40] Zhao J, Pati S, Redell JB, Zhang M, Moore AN, Dash PK: Caffeic Acid phenethyl ester protects blood-brain barrier integrity and reduces contusion volume in rodent models of traumatic brain injury. *J. Neurotrauma* 2012, 29:1209-1218.
[41] Kerman M, Kanter M, Coskun KK, Erboga M, Gurel A: Neuroprotective effects of caffeic acid phenethyl ester on experimental traumatic brain injury in rats. *J. Mol. Histol* 2012, 43:49-57.
[42] Feng Y, Lu YW, Xu PH, Long Y, Wu WM, Li W, Wang R: Caffeic acid phenethyl ester and its related compounds limit the functional alterations of the isolated mouse brain and liver mitochondria submitted to in vitro anoxia-reoxygenation: relationship to their antioxidant activities. *Biochim. Biophys. Acta* 2008, 1780:659-672.
[43] Suuronen T, Huuskonen J, Pihlaja R, Kyrylenko S, Salminen A: Regulation of microglial inflammatory response by histone deacetylase inhibitors. *J. Neurochem.* 2003, 87:407-416.
[44] DeVivo MJ: Causes and costs of spinal cord injury in the United States. *Spinal Cord* 1997, 35:809-813.
[45] NSCISC: National Spinal Cord Injury Statistical Center , University of Alabama at Birmingham (http://www.spinalcord.uab.edu/).
[46] Wickelgren I: Neuroscience. Animal studies raise hopes for spinal cord repair. *Science* 2002, 297:178-181.
[47] Poulsen DJ, Harrop JS, During MJ: Gene therapy for spinal cord injury and disease. *J. Spinal Cord Med.* 2002, 25:2-9.
[48] Olson L: Medicine: clearing a path for nerve growth. *Nature* 2002, 416:589-590.
[49] Bunge MB, Pearse DD: Transplantation strategies to promote repair of the injured spinal cord. *J. Rehabil Res. Dev.* 2003, 40:55-62.
[50] McDonald JW, Becker D, Holekamp TF, Howard M, Liu S, Lu A, Lu J, Platik MM, Qu Y, Stewart T, Vadivelu S: Repair of the injured spinal cord and the potential of embryonic stem cell transplantation. *J. Neurotrauma* 2004, 21:383-393.
[51] Bareyre FM, Kerschensteiner M, Raineteau O, Mettenleiter TC, Weinmann O, Schwab ME: The injured spinal cord spontaneously forms a new intraspinal circuit in adult rats. *Nat. Neurosci.* 2004, 7:269-277.
[52] Rosenzweig ES, McDonald JW: Rodent models for treatment of spinal cord injury: research trends and progress toward useful repair. *Curr. Opin. Neurol.* 2004, 17:121-131.
[53] Silva NA, Sousa N, Reis RL, Salgado AJ: From basics to clinical: a comprehensive review on spinal cord injury. *Prog. Neurobiol.* 2014, 114:25-57.
[54] Ilhan A, Koltuksuz U, Ozen S, Uz E, Ciralik H, Akyol O: The effects of caffeic acid phenethyl ester (CAPE) on spinal cord ischemia/reperfusion injury in rabbits. *Eur. J. Cardiothorac. Surg.* 1999, 16:458-463.
[55] Gocmez C, Celik F, Kamasak K, Kaplan M, Uzar E, Arikanoglu A, Evliyaoglu O: Effects of intrathecal caffeic acid phenethyl ester and methylprednisolone on oxidant/antioxidant status in traumatic spinal cord injuries. *J. Neurol. Surg. A. Cent Eur. Neurosurg.* 2015, 76:20-24.
[56] Uzar E, Sahin O, Koyuncuoglu HR, Uz E, Bas O, Kilbas S, Yilmaz HR, Yurekli VA, Kucuker H, Songur A: The activity of adenosine deaminase and the level of nitric oxide

in spinal cord of methotrexate administered rats: protective effect of caffeic acid phenethyl ester. *Toxicology* 2006, 218:125-133.

[57] Kasai M, Fukumitsu H, Soumiya H, Furukawa S: Caffeic acid phenethyl ester reduces spinal cord injury-evoked locomotor dysfunction. *Biomed. Res.* 2011, 32:1-7.

[58] Moser H. SKD, Watkins P. A., Powers J. and Moser A. B.: *X-linked adrenoleukodystrophy, in The Metabolic and Molecular Basis of Inherited Disease.* McGraw-Hill, New York.: Scriver C., ed.; 2001.

[59] Singh I, Moser AE, Goldfischer S, Moser HW: Lignoceric acid is oxidized in the peroxisome: implications for the Zellweger cerebro-hepato-renal syndrome and adrenoleukodystrophy. *Proc. Natl. Acad. Sci. USA* 1984, 81:4203-4207.

[60] Mosser J, Douar AM, Sarde CO, Kioschis P, Feil R, Moser H, Poustka AM, Mandel JL, Aubourg P: Putative X-linked adrenoleukodystrophy gene shares unexpected homology with ABC transporters. *Nature* 1993, 361:726-730.

[61] Gilg AG, Singh AK, Singh I: Inducible nitric oxide synthase in the central nervous system of patients with X-adrenoleukodystrophy. *J. Neuropathol. Exp. Neurol.* 2000, 59:1063-1069.

[62] Powers JM, Pei Z, Heinzer AK, Deering R, Moser AB, Moser HW, Watkins PA, Smith KD: Adreno-leukodystrophy: oxidative stress of mice and men. *J. Neuropathol. Exp. Neurol.* 2005, 64:1067-1079.

[63] Vargas CR, Wajner M, Sirtori LR, Goulart L, Chiochetta M, Coelho D, Latini A, Llesuy S, Bello-Klein A, Giugliani R, et al: Evidence that oxidative stress is increased in patients with X-linked adrenoleukodystrophy. *Biochim. Biophys. Acta* 2004, 1688:26-32.

[64] Uto T, Contreras MA, Gilg AG, Singh I: Oxidative imbalance in nonstimulated X-adrenoleukodystrophy-derived lymphoblasts. *Dev. Neurosci.* 2008, 30:410-418.

[65] Baarine M, Beeson C, Singh A, Singh I: ABCD1 deletion-induced mitochondrial dysfunction is corrected by SAHA: implication for adrenoleukodystrophy. *J. Neurochem.* 2014.

[66] Singh J, Khan M, Singh I: Silencing of Abcd1 and Abcd2 genes sensitizes astrocytes for inflammation: implication for X-adrenoleukodystrophy. *J. Lipid Res.* 2009, 50:135-147.

[67] Baarine M, Andreoletti P, Athias A, Nury T, Zarrouk A, Ragot K, Vejux A, Riedinger JM, Kattan Z, Bessede G, et al: Evidence of oxidative stress in very long chain fatty acid--treated oligodendrocytes and potentialization of ROS production using RNA interference-directed knockdown of ABCD1 and ACOX1 peroxisomal proteins. *Neuroscience* 2012, 213:1-18.

[68] Baarine M, Ragot K, Athias A, Nury T, Kattan Z, Genin EC, Andreoletti P, Menetrier F, Riedinger JM, Bardou M, Lizard G: Incidence of Abcd1 level on the induction of cell death and organelle dysfunctions triggered by very long chain fatty acids and TNF-alpha on oligodendrocytes and astrocytes. *Neurotoxicology* 2012, 33:212-228.

[69] Singh J, Khan M, Pujol A, Baarine M, Singh I: Histone deacetylase inhibitor upregulates peroxisomal fatty acid oxidation and inhibits apoptotic cell death in abcd1-deficient glial cells. *PLoS One* 2013, 8:e70712.

[70] Khan M, Pahan K, Singh AK, Singh I: Cytokine-induced accumulation of very long-chain fatty acids in rat C6 glial cells: implication for X-adrenoleukodystrophy. *J. Neurochem.* 1998, 71:78-87.

[71] Khan M, Singh J, Gilg AG, Uto T, Singh I: Very long-chain fatty acid accumulation causes lipotoxic response via 5-lipoxygenase in cerebral adrenoleukodystrophy. *J. Lipid Res.* 2010, 51:1685-1695.

[72] Khan M, Singh J, Singh I: Plasmalogen deficiency in cerebral adrenoleukodystrophy and its modulation by lovastatin. *J. Neurochem.* 2008, 106:1766-1779.

[73] Bizzozero OA, Reyes S, Ziegler J, Smerjac S: Lipid peroxidation scavengers prevent the carbonylation of cytoskeletal brain proteins induced by glutathione depletion. *Neurochem Res* 2007, 32:2114-2122.

[74] Ilhan A, Akyol O, Gurel A, Armutcu F, Iraz M, Oztas E: Protective effects of caffeic acid phenethyl ester against experimental allergic encephalomyelitis-induced oxidative stress in rats. *Free Radic. Biol. Med.* 2004, 37:386-394.

[75] Ma Z, Wei X, Fontanilla C, Noelker C, Dodel R, Hampel H, Du Y: Caffeic acid phenethyl ester blocks free radical generation and 6-hydroxydopamine-induced neurotoxicity. *Life Sci.* 2006, 79:1307-1311.

[76] Noelker C, Bacher M, Gocke P, Wei X, Klockgether T, Du Y, Dodel R: The flavanoide caffeic acid phenethyl ester blocks 6-hydroxydopamine-induced neurotoxicity. *Neurosci Lett* 2005, 383:39-43.

[77] Fontanilla CV, Ma Z, Wei X, Klotsche J, Zhao L, Wisniowski P, Dodel RC, Farlow MR, Oertel WH, Du Y: Caffeic acid phenethyl ester prevents 1-methyl-4-phenyl-1,2,3,6-tetrahydropyridine-induced neurodegeneration. *Neuroscience* 2011, 188:135-141.

[78] Barber SC, Higginbottom A, Mead RJ, Barber S, Shaw PJ: An in vitro screening cascade to identify neuroprotective antioxidants in ALS. *Free Radic. Biol. Med.* 2009, 46:1127-1138.

[79] Fontanilla CV, Wei X, Zhao L, Johnstone B, Pascuzzi RM, Farlow MR, Du Y: Caffeic acid phenethyl ester extends survival of a mouse model of amyotrophic lateral sclerosis. *Neuroscience* 2012, 205:185-193.

[80] Lee JM, Calkins MJ, Chan K, Kan YW, Johnson JA: Identification of the NF-E2-related factor-2-dependent genes conferring protection against oxidative stress in primary cortical astrocytes using oligonucleotide microarray analysis. *J. Biol. Chem.* 2003, 278:12029-12038.

In: Caffeic Acid
Editor: Leanna Vaughn
ISBN: 978-1-63483-121-5

Chapter 7

CAFFEIC ACID AND ITS DERIVATIVES AS INHIBITORS OF LIPOXYGENASES IN DIFFERENT ORGANISMS

Miroslava Cuperlovic-Culf[1,2] and Mohamed Touaibia[2]
[1]National Research Council of Canada, Moncton, Canada
[2]Department of Chemistry and Biochemistry, Université de Moncton, Moncton, Canada

ABSTRACT

Hydroxycinnamic acids including caffeic acid are major phenolic secondary metabolites in plants involved in growth and metabolism regulation as well as response to pathogen challenges. Caffeic acid is broadly distributed in several species of plant kingdom and is widely consumed in human diet. Caffeic acid and derivatives have been extensively studied revealing a broad spectrum of biological activities in different species. In addition to the antioxidant activity of caffeic acid, this molecule has been indicated as an inhibitor of number of proteins conserved across distinct species. This chapter will focus on the effect of caffeic acid and several targeted derivatives on lipoxygenases. Lipoxygenases (LOXs) are enzymes involved in peroxidation of lipids to bioactive lipid mediators such as leukotriene. LOXs are significant and highly conserved enzymes in eukaryotes with known function in metabolism of plant, fungal and mammalian cells. LOX inhibitory potential of caffeic acid has therefore major relevance to public health as well as agriculture and environment.

This chapter is aimed at summarizing the current knowledge of the inhibition of LOXs by caffeic acid and its application as well as potential in human health and agriculture. In addition to reviewing literature on LOX inhibition we will provide analysis of the LOX protein conservation in highly distinct species and present computational illustration of pan-eukaryotic LOX inhibition by caffeic acid and its derivatives.

INTRODUCTION

Caffeic acid (**1**) (**Scheme** 1) is one of the major representatives of hydroxycinnamic acids and phenolic acid widely distributed in plants. These plant-derived phenylpropanoids are the largest group of secondary metabolites produced in higher plants with major role in the protection against biotic or abiotic stresses such as infections or herbivores attacks, wounding, UV irradiation, exposure to ozone or pollutants. Phenylpropanoids are amongst the resistance related metabolites determined in a response of barley [Bollina, 2010] and wheat [Gunnaiah, 2014] against fusarium head blight as well as rust [Bolton, 2008] and sugar cane against Sporisorium scitamineaum fungus [Santiago, 2010]. Phenylpropanoids are also synthesized and released as part of plant response to insect feeding [Chrzanoski, 2008] as well as infection by the *Pseudomonas syringae* pv. *Glycinea* [Zabala, 2006] to name just a few examples. Phenolic acids also act as signaling molecules in plant-microbe symbioses [Mandal, 2010]. Roles of phenolics in the resistance mechanisms of plants against fungal pathogens and insects have been reviewed by Lattanzio et al. [Lattanzio, 2006].

Phenylpropanoids including caffeic acid (**1**) are produced in plants in the Phenylpropanoid biosynthesis from phenylalanine deriving from Shikimate pathway (Figure 1). Many phenolic acids shown in Figure 1 have biological activity. In addition, through *p*-coumarate, phenolic acids are entering flavonoid and stilbenoids biosynthesis pathways, producing number of active compounds relevant for plant as well as human health. Ester of caffeic acid, Caffeic acid phenylethyl ester – CAPE (**2**) (**Scheme 1**), is one of the most active compounds in propolis, illustrating the diversity of biological activities of phenolic acids.

Benefits of phenylpropanoids to human health have been extensively explored. Phenylpropanoids, including caffeic acid (**1**) and CAPE (**2**), act as inhibitors of inflammatory response, have anti-apoptotic activity, regulate reactive oxygen species (ROS), inhibit NF-kB activity, regulate cell cycle arrest, inhibit AKT activation and regulate ERK activation [review in Kim, 2011]. CAPE have been found to have anti-cancer, anti-inflammatory and immune-regulatory activities [Grunberger, 1988; Natarajan,1996]. Caffeic acid (**1**), CAPE (**2**) and their derivatives are potent antioxidants, effective scavengers of variety of free radicals and superoxide anion radicals [Calikoglu, 2003; Gurel, 2004]. CAPE (**2**) is a specific inhibitor of the nuclear transcription factor, NF-kB and also significantly suppresses the lipoxygenase pathway of arachidonic acid metabolism during inflammation [Natarajan, 1996; Sud'ina, 1993; Mirzoeva, 1996]. In addition, these molecules have high reducing power and chelate on ferrous ions [Gulcin, 2005]. Caffeic acid (**1**) was indicated as α-tocopherol protectant in low-density lipoprotein (LDL) [Laranjinha, 1995]. Caffeic acid (**1**) and its derivatives have been proven as good substrates of polyphenol oxidase and may under particular conditions be oxidized in plant tissues or plant products [Kerr, 1998; Bassil, 2005; Gulcin, 2006]. In addition to general anti-oxidant activity, redox activity of caffeic acid and derivatives makes them very interesting specific inhibitors of metaloenzymes. In for example lipoxygenases these compounds can keep the active site iron in the ferrous state thereby uncoupling the catalytic cycle of the enzyme. This more specific function of caffeic acid (**1**) as protein inhibitor will be reviewed here.

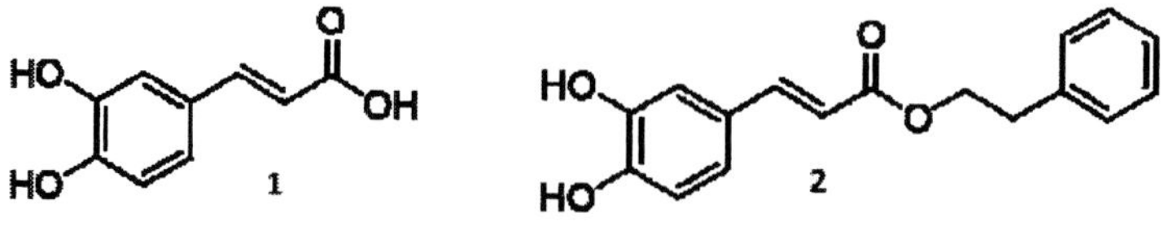

Scheme 1.

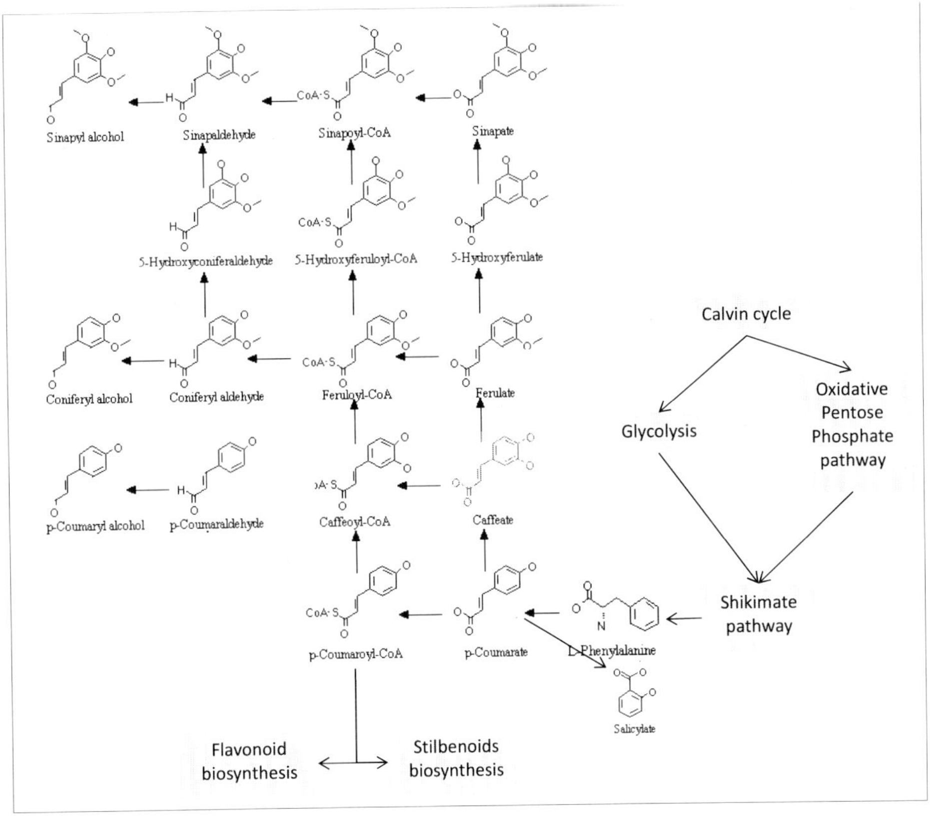

Figure 1. Phenylpropanoid biosynthesis pathway for production of phenylpropanoids including caffeic acid in plants from phenylalanine derived from Shikimate pathway. Through p-coumarate, phenolic acids enter flavonoid and stilbenoids biosynthesis pathways.

PROTEIN INHIBITION BY CAFFEIC ACID AND DERIVATIVES

Analysis of published bioactivity of caffeic acid (**1**) and CAPE (**2**) performed through PubChem database (https://pubchem.ncbi.nlm.nih.gov/) shows number of proteins that have been determined as targets for these two phenylpropanoids in at least one published essay (Figure 2 and Table 1).

According to Gene Ontology analysis using AmiGO2 (amigo. geneontology.org) majority of these proteins are involved in metabolism including bicarbonate transport, one-carbon metabolic process, ion (organic anion) transport and lipoxygenase pathway. Significant proportions of these proteins are metaloenzymes containing either zinc or iron with cation binding activity. As previously shown caffeic acid (**1**) and derivatives function as free radical scavengers and antioxidants and in metaboloenzymes they function by reducing the catalytically-active metal ion. However, in inhibiting metaloenzymes these molecules have to also show specific structural features in order to be able to enter enzymes active site [Boudreau, 2012]. Interestingly caffeic acid (**1**) and CAPE (**2**) target mostly proteins that are ubiquitously present in variety of organisms such as lipoxygenase and carbonic anhydrase enzymes. Majority of essays have explored the effects of these compounds on mammalian cells and protein targets [Boudreau, 2012; Odaci, 2007; Feliciano, 2009; Parlakpinar, 2005; Gokalp, 2006; Suud'ina, 1993; Mirzoeva, 1996; Michaluart, 1999; Boudreau, 2009], however several publications show inhibition of lipoxygenase by caffeic acid (**1**) or CAPE (**2**) in non-mammalian cells [Koukoulitsa, 2007]. For caffeic acid (**1**) to function as inhibitor of these proteins in different species proteins' active sites have to be sufficiently similar and this can be explored using computational methods and will be shown below. Similarity of protein active sites and possibility for the development of specific multi-species protein inhibitors can have wide ranging applications. Anti-inflammatory effect of lipoxygenase inhibitors combined with antifungal activities was shown as highly valuable in the development of treatment against fungal infections [Hoobler, 2013]. Great example of an application of plant metabolite as highly effective antifungal agent in agriculture has been recently presented by Piotrowski and co-workers [Piotrowski, 2015]. Development of agricultural-pest treatment methods that are biodegradable and at the same time possibly beneficial for higher animals and plants is needed.

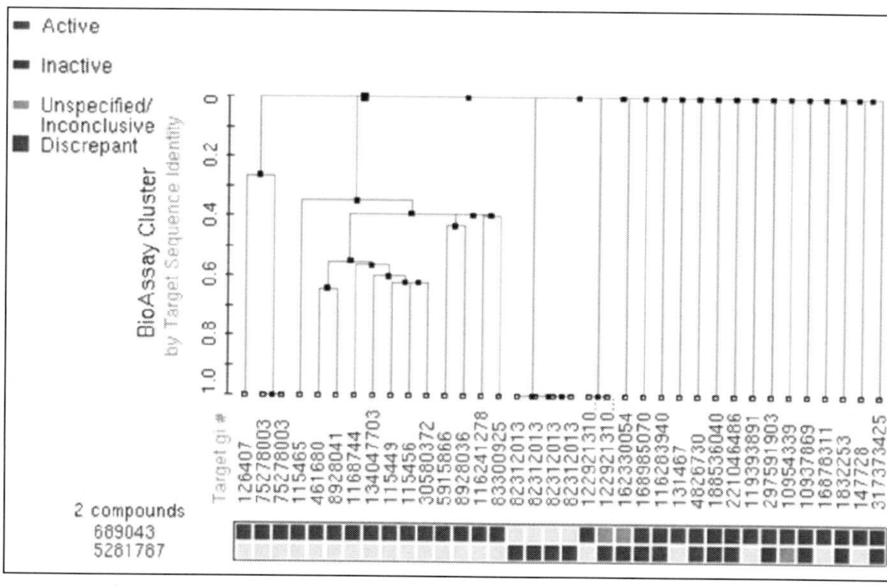

Figure 2. Proteins listed in PubChem as targets of CAPE (CID: 5281787) and/or caffeic acid (CID:689043). Figure shows only confirmed active targets according to information provided in the PubChem database information. Figure shows protein targets GI numbers. Protein names are listed in Table 1.

Table 1. Proteins listed in PubChem as targets of CAPE (CID: 5281787) and/or caffeic acid (CID:689043). List shows only confirmed targets according to information provided in the PubChem database information

Target gi	Protein Name
1832253	15-lipoxygenase, partial [Homo sapiens]
75278003	5-lipoxygenase
317373425	Arachidonate 15-lipoxygenase B
126407	Arachidonate 5-lipoxygenase

Target gi	Protein Name
116283940	ATAD5 protein, partial [Homo sapiens]
115449	Carbonic anhydrase 1
5915866	Carbonic anhydrase 12
30580372	Carbonic anhydrase 13
8928036	Carbonic anhydrase 14
115456	Carbonic anhydrase 2
134047703	Carbonic anhydrase 3
115465	Carbonic anhydrase 4
461680	Carbonic anhydrase 5A, mitochondrial
8928041	Carbonic anhydrase 5B, mitochondrial
116241278	Carbonic anhydrase 6
1168744	Carbonic anhydrase 7
83300925	Carbonic anhydrase 9
221046486	Chain A, Human 2-Oxoglutarate Oxygenase Loc390245
162330054	Chain A, Jmjd2a Tandem Tudor Domains
122921310	Chain A, The Structure Of Wild-Type Human Hadh2
168985070	euchromatic histone-lysine N-methyltransferase 2 [Homo sapiens]
82312013	Integrase
297591903	large T antigen [Simian virus 40]
119393891	lysosomal alpha-glucosidase preproprotein [Homo sapiens]
16878311	MPI protein [Homo sapiens]
188536040	nuclear receptor ROR-gamma isoform 1 [Mus musculus]
10954339	phosphopantetheinyl transferase [Bacillus subtilis]
147728	RNA polymerase beta subunit (EC 2.7.7.6), partial [Escherichia coli]
4826730	serine/threonine-protein kinase mTOR [Homo sapiens]
131467	Tyrosine-protein phosphatase non-receptor type 1; AltName: Full=Protein-tyrosine phosphatase 1B; Short=PTP-1B [Homo sapiens]
10937869	survival motor neuron protein isoform d [Homo sapiens]

Both of the major known targets of caffeic acid (**1**) and CAPE (**2**) - carbonic anhydrases and lipoxygenases have significant roles in the metabolism of living organisms. Carbonic anhydrases regulate CO_2 levels in living organisms while lipoxygenases oxygenate polyunsaturated fatty acids containing a series of cis double bonds. Carbonic anhydrases have been found in all living organisms. Lipoxygenase are widely distributed in eukaryotes and are found to be ubiquitous in plants and animals. Prokaryotes do not have polyunsaturated fatty acids however in several bacteria lipoxygenases have been found and analysed. It has been suggested that *P. aeruginosa* uses its lipoxygenase as an extracellular mediator that can extract and modify unsaturated phospholipids from eukaryotic membranes, playing a role in interaction between bacteria and the host cells. Both groups of proteins have a major role in cells' and organisms' metabolism as well as response to pathogens and diseases and therefore possible role of caffeic acid (**1**) and its derivatives as dual inhibitor of carbonic anhydrases and lipoxygenases in different organisms can have wide reaching applications. In the following we will explore similarity and differences between lipoxygenases in different organisms and computationally determine inhibitory power of caffeic acid (**1**), CAPE (**2**) as well as some of the published derivatives against lipoxygenases of different species. For this analysis we will use only experimentally determined crystal strictures of lipoxygenases from mammalian cells, plant cells, bacteria and coral.

LIPOXYGENASES

Lipoxygenases are a family of monomeric non-heme, non-sulfur iron dioxygenases, which catalyze the conversion of polyunsaturated fatty acids into conjugated hydroperoxides. Polyunsaturated fatty acids are important components of lipid membranes as well as substrates for the biosynthesis of signaling molecules such as prostaglandins, leukotriens and endocannabinoids (reviewed in Horn, 2015). Lipoxygenases are found in plants, fungi and animals and their roles in some of these species have been extensively studied and reviewed [Chedea, 2011, Brodhum, 2011, Brash, 1999, Feussner, 2002, Brodhum, 2011, Horn, 2015, Schneider, 2011; Kuhn, 2014; Haeggstrom, 2011; Andreou, 2009; Hansen, 2013]. Lipoxygenase catalysis is shown in Figure 3 [based on Brash, 1999]. Keeping catalytic core Fe in inactive, Fe^{2+} state inhibits the catalysis.

Briefly, in plants lipoxygenases have a role in regulation of development and environmental response. First lipoxygenases have been discovered in plants [Andre, 1932] almost hundred years ago and since then oxylipins formed through the action of variety of lipoxygenases have been determined to play a highly significant role in stress response [Grebner, 2013], growth and development [Vellosillo, 2007], germination [Feussner, 2001], tuber formation [Kolomiets, 2001], sex determination [Acosta, 2009], senescence [Prakash, 1990], fruit ripening [Zhang, 2006] and aroma production [Shen, 20014]. In these systems lipoxygenases are using mostly C18 fatty acids as precursors [Feussner, 2002]. Lipoxygenases are implicated in the production of signaling compounds such as jasmonates, antimicrobial and antifungal compounds such as leaf aldehydes as well as plant-specific volatiles including leaf alcohols [Feussner, 2002].

Several lipoxygenases have been identified in fungi, with the most unusual being the manganese lipoxygenase which, unlike all the other lipoxygenases that have mononuclear iron core, contains catalytic manganese center [reviewed in Brodhum, 2011]. Oxylipins produced as result of fungal lipoxygenases have suggested roles in host-fungus crosstalk and fungal pathogeneses. Further analysis of the role of fungal lipoxygenases and the significance of their inhibition in fungi is needed.

Mammalian lipoxygenases have been extensively studies with number of reports describing functions and properties of various lipoxygenase isozymes in these systems. In human cells six lipoxygenase isoforms have been described and sequences [reviewed in Horn, 2015] and number of biological roles has been described for these enzymes. In mammalian cells, C20 fatty acids are predominantly used as precursors with 5-Lipoxygenase being the key enzyme in the conversion of arachidonic acid (AA) to the leukotrienes (LTs), potent mediators of inflammation [Peters-Golden, 2007; Romano, 2003]. Elevated levels of LTs have been associated with several inflammatory and allergic disorders and are found in various pathologic tissues [Baron, 2008; Praveen, 2006]. 5-lipoxygenase mutations was linked to the development of bronchial [Bizzintino, 2009] and allergy-induced asthma [Funk, 2002]. This enzyme has also been linked to pathogenesis of atherosclerosis [Rydberg, 2004] amongst several other diseases and disorders.

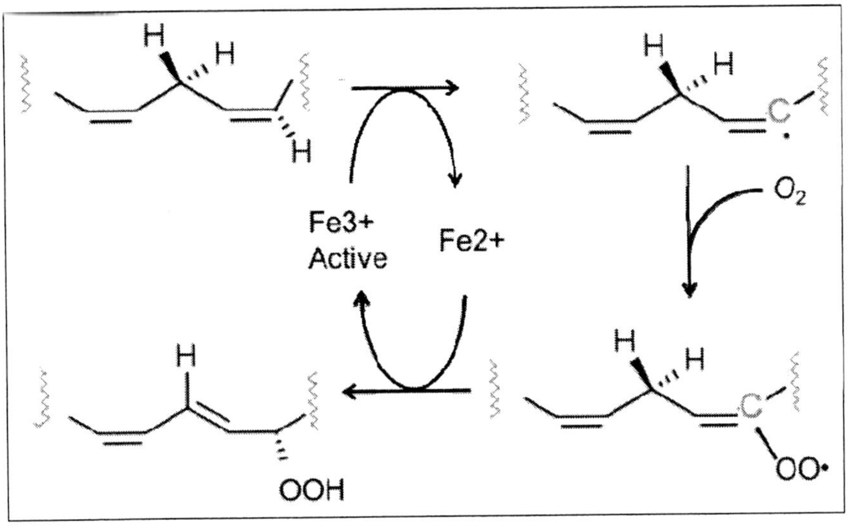

Figure 3. General lipoxygenase catalysis process [based on Brash, 1999].

Crystal structures for a number of lipoxygenases from variety of species have been presented and made available at RCSB Protein Data Bank (http://www.rcsb.org/pdb/home/home.do). These structures include complexes between lipoxygenases and inhibitors as well as fatty acid. Experimental data provides an opportunity for comparison of structures of lipoxygenases in different species. Availability of information regarding the shape and characteristics of the binding site provided an opportunity for the development of either pan-species lipoxygenase inhibitors or specific inhibitors for different species and lipoxygenase isoforms. With our interest primarily in the effect of caffeic acid and its derivatives we will explore similarities between active sites of lipoxygenases of different species and the possibility for their inhibition with caffeic acid, CAPE and related molecules developed in our group [Boudreau, 2012].

Lipoxygenase activity of proteins across species does not require closely related sequences. For example, soybean (Glycin max) vegetative lipoxygenase isoform VLX-B (crystal structure is available as PDB 2IUJ) [Youn, 2006] shows only 22% sequence similarity to human 15-lipoxygenase-2 (crystal structure determined by Kobe, 2014 and available at pdb:4NRE) and only 70% similarity to soybean lipoxygenase-3 (crystal structure ID PDB: 1JNQ) [Skrzypczak-Jankun, 2003]. With such low sequence similarity it is interesting to investigate structural and active site similarities allowing these proteins to perform similar functions. Crystal structures for several of mammalian, soybean, coral and bacterial lipoxygenases are available. Comparison of overall structures of lipoxygenases for Pseudomonas aeruginosa (PDB: 4G33) [Garreta, 2013], soybean (lipoxygenase-3, PDB: 1HU9) [Skrzypczak-Jankun, 2003], lipoxygenases 8R from Caribbean sea whip coral Plexaura homomalla (PDB: 2FNQ) [Oldham, 2005] and human stable lipoxygenase-5 (PDB:3O8Y) [Gilbert, 2011] is shown in Figure 4. All lipoxygenases consist of two domains, a smaller N-terminal β-barrel, membrane-associating C2 domain, and a larger C-terminal domain mostly composed of long α-helixes, and containing the redox-active iron. In the compared structures these domains are structurally very similar.

The major difference between these structures is in the cluster of helices acting as a lid to the active center in *Pseudomonas aeruginosa* (indicated with red arrow on Figure 4). Although this unique feature of bacterial lipoxygenase can affect the access to the active site, it does not change the active site. Comparison of active sites and the iron coordination environment between number of different lipoxygenases is shown in Figure 5 with distances in the Fe-coordinated core listed in Table 2

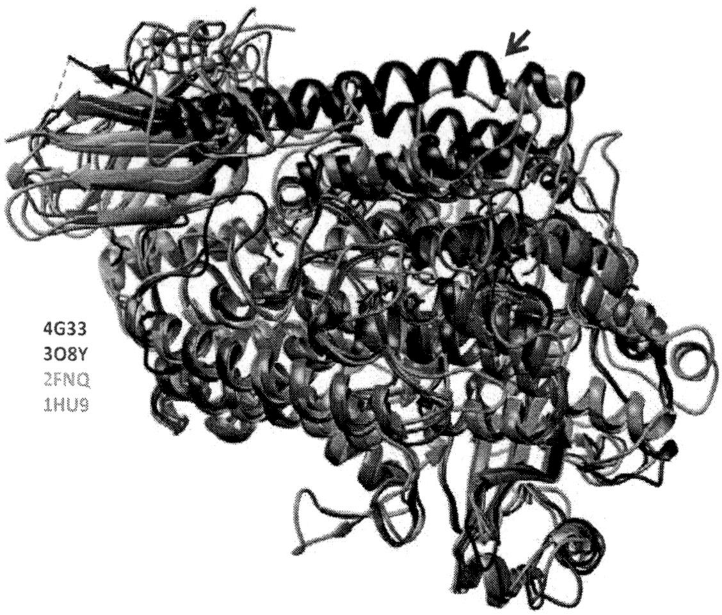

Figure 4. Comparison of overall structures of lipoxygenases for Pseudomonas aeruginosa (red PDB: 4G33) [Garreta, 2013], soybean (green lipoxygenase-3, PDB: 1HU9) [Skrzypczak-Jankun, 2003], lipoxygenases 8R from Caribbean sea whip coral Plexaura homomalla (orange PDB: 2FNQ) [Oldham, 2005] and human stable 5-lipoxygenase (blue PDB:3O8Y) [Gilbert, 2011]. Region that is significantly different in bacterial lipoxygenase is indicated with a red arrow.

Distances between iron and ligands in the active sites shown in Table 2 are highly comparable to the detailed analysis of structural differences between several lipoxygenase isoforms presented by Xu et al. [Xu, 2012]. Although there are some differences in exact distances in the active sites, overall when considering the error in the crystal structure measurement and the structure flexibility in solution these distances are highly similar and suggest possibility for pan-species inhibitors of lipoxygenases.

Table 2. Comparison of the iron coordination environments for lipoxygenases from four different species

SOYBEAN						HUMAN		Ps. aeruginosa		Coral	
1HU9		1IK3		1RRL		3O8Y		4G33		2FNQ	
Fe-Ile 857	2.071	Fe-Ile 857	2.050	Fe-Ile 857	2.136	Fe-Ile 673	2.235	Fe-Ile 685	2.292	Fe-Ile1066	1.691
Fe-His518	2.227	Fe-His518	2.232	Fe-His523	2.130	Fe-His550	2.208	Fe-His382	2.206	Fe-His757	2.318
Fe-His709	2.242	Fe-His709	2.279	Fe-His518	2.186	Fe-His372	2.100	Fe-His555	2.164	Fe-His943	2.664
Fe-His523	2.224	Fe-His523	2.241	Fe-His709	2.161	Fe-His367	2.244	Fe-His377	2.115	Fe-His762	2.333
Fe-Asn713	2.243	Fe-Asn713	2.281	Fe-Asn713	3.406	Fe-Asn554	3.225	Fe-Asn559	2.241	Fe-Asn947	2.567
Fe-HM	2.009	Fe-FA	2.008	Fe-HOH	2.333	H2O-Ile673	2.997	Fe-HOH	2.146		

* FA - 13(S)-HYDROPEROXY-9(Z),11(E)-OCTADECADIENOIC ACID; HM-CURCUMIN.

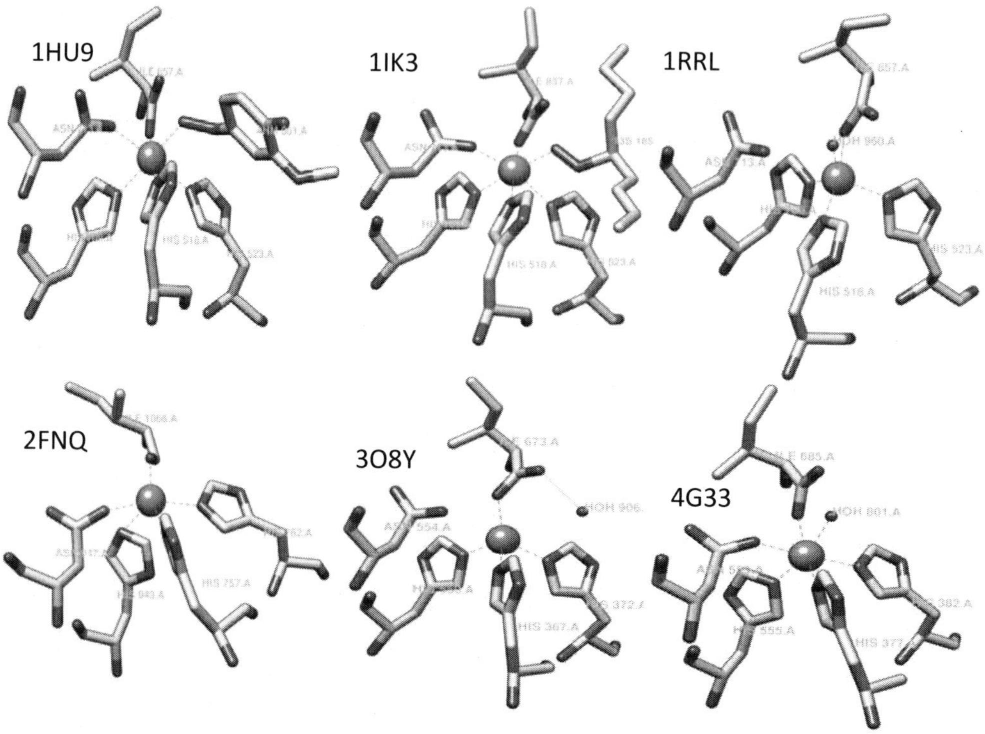

Figure 5. Comparison of active sites and the iron coordination environment between studied lipoxygenases. Distances listed in Table 2.

Lipoxygenase Inhibition

Targeted inhibition of lipoxygenases is a promising strategy for treatment of inflammatory and allergic ailments, including asthma and possibly cancer and neuronal vulnerability in aging. At the same time inhibition of lipoxygenase could provide a novel approach for protection of plants against biotic and abiotic challenges. Finally, inhibition of lipoxygenases has a potential for regulation and control of pathogenic organisms such as fungi and bacteria. Structural similarities across organisms suggest that similar molecules can be used for targeting of lipoxygenases of different species. Caffeic acid possibly provides a scaffold for this development.

Caffeic acid (**1**) and its derivatives including CAPE (**2**) have a significant inhibitory effect on human 5-lipoxygenase [Boudreau, 2012; Du, 2005]. Computational analysis of Koukoulitsa et al. [Koukoulitsa, 2007] suggested caffeic acid (**1**) and related molecules as soybean lipoxygenase inhibitors. Several types of low molecular weight inhibitors that potently suppress leukotriene synthesis as well as leukotriene receptor antagonists have been developed in the past years [Pergola, 2010]. This therapeutic target has been validated by the approval of leukotriene receptor antagonists for use as anti-asthma drugs, while zileuton (**4**) is the only inhibitor of 5- lipoxygenase approved for the treatment of inflammatory diseases. However, zileuton's clinical use has been limited because of a poor pharmacokinetic profile and liver toxicity [Berger, 2007; Garcia-Marcos, 2003]. Therefore there is a need for new effective inhibitors of 5- lipoxygenase and the search for alternative and potent 5-lipoxygenase inhibitors with fewer side effects is ongoing. Among the known promising inhibitors of 5- lipoxygenase are a variety of polyhydroxylated natural products such as caffeic acid (**1**) [Koshihara, 1984] and flavonoids [Baumann, 1980]. However, the rational design of efficient inhibitors requires structural and functional information about the target itself. The recently obtained X-ray structure of the 5- lipoxygenase enzyme [Gilbert, 2011] as well as number of other lipoxygenases will greatly help the optimization of new inhibitors,

Both ester and amide analogues of CAPE (**2**) were designed and tested by Boudreau et al. (Schema 2). CAPE (**2**) was a more potent inhibitor than zileuton (**4**) for 5- lipoxygenase leukotriene biosynthesis in stimulated human PMNL, although its inhibition of leukotriene biosynthesis in whole blood was similar to that of zileuton (**4**) [Boudreau, 2012]. The phenethyl ester group of CAPE (**2**) was essential for effective inhibition since caffeic acid (**1**) and cinnamic acid did not show significant activity. Contrary to the belief that polyphenols inhibit 5-lipoxygenase exclusively through their antioxidant properties, these results clearly show that the presence or absence of specific functional groups influence 5- lipoxygenase inhibition independent of their impact on anti-oxidant activity. While amide (**3**) (Scheme 2) inhibited leukotriene biosynthesis and 5-lipoxygenase activity, the presence of the amide linkage reduced its potency compared to the ester CAPE (**2**).

Based on the biological properties of selenium [Nordberg, 2001; Schewe, 1995], Lin and co-workers described the synthesis and biological evaluation of the CAPE (**2**) selenium-containing analog (Scheme 3). Compound (**5**) was found to be a very effective antioxidant and 5- lipoxygenase inhibitor with activity comparable to or better than CAPE (**2**) (IC_{50} = 0.22 µM; K_i 0.30 µM) [Lin, 2005].

In an application of clusters or dendrimers, Doiron and co-workers applied the versatile "click chemistry" conditions to an azido or alkyne-functionalized caffeoyl derivatives for novel dimeric analogues synthesis (**Scheme 4**) [Doiron, 2009]. In dose-response studies, some compounds showed concentration-dependent inhibition of 5-lipoxygenase product synthesis with IC_{50} values below 1 µM (**6a** IC_{50} = 0.68 µM; **6c** IC_{50} = 0.74 µM) [Doiron, 2009] that were comparable to the inhibitory activity of zileuton (IC_{50} = 0.5-1 µM) [Carter, 1991]. On the basis of corrected values on a per caffeoyl residue, dimer (**6c**) (**Schema 4**), readily surpassed the activity of caffeic acid (**1**) (IC_{50} = 25 µM) by more than 10-fold.

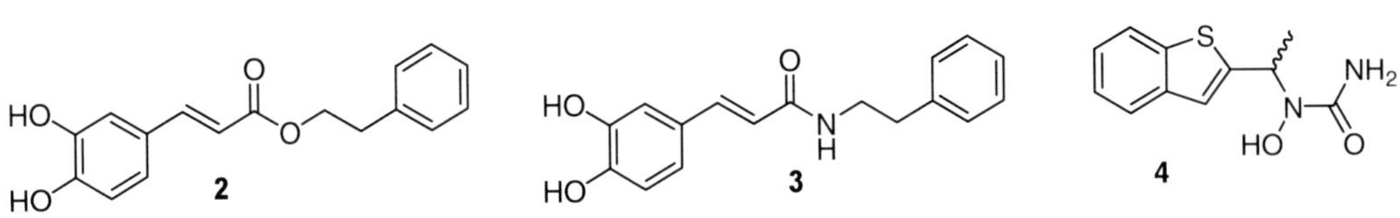

Scheme 2.

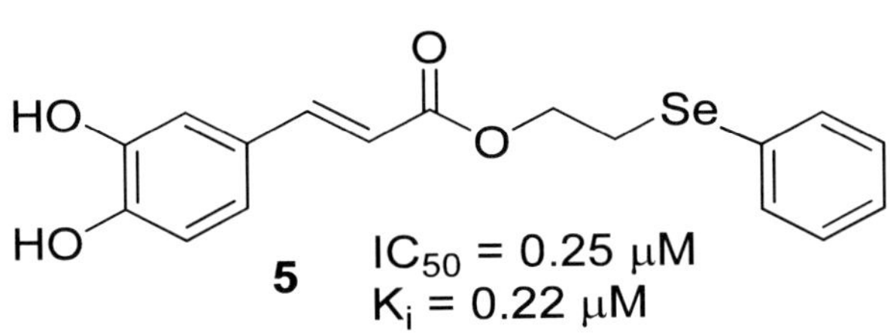

Scheme 3.

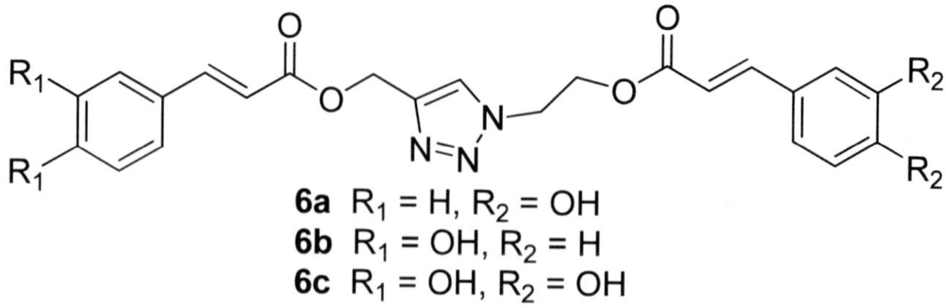

Scheme 4.

The inhibitory capacities of these compounds were then evaluated in intact HEK293 cells [Boudreau, 2012]. These cells have the cellular machinery required for leukotriene biosynthesis. Compound **6b** caused almost complete 5-lipoxygenase inhibition at 10 mM, and was more effective than the known 5- lipoxygenase inhibitor zileuton (**4**) [Carter, 1991]. Inhibition tests with diacetylated precursors of molecules (**6a**) and (**6b**) resulted in the complete inhibition of 5-lipoxygenase metabolites in intact cells. These results could suggest that these molecules infiltrate into intact cells more efficiently than **6a** and **6b**, resulting in a more efficient inhibition of 5-lipoxygenase following esterase action, or a possible interference with 5-lipoxygenase associated proteins (e.g., coactosin-like protein or FLAP).

From known 5-lipoxygenase inhibitor (**6b**) as a lead, cinnamic esters and amides as well as caffeic esters and amides (**7** and **8;** Scheme 5) were synthesized by Cu(I)-catalyzed [1,3]-dipolar cycloaddition with the appropriate azide precursors and terminal alkynes (Scheme 6). The clicked caffeic esters (**7** and **8**) bearing a *para*-methylphenyl moiety or a *para*-fluorophenyl moiety were four times more potent than zileuton (**4**) [Doiron, 2014]. The corresponding clicked caffeic amides were equipotent to zileuton (**4**). In thapsigargin-stimulated human PMNL, the same esters were equivalent to zileuton (**4**) while the corresponding amides were less active than zileuton (Schema 2.**4**).

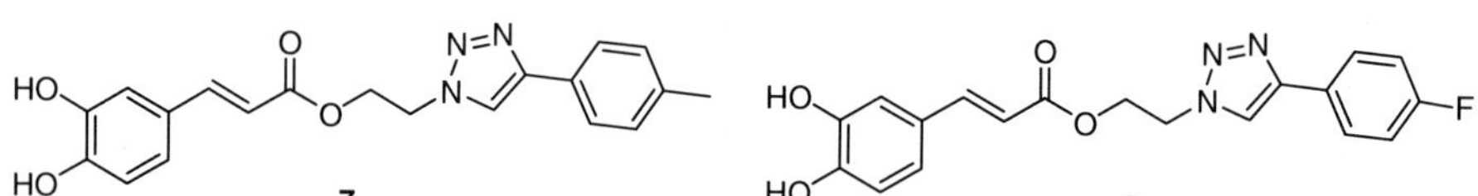

7 **8**

Scheme 5.

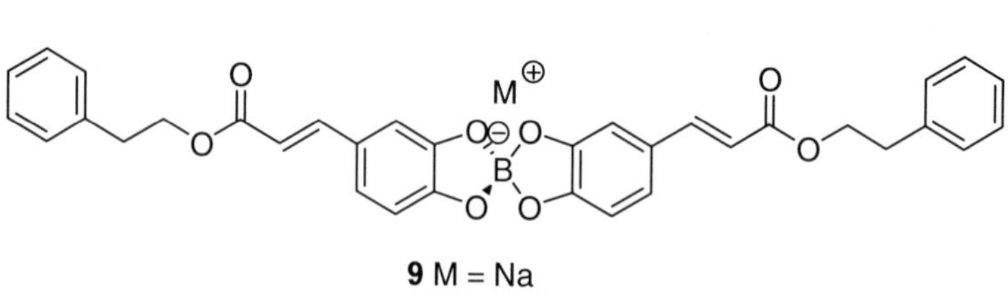

Scheme 6.

Hébert and co-workers described arylspiroborate sodium (**9**) and ammonium (**10**) salts derived from CAPE (.**2**) that have similar 5-lipoxygenase products biosynthesis inhibitory activity in intact HEK293 cells as zileuton (**4**) and CAPE (**2**) [Hebert, 2015]. This result clearly shows the effectiveness of the synthesized boron salts as transporter of CAPE (**2**), where a total or even partial hydrolysis occurs for the release of CAPE (**2**).

Novel, caffeic acid derivatives have been explored primarily as 5-lipoxygenase inhibitors; however similarity between lipoxygenases (Figures 4 and 5) suggest that they can be investigated in the development of pan-lipoxygenase inhibitor. Development of multi-species lipoxygenase inhibitors can have therapeutic significance as shown by Hoobler et al. [Hoobler, 2013] or as environmentally friendly plant pathogen methods with multifaceted effects. Comparison between studied structures shows that in all four lipoxygenases studied from highly distinct organisms there is room in the active site for one of known natural product inhibitors curcumin (Figure 6) which was presented in the solybean lipoxygenase structure 1HU9. Analysis of binding potential for related molecules such as caffeic acid (**1**) and its derivatives can be based on the measured interaction between curcumin and soybean lipoxygenase.

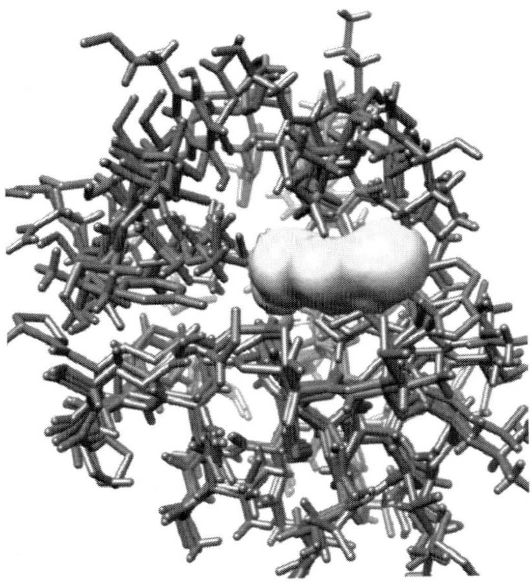

Figure 6. Comparison between active sites of 3O8Y, 1HU9, 4G33, 2FNQ (gray) with curcumin measured in the active site of 1HU9 (yellow surface).

Table 3. Crystal structures used in the docking analysis

Species	Name	Fe charge	Resolution (A)	ID	Reference
Human	LOX-5	2	2.39	3O8Y	Gilbert, 2011
Soybean	LOX-3 + curcumin	3	2.20	1HU9	Skrzypczak-Jankun, 2003
Pseudomonas aeruginosa.		2	2.03	4G33	Garreta, 2013
Coral	8R-Lipoxygenase	2	3.20	2FNQ	Oldham, 2005

Molecular docking is a popular method used in structure-based rational drug design for evaluation of complex formation between small ligand and large biomolecules. Docking computational analysis simulates interaction between molecules and shows relative binding energy between ligand and targets. Comparison of docking analysis for number of caffeic acid analogs and lipoxygenases from different organisms is shown below. Analysis was performed on lipoxygenases with experimentally determined crystal structures and this includes proteins from human, soybean, coral and pseudomonas aeruginosa (more detailed information regarding protein structures used is shown in Table 3). Caffeic acid (**1**), CAPE (**2**) and compound (**3**) were used as ligand. Docking was performed using AutoDock 4 (Scripps Research Institute, USA). AutoDock4 performs a rapid energy evaluation through a pre-calculated grid of affinity potentials. The macromolecule is kept rigid while torsional flexibility is allowed for the ligand molecule. The grid box for all calculations included active site based on X-ray and literature information. Initial structures of all ligands were drawn in ChemDraw. Following the addition of hydrogen atoms, all ligand structures were optimized using Gaussian 97 software using Semi-Empirical and Density Functional optimization model (using 631Gd,p basis set). Atom charges in ligands were calculated using the Gasteiger method in Autodock4 and compared to charges calculated using ChelpG functionality in Gaussian 97. Resulting docking binding energies are shown in Table with obtained positions for ligands shown in Figures 7 and 8.

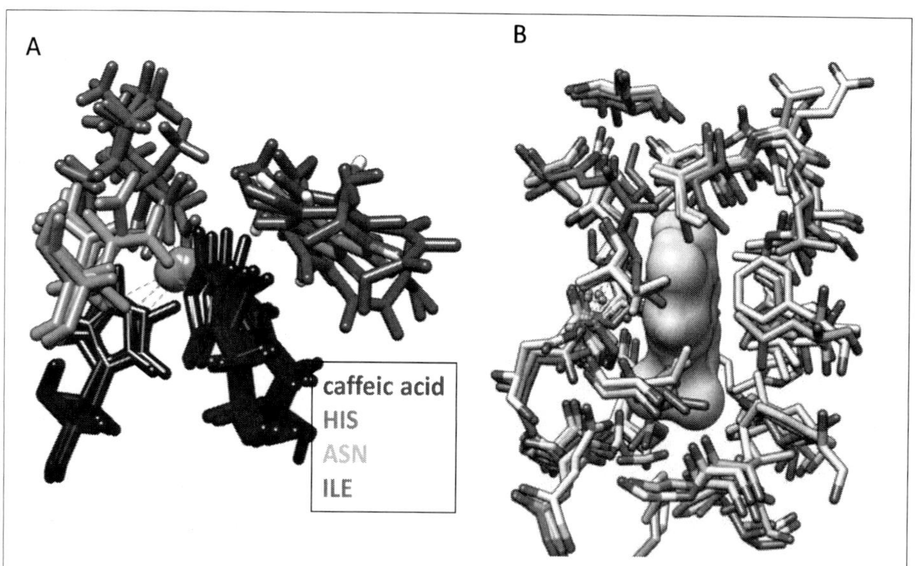

Figure 7. Docking of caffeic acid (1) to 3O8Y, 1HU9, 4G33, 2FNQ. A. Optimal orientation of caffeic acid (1) in active sites of all four proteins and the coordinated amino acids in the active sites. B. Surface presentation of caffeic acid (1) with adjacent amino acids in the active sites of all proteins.

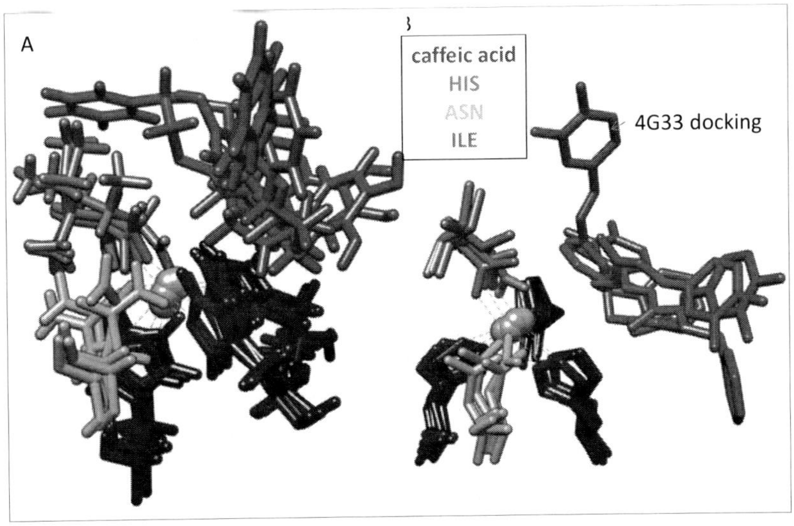

Figure 8. A. Orientation of CAPE (2) following docking to 3O8Y, 1HU9, 4G33, 2FNQ. B Orientation of compound (3) following docking to 3O8Y, 1HU9, 4G33, 2FNQ.

Table 4. Docking binding energies determined for studied targets and ligands using AutoDock Vina. The grid size was: 15; 16; 16 and it was centred around iron core and location of curcuumin in 1HU9 structure

Species	ID	Docking binding energy (kcal/mol)		
		Caffeic Acid (1)	CAPE (2)	Compound (3)
Human	3O8Y	-6.3	-8.1	-7.8
Soybean	1HU9	-6.1	-5.1	-3.6
Pseudomonas aeruginosa.	4G33	-6.8	-7.9	-8.3
Coral	2FNQ	-5.5	-4.4	-1.2

Both orientations and docking energies are highly similar in the analysis of binding of caffeic acid (**1**) to the studied proteins (Table 4 and Figure 7). Similar binding characteristics shown here are in agreement with pan-species activity of caffeic acid (**1**). In all cases caffeic acid (**1**) is positioned at the active site, in the vicinity of iron as well as coordinated amino acids. Structures of all four proteins provide empty cavity for caffeic acid (**1**) to approach their active core iron. In the case of larger molecules of CAPE (**2**) and compound (**3**) strong binding is observed for human and *Pseudomonas aeruginosa* lipoxygenases. However, binding to lipoxygenases of coral and soybean lead to lower binding energies suggesting steric issues in the vicinity of the active site. Different orientation of these compound is visible in Figure 8. Thus, derivatives of caffeic acid can be explored as either pan-species or species specific lipoxygenase inhibitors.

CONCLUSION

Structures of lipoxygenases across highly distinct organisms show sufficient similarity explaining pan-species activity of caffeic acid (**1**) as a lipoxygenase inhibitor. The inhibitory activity of caffeic acid (**1**), CAPE (**2**) and compound (**3**) was investigated using docking analysis. Caffeic acid (**1**) binds to all distinct lipoxygenases with similar docking energy suggesting that it can be used as multi-species inhibitors or as a precursor for the development of universal inhibitory molecule. This study explains pan-species activity of caffeic acid (**1**) through the inhibition of lipoxygenases showing similar structures of active sites in different systems. At the same time, structural differences between lipoxygenases of different species are sufficient to provide distinct binding energies for some caffeic acid derivatives and therefore caffeic acid (**1**) can also provide a precursor for species and isoform specific inhibitor development. Further theoretical and experimental efforts are needed in both of these highly distinct and useful directions.

REFERENCES

Acosta, I. F., Laparra, H., Romero, S. P., Schmelz, E., Hamberg, M., Mottinger, J. P. et al. (2009). Tasselseed1 is a lipoxygenase affecting jasmonic acid signaling in sex determination of maize. *Science*, *323*, 262–5.

Andre, D. & Hou, K. (1932). The presence of a lipoid oxidase in soybean. Glycine soya. *C R Acad Sci.*, *194*, 645–7.

Andreou, A. & Feussner I. (2009). Lipoxygenases – structure and reaction mechanism. *Phytochemistry*, *70*, 1504–10.

Arakawa, R., Yamaguchi, M., Hotta, H., Osakai, T. & Kimoto, T. (2004). Product analysis of caffeic acid oxidation by on-line electrochemistry/electrospray ionization mass spectrometry. *Journal Am Society for Mass Spectrometry*, *15*, 1228–1236.

Baranowski, J. D., Davidson, P. M., Nagel, C. W. & Brannen, R. L. (1980). Inhibition of Pseudomonas fluorescens by hydroxycinnamic acids and their alkyl esters, *J. Food Sci.*, *45*, 592-594.

Baron, J. A., Sandler, R. S., Bresalier, R. S., Lanas, A., Morton, D. G., Riddell, R., Iverson, E. R. & DeMets, D. L. (2008). Cardiovascular events associated with rofecoxib: final analysis of the APPROVe trial, *Lancet*, *372*, 1756-1764.

Baumann, J., Von Bruchhausen, F. & Wurm, G. (1980). Flavonoids and related compounds as inhibition of arachidonic acid peroxidation, *Prostaglandins*, *20*, 627-639.

Berger, W., De Chandt, M. T. & Cairns, C. B. (2007). Zileuton: clinical implications of 5-Lipoxygenase inhibition in severe airway disease, *Int. J. Clin. Pract.*, *61*, 663-676.

Bizzintino, J. A., Khoo, S. K., Zhang, G., Martin, A. C., Rueter, K., Geelhoed, G. C. et al. (2009). Leukotriene pathway polymorphisms are associated with altered cysteinyl leukotriene production in children with acute asthma. *Prostaglandins Leukot Essent Fatty Acids*, *81*, 9–15.

Bollina, V., Kumaraswamy, G. K., Kushalappa, A. C., Choo, T. M., Dion, Y., Rioux, S. & Hamzehzarghani, H. (2010). Mass spectrometry-based metabolomics application to identify quantitative resistance-related metabolites in barley against Fusarium head blight, *Mol Plant Pathology*, *11*, 769–782.

Bolton, M. D., Kolmer, J. A., Xu, W. W. & Garvin, D. F. (2008). Lr34 -Mediated Leaf Rust Resistance in Wheat: Transcript Profiling Reveals a High Energetic Demand Supported by Transient Recruitment of Multiple Metabolic Pathways, *21*, 1515–1527.

Boudreau, L. H., Maillet, J., LeBlanc L. M., Jean-Francois, J., Touaibia, M., Flamand, N. & Surette, M. E. (2012). Caffeic acid phenethyl ester and its amide analogue are potent inhibitors of leukotriene biosynthesis in human polymorphonuclear leukocytes, *Plos One*, *7*(2), e31833.

Boudreau, L. H., Picot, N., Doiron, J., Villebonet, B., Surette, M. E., Robichaud, G. A. & Touaibia, M. (2009). Caffeoyl and cinnamoyl clusters with anti-inflammatory and anti-cancer effects. Synthesis and structure–activity relationship. *New J Chem*, *33*, 1932–1940

Brash, A. R. (1999). Lipoxygenases: Occurrence, Functions, Catalysis, and Acquisition of Substrate, *J Biol Chem*, *274*, 23679–23683.

Brodhun, F. & Feussner, I. (2011). Oxylipins in fungi. *The FEBS Journal*, *278*, 1047–63.

Carter, G. W., Young, P. R., Albert, D. H., Bouska, J., Dyer, R., Bell, R. L., Summers, J. B. & Brooks, D. W. (1991). 5-lipoxygenase inhibitory activity of zileuton, *J. Pharmacol. Exp. Ther.*, *256*, 929-937.

Chao, C-Y., Mong, M-C., Chan, K-C. & Yin, M-C. (2010). Anti-glycative and anti-inflammatory effects of caffeic acid and ellagic acid in kidney of diabetic mice. *Mol Nutr Food Res*, *54*, 388–395.

Chedea, V. S. & Jisaka, M. (2011). Inhibition of Soybean Lipoxygenases – Structural and Activity Models for the Lipoxygenase Isoenzymes Family. Recent Trends for Enhancing the Diversity and Quality of Soybean Products, InTech, editor D. Krezhova, 109–130.

Chrzanowski, G. & Leszczyński, B. (2008). Induced accumulation of phenolic acids in winter triticale (Triticosecale Wittm.). under insects feeding, *Herb. Pol.*, *54*, 33–40.

Chuu, C-P., Lin, H-P., Ciaccio, M. F., Kokontis, J. M., Hause, R. J., Hiipakka, R. A., Liao, S. & Jones, R. B. (2012). Caffeic acid phenethyl ester suppresses the proliferation of human prostate cancer cells through inhibition of p70S6K and akt signaling networks. *Cancer Prev Res*, *5*, 788–797.

Doiron, J., Boudreau, L. H., Picot, N., Villebonet, B., Surette, M. E. & Touaibia, M. (2009). Synthesis and 5-lipoxygenase inhibitory activity of new cinnamoyl and caffeoyl clusters, *Bioorg. Med. Chem. Lett.*, *19*, 1118-1121.

Doiron, J. A; Métayer, B., Richard, R. R., Desjardins, D., Boudreau, L. H; Levesque, N. A., Jean-François, J., Poirier, S. J., Surette, M. E. & Touaibia, M. (2014). Clicked Cinnamic/Caffeic Esters and Amides as Radical Scavengers and 5-Lipoxygenase Inhibitors, *Int. J. Med. Chem.*, Volume *2014*, 931756.

Du, L., Zhang, Z., Luo, X., Chen, K., Shen, X. & Jiang, H. (2006). Binding investigation of human 5-lipoxygenase with its inhibitors by SPR technology correlating with molecular docking simulation. *J Biochemistry*, *139*, 715–723.

Feliciano, R. P., Bravo, M. N., Pires, M. M., Serra, A. T., Duarte, C. M., Boas, L. V. & Bronze, M. R. (2009). Phenolic Content and Antioxidant Activity of Moscatel Dessert Wines from the Setubal Region in Portugal. *Food Anal Methods*, *2*, 149–161

Feussner, I. & Wasternack, C. (2002). The lipoxygenase pathway. *Annual Review of Plant Biology*, *53*, 275–297.

Feussner, I., Kühn, H. & Wasternack, C. (2001). Lipoxygenase-dependent degradation of storage lipids. *Trends Plant Sci*, *6*, 268–73.

Funk, C. D., Chen, X. S., Johnson, E. N. & Zhao, L. (2002). Lipoxygenase genes and their targeted disruption. *Prostaglandins Other Lipid Mediat*, 68–69, 303–12.

Garcia-Marcos, L., Schuster, A. & Perez-Yarza, E. G. (2003). Benefit-risk assessment of antileukotrienes in the management of asthma, *Drug Saf.*, 26, 483-518.

Garreta, A., Val-Moraes, S. P., García-Fernández, Q., Busquets, M., Juan, C., Oliver, A., Ortiz, A., Gaffney, B. J., Fita, I., Manresa, À. & Carpena, X. (2013). Structure and interaction with phospholipids of a prokaryotic lipoxygenase from Pseudomonas aeruginosa. *FASEB J. 27*, 4811-21.

Gilbert, N. C., Sue G. Bartlett, S. G., Waight, M. T., David B. Neau, D. B., William E. Boeglin, W. E., Alan R. Brash, A. R., Marcia E. & Newcomer, M. E. (2011). The structure of human 5-lipoxygenase, *Science*, *331*, 217-219.

Gokalp, O., Uz, E., Cicek, E., Yilmaz, H. R., Ozer, M. K. et al. (2006). Ameliorating role of caffeic acid phenethyl ester (CAPE). against isoniazid-induced oxidative damage in red blood cells. *Mol Cell Biochem*, *290*, 55–59.

Grebner, W., Stingl, N. E., Oenel, A., Mueller, M. J. & Berger, S. (2013). Lipoxygenase 6-dependent oxylipin synthesis in roots is required for abiotic and biotic stress resistance of Arabidopsis. *Plant Physiol.*, *161*, 2159–70.

Gunnaiah, R. & Kushalappa, A. C. (2014). Metabolomics deciphers the host resistance mechanisms in wheat cultivar Sumai-3, against trichothecene producing and non-producing isolates of Fusarium graminearum. *Plant Physiology and Biochemistry*, *83C*, 40–50.

Haeggström, J. Z. & Funk, C. D. (2011). Lipoxygenase and leukotriene pathways: biochemistry, biology, and roles in disease. *Chem Rev. 111*, 5866–98.

Hansen, J., Garreta, A., Benincasa, M., Fusté, M. C., Busquets, M. & Manresa, A. (2013). Bacterial lipoxygenases, a new subfamily of enzymes? A phylogenetic approach. *Appl Microbiol Biotechnol*, *97*, 4737–47.

Hébert, M. J. G., Flewelling, A. J., Clark, T. N., Levesque, N. A., Jean-François, J., Surette, M. E., Gray, C. A., Vogels, C. M., Touaibia, M. & Westcott, S. A. (2015). Synthesis and biological activity of arylspiroborate salts derived from caffeic Acid phenethyl ester, *Int. J. Med. Chem.*, 2015, 418362.

Horn, T., Adel, S., Schumann, R., Sur, S., Kakularam, K. R., Polamarasetty, A. & Heydeck, D. (2015). Evolutionary aspects of lipoxygenases and genetic diversity of human leukotriene signaling. *Progress in Lipid Research*, *57*, 13–39.

Kim, W., Seong, K. M. & Youn, B. (2011). Phenylpropanoids in radioregulation: double edged sword. *Exp Mol Med.*, *43*, 323-33.

Kobe, M. J., Neau, D. B., Mitchell, C. E., Bartlett, S. G. & Newcomer, M. E. (2014). The structure of human 15-lipoxygenase-2 with a substrate mimic. *J. Biol. Chem.*, *289*, 8562-8569.

Kolomiets, M. V., Hannapel, D. J., Chen, H., Tymeson, M. & Gladon, R. J. (2001). Lipoxygenase is involved in the control of potato tuber development. *Plant Cell*, *13*, 613–26.

Koshihara, Y., Neichi, T., Murota, S. I., Lao, A. N., Fujimoto, Y. & Tatsuno, T. (1984). Caffeic acid is a selective inhibitor for leukotriene biosynthesis, *Biochim. Biophys. Acta*, *792*, 92-97.

Koukoulitsa, C., Hadjipavlou-Litina, D., Geromichalos, G. D. & Skaltsa, H. (2007) Inhibitory effect on soybean lipoxygenase and docking studies of some secondary metabolites, isolated from Origanum vulgare L. ssp. hirtum. *Journal of Enzyme Inhibition and Medicinal Chemistry*, *22*, 99–104.

Kudugunti, S. K., Vad, N. M., Whiteside, A. J., Naik, B. U., Yusuf, M. A., Srivenugopal, K. S. & Moridani, M. Y. (2010). Biochemical mechanism of Caffeic Acid Phenyl-ethyl Ester (CAPE). selective toxicity towards melanoma cell lines. *ChemBiol Interact*, *188*, 1–14.

Kuhn, H., Banthiya, S. & van Leyen, K. (2015). Mammalian lipoxygenases and their biological relevance. *Biochim Biophys Acta.*, *1851*, 308-330.

Lattanzio, V. (2006). Role of phenolics in the resistance mechanisms of plants against fungal pathogens and insects. *Phytochemistry: Advances in Research*, *661*, 23–67.

Liao, H. F., Chen, Y. Y., Liu, J. J., Hsu, M. L., Shieh, H. J., Liao, H. J., Shieh, C. J., Shiao, M. S. & Chen, Y. J. (2003). Inhibitory Effect of Caffeic Acid Phenethyl Ester on Angiogenesis, Tumor Invasion, and Metastasis, *J. Agric. Food Chem.*, *51*, 7907-7912.

Lin, C. F., Chang, T. C., Chiang, C. C., Tsai, H. J. & Hsu, L. Y. (2005). Synthesis of selenium-containing polyphenolic acid esters and evaluation of their effects on antioxidation and 5-lipoxygenase inhibition, *Chem. Pharm. Bull.*, *53* (11). 1402-1407.

Mandal, S. M., Chakraborty, D. & Dey, S. (2010). Phenolic acids act as signaling molecules in plant-microbe symbioses. *Plant Signaling & Behavior*, *5*, 359–368.

Michaluart, P., Masferrer, J. L., Carothers, A. M., Subbaramaiah, K., Zweifel, B. S., Koboldt, C., Mestre, J. R., Grunberger, D., Sacks, P. G., Tanabe, T. & Dannenberg, A. J. (1999). Inhibitory Effects of Caffeic Acid Phenethyl Ester on the Activity and Expression of Cyclooxygenase-2 in Human Oral Epithelial Cells and in a Rat Model of Inflammation, *Cancer Res.*, *59*, 2347-2352.

Mirzoeva, O. K. & Calder, P. C. (1996). The effect of propolis and its components on eicosanoid production during the inflammatory response. *Prostaglandins Leukot Essent Fatty Acids*, *55*, 441–449.

Natarajan, K., Singh, S., Burke, T. R., Grunberger, D. & Aggarwal, B. B. (1996). Caffeic acid phenethyl ester is a potent and specific inhibitor of activation of nuclear transcription factor NF-kappa B. *Proc Natl Acad Sci*, *93*, 9090-9095.

Newcomer, M. E. & Gilbert, N. C. (2010). Location, location, location: compartmentalization of early events in leukotriene biosynthesis. *J Biol Chem*, *285*, 25109–14.

Nordberg, J. & Arner, E. S. (2001). Reactive oxygen species, antioxidants, and the mammalian thioredoxin system, *Free Radic. Biol. Med.*, *31*, 1287-1312.

Odaci, D., Timur, S., Pazarlioglu, N., Montereali, M. R., Vastarella, W., Pilloton, R. & Telefoncu, A. (2007). Determination of phenolic acids using Trametes versicolor laccase. *Talanta*, *71*, 312–317.

Oldham, M. L., Brash, A. R. & Newcomer, M. E. (2005). Insights from the X-ray crystal structure of coral 8R-lipoxygenase: calcium activation via a C2-like domain and a structural basis of product chirality. *J Biol Chem.*, *280*, 39545-52.

Parlakpinar, H., Tasdemir, S., Polat, A., Bay-Karabulut, A., Vardi. N., et al. (2005). Protective role of caffeic acid phenethyl ester (cape). on gentamicin-induced acute renal toxicity in rats. *Toxicology*, *207*, 169–177.

Pati, S., Losito, I., Palmisano, F. & Zambonin, P. G. (2006). Characterization of caffeic acid enzymatic oxidation by-products by liquid chromatography coupled to electrospray ionization tandem mass spectrometry. *Journal of Chromatography A*, *1102*, 184–192.

Pergola, C. & Werz, O. (2010). 5-Lipoxygenase inhibitors: a review of recent developments and patents, *Expert Opin. Ther. Patents*, *20*, 355-375.

Peters-Golden, M. & Henderson, W. R. (2007). Leukotrienes, *N. Engl. J. Med*, *357*, 1841-1854.

Prakash, T. R., Swamy, P. M., Suguna, P. & Reddanna, P. (1990). Characterization and behaviour of 15-lipoxygenase during peanut cotyledonary senescence. *Biochem Biophys Res Commun*, *172*, 462–70.

Praveen Rao, P. N., Chen, Q.-H. & Knaus, E. E. (2006). Synthesis and structure-activity relationship studies of 1,3-diarylprop-2-yn-1-ones: dual inhibitors of cyclooxygenases and lipoxygenases, *J. Med. Chem.*, *49*, 1668-1683.

Romano, M. & Claria, J. (2003). Cyclooxygenase-2 and 5-lipoxygenase converging functions on cell proliferation and tumor angiogenesis: implications for cancer therapy, *FASEB J*, *17*, 1986-1995.

Rydberg, E. K., Krettek, A., Ullstrom, C., Ekstrom, K., Svensson, P. A., Carlsson, L. M., et al. (2004). Hypoxia increases LDL oxidation and expression of 15-lipoxygenase-2 in human macrophages. *Arterioscler Thromb Vasc Biol*, *24*, 2040–5.

Sato, Y., Itagaki, S., Kurokawa, T., Ogura, J., Kobayashi, M., Hirano, T., Sugawara, M. & Iseki, K. (2011). In vitro And in vivo antioxidant properties of chlorogenic acid and caffeic acid. *Int J Pharm*, *403*, 136–138.

Schewe T. (1995). Molecular actions of ebselen--an antiinflammatory antioxidant, *Gen. Pharmacol.*, *26*, 1153-1169.

Schneider, C. & Pozzi, A. (2011). Cyclooxygenases and lipoxygenases in cancer. *Cancer Metastasis Reviews*, *30*, 277–94.

Skrzypczak-Jankun, E., Zhou, K. & Jankun, J. (2003). Inhibition of lipoxygenase by (-)-epigallocatechin gallate: X-ray analysis at 2. 1 A reveals degradation of EGCG and shows soybean LOX-3 complex with EGC instead. *INT. J. MOL. MED.*, *12*, 415-420

Sud'ina, G. F., Mirzoeva, O. K., Pushkareva, M. A., Korshunova, G. A., Sumbatyan, N. V. & Varfolomeev, S. D. (1993). Caffeic acid phenethyl ester as a lipoxygenase inhibitor with antioxidant properties. *FEBS Letters*, *329*, 21-24.

Vellosillo, T., Martínez, M., López, M. A., Vicente, J., Cascón, T. & Dolan, L. et al. (2007). Oxylipins produced by the 9-lipoxygenase pathway in Arabidopsis regulate lateral root development and defense responses through a specific signaling cascade. *Plant Cell*, *19*, 831–46.

Xu, S., Mueser, T. C., Marnett, L. J. & Funk, M. O. (2012). Crystal structure of 12-Lipoxygenase catalytic-domain-inhibitor complex identifies a substrate-binding channel for catalysis. *Structure*, *20*, 1490–1497.

Youn, B., Sellhorn, G. E., Mirchel, R. J., Gaffney, B. J., Grimes, H. D. & Kang, C. (2006). Crystal structures of vegetative soybean lipoxygenase VLX-B and VLX-D, and comparisons with seed isoforms LOX-1 and LOX-3. *Proteins*, *65*, 1008.

Zabala, G., Zou, J., Tuteja, J., Gonzalez, D. O., Clough, S. J. & Vodkin, L. O. (2006). Transcriptome changes in the phenylpropanoid pathway of Glycine max in response to Pseudomonas syringae infection. *BMC Plant Biology*, *6*, 26.

Zhang, B., Chen, K., Bowen, J., Allan, A., Espley, R., Karunairetnam, S., et al. (2006). Differential expression within the LOX gene family in ripening kiwifruit. *J Exp Bot*, *57*, 3825–36.

In: Caffeic Acid
Editor: Leanna Vaughn

ISBN: 978-1-63483-121-5

Chapter 8

THE ROLE OF LIQUID CHROMATOGRAPHY-MASS SPECTROMETRY IN THE ANALYSIS OF CAFFEIC ACID

***Paolo Lucci*[1]*, Oscar Núñez*[2,3,*] *and Javier Saurina*[2]**

[1]Department of Nutrition and Biochemistry, Faculty of Sciences, Pontificia Universidad Javeriana, Bogotá, Colombia
[2]Department of Analytical Chemistry, University of Barcelona. Martí i Franquès, Barcelona, Spain
[3]Serra Húnter Fellow, Generalitat de Catalunya, Spain

ABSTRACT

Phenolic acids, one of the major classes of polyphenols, are widely distributed in the human diet, particularly in fruit, vegetables and beverages. Caffeic acid is a phenolic acid naturally occurring in many plant foods such as carrots, cabbage, tomatoes and several berries. It is also present in multiple beverages such as coffee, fruit juices and wine. Moreover, caffeic acid is one of the major active components in many traditional Chinese medicines. As a natural product, it possesses several biological and physiological activities. These activities include antioxidant, anti-mutagenic, anti-tumor and anti-obesity effects. For these reasons, caffeic acid is also being developed into drug for sale, for instance for its use in the development of asthmatic and allergenic drugs.

Although caffeic acid as well as other polyphenols are not essential nutrients, they are secondary metabolites synthesized by plants with bioactive properties, also known as phytochemicals, that may have an important effect on human health. For this reason, the interest in the analysis of this kind of compounds for the characterization of plant extracts is raising in the last years. Liquid chromatography coupled to mass spectrometry or to tandem mass spectrometry (LC-MS(/MS)) is one of the most frequently used techniques for that purpose, and the number of LC-MS(/MS) methods published in the literature dealing with the analysis of polyphenols, including caffeic acid, not only for their quantitative determination in food products and beverages but also for the characterization of foodstuffs and pharmaceutical preparations as a way to guarantee authentication regarding the natural product of origin is increasing. Additionally, LC-

[*] E-mail address: oscar.nunez@ub.edu

MS(/MS) techniques has also been used for the analysis of caffeic acid and its active metabolites in biological fluids to carry out pharmacological studies.

Recently, high resolution mass spectrometry (HRMS) is acquiring also an important role in the analysis of caffeic acid, and polyphenols in general, and today several LC-HRMS methods either using time-of-flight (TOF) or Orbitrap mass analyzers are described for the quantitative determination of these compounds as well as a fingerprint analysis for characterization, for instance in the case of herbal medicines.

In this chapter the role of liquid chromatography-mass spectrometry in the analysis of caffeic acid will be discussed. LC-MS(/MS) chromatographic conditions, ionization sources, and MS and HRMS analyzers frequently used, as well as strategies for the structural characterization and the qualitative and quantitative analysis of caffeic acid, its active metabolites, and other phytochemicals will be discussed by means of relevant applications. Coverage of all kind of applications is beyond the scope of the present contribution, so we will focus on the most relevant applications published in the last years.

INTRODUCTION

Since several years ago, researchers and food manufactures, as well as the public in general, are becoming very interested in polyphenolic compounds. Polyphenols are aromatic secondary metabolites that are ubiquitously spread through the plant kingdom. This family of compounds comprise more than 8,000 substances with highly diverse structures and molecular masses ranging from small molecules (<100 Da) such as phenolic compounds to big molecules (>30,000 Da) for highly polymerized polyphenolic compounds. The main reason for the interest in this family of compounds is the recognition of their antioxidant properties, their great abundance in our diet, and their probable role in the prevention of various diseases [1, 2]. Furthermore, polyphenols, which also constitute the active substances found in many medicinal plants, modulate the activity of a wide range of enzymes and cell receptors [3].

One of the major classes of polyphenols is phenolic acids that are widely distributed in the human diet, particularly in fruit, vegetables and beverages. Caffeic acid is a phenolic acid classified in the family of hydroxycinnamic acids, which consists of both phenolic and acrylic functional groups, and naturally occurring in many vegetables such as red beetroots, broccoli, Brussels sprouts, cabbage, carrots, celery, garlic, onions, lettuces, radish, potatoes and parsley, among others. Caffeic acid has also been detected in many fruits such as lemon, orange, apricots, coconut, melon, tomato, and berries. Consequently, it is also present in fruit juices, but also in other beverages such as coffee and wine. It is also the major active component in many plants used to make traditional medicines, such as *Davallia mariesii Moore* (Korean medicine for the cold, neuralgia, and stomach cancer; Chinese medicine for lumbago, rheumatism, toothache, and tinnitus); *Carissa spinarum L.* (Indian purgative and rheumatism treatment); *Ixora javanica* (Indian antitumor agent, gastric sedative, intestinal antiseptic, and astringent); *Centaurium umbellatum Gil.* (medicinal plant used in many countries), etc. In general, caffeic acid is present in all plants because it is a key intermediate in the biosynthesis of lignin, one of the major components of biomass and its residues [4]. Caffeic acid is biosynthesized by hydroxylation of coumaroyl ester of quinic acid. This hydroxylation produces the caffeic acid ester of shikimic acid, which converts to chlorogenic acid. It is also the precursor to ferulic acid, coniferyl alcohol, and sinapyl alcohol, all of which are significant building blocks in lignin [4]. Its transformation to ferulic acid is catalyzed by the enzyme caffeic acid-O-methyltransferase. Moreover, caffeic acid and its derivative caffeic acid phenethyl ester (CAPE) are produced in many kinds of plants.

As a natural product, caffeic acid possesses several biological and physiological activities, including antioxidant [5], anti-mutagenic [6], anti-tumor [7], and anti-obesity effects [8]. Lately, caffeic acid is also being developed into drug for sale, for instance for its use in the development of asthmatic and allergenic drugs. Regarding the studies of carcinogenicity of caffeic acid, mixed results have been published. Some studies have shown that it inhibits carcinogenesis, but other experiments have shown carcinogenic effects.

Liquid chromatography coupled to mass spectrometry or tandem mass spectrometry (LC-MS(/MS)) has attracted great attention for the analysis of caffeic acid among other polyphenols in a great variety of matrices [9, 10]. Both quantitative determination in foodstuffs and beverages and characterization of food products and pharmaceutical preparations by LC-MS/MS techniques as a way to guarantee authentication regarding the natural product of origin is increasing. Mass spectrometry techniques are also chosen in the analysis of caffeic acid and its main active metabolites in biological fluids to perform pharmacological studies [11, 12]. Recently, high resolution mass spectrometry (HRMS) is also playing an important role to help in the determination and characterization of polyphenolic compounds if foodstuffs and biological matrices [13-16].

LC-MS in the Analysis of Caffeic Acid in Biological Fluids

Liquid chromatography coupled to tandem mass spectrometry (LC-MS/MS) is one of the techniques of choice in pharmacokinetic studies of biological active compounds in a great variety of biological fluids. Among biological active compounds related to caffeic acid, caffeic acid phenethyl ester (CAPE), a polyphenolic ester firstly reported in 1987, is an important bioactive compound present in propolis. CAPE has attracted increasing attention due to its important pharmacological activity as anticancer/tumor, antiviral, anti-inflammatory, antioxidant, cytoprotection, and vasorelaxant effects [17-20]. CAPE can be easily hydrolyzed [21], and the main hydrolyzed metabolite is caffeic acid, a natural product also possessing, as previously commented, several biological and physiological activities [22], including antioxidant, anti-mutagenic, anti-tumor, and anti-obesity effects [5-8]. Because of these interesting properties of caffeic acid, the major metabolite of CAPE *in vivo*, it is necessary to analyze the bioactivities of caffeic acid when studying CAPE. Thus, pharmacokinetic parameters of both CAPE and caffeic acid in animal or human body should be determined when CAPE is studied as a therapeutic agent. Tang et al. [11] proposed a LC-MS/MS method for the simultaneous determination of CAPE and caffeic acid in dog plasma. Electrospray in negative ionization mode was proposed. The most abundant product ion observed was *m/z* 135 for both CAPE and caffeic acid. This ion is the decarboxylated product from the caffeate ion (*m/z* 179) which is the precursor ion of caffeic acid and the ester-cleaved product ion of CAPE. The other major product ion observed at *m/z* 161 may be due to the dehydratation from the caffeate ion (*m/z* 179). CAPE and caffeic acid were separated by reversed-phase chromatography, and a quantification range in multiple reaction monitoring (MRM) mode from 10.0 to 10,000.0 ng/mL was established, with good intra- and inter-batch accuracies (relative standard deviations (RSD) values lower than 10.5%). CAPE and caffeic acid were proven to be stable in stabilizer-treated dog blood and treated plasma after protein precipitation during the sampling and pretreatment period.

Rubió et al. [23] described a selective and sensitive method, based on microelution solid-phase extraction (µSPE) plate and ultra-high performance liquid chromatography coupled to tandem mass spectrometry (UHPLC-MS/MS) for the determination of plasma metabolites of the bioactive compounds of thyme. Thyme comes originally from the regions around the Mediterranean and is used as cough medicine. It has also been commonly used as a culinary herb for adding flavor. Several classes of bioactive compounds, such as phenolic acids (caffeic acid, homovanillic acid, vanillic acid, rosmarinic acid…), flavonoids (naringenin, thymusin, quercetin-glucoside…), terpenoids (thymol, carvacrol, borneol…) and essential oils are present in thyme [24]. Although thyme extracts and essential thyme oils have been shown to posses various pharmacodynamic activities, there have been no *in vivo* studies of the oral administration of thyme extract with all its bioactive compounds to clarify its bioavailability and metabolism. For that purpose, and in order to achieve method validation, standards of the more representative components of the phenolic and monoterpene fractions of thyme were spiked in plasma samples and then the quality parameters of the method were determined. Extraction recoveries higher than 75% for targeted analytes with a matrix effect lower than 18% were obtained. LODs ranged from 1 to 65 µg/L, except for the thymol sulfate metabolite which was 240 µg/L. The authors applied the method to the analysis of rat plasma obtained at different times, from 0 to 6 h, after an acute intake of thyme extract (5g/kg body weight). Different thyme metabolites were identified and were mainly derived from rosmarinic acid (coumaric acid sulfate, caffeic acid sulfate, ferulic acid sulfate, hydroxyphenylpropionic acid sulfate, dihydroxyphenylpropionic acid sulfate and hydroxybenzoic acid) and thymol (thymol sulfate and thymol glucuronide).

In a similar work, Su et al. [12] developed a sensitive and rapid UHPLC-MS/MS method for the determination of ferulic acid, caffeic acid, vanillic acid, and paeoniflorin in rat plasma. Separation was carried out by reversed-phase chromatography using a mobile phase consisting of acetonitrile and water (containing 0.1% formic acid), and electrospray in positive and negative ionization mode and MRM acquisition mode was used to identify and quantify these active components. The method was successfully applied to pharmacokinetic study of all three aromatic acids and the monoterpene in rat plasma.

Recently, Wang et al. [25] developed a simple, sensitive and selective LC-ESI-MS/MS method for the simultaneous determination and the pharmacokinetic study of caffeic acid and its major pharmacologically active metabolites in rat plasma. Caffeic acid is extensively metabolized in rats via conjugative (phase II) pathways, including methylated, glucuronide and sulfated caffeic acid matabolites. Attention must be pay to the role of active metabolites when dealing with new drug studies, because sometimes the pharmacological activities of metabolites are related to their safety and efficacy, and they may make a very important contribution to the pharmacodynamics of a given compound. Ferulic acid and isoferulic acid are the two major bioactive metabolites of caffeic acid, therefore analytical methodologies for their simultaneous quantification with caffeic acid are necessary. For that purpose, a reversed-phase LC-MS/MS method was proposed to determine and to evaluate pharmacokinetic profiles of caffeic, ferulic and isoferulic acids in rats after oral and intravenous administration of caffeic acid [25]. All analytes formed deprotonated molecules [M-H]$^-$ as major ion peaks, as expected. Among fragment product ions observed, those at *m/z* 135 for caffeic acid (loss of carboxylic group) and for ferulic and isoferulic acids (loss of the carboxylic group and the methyl from the methoxy group) were the major peaks in the product ion spectra, and then selected as transitions for quantitation purposes. Limits of quantification (LOQ) values down to 5.0 ng/mL were obtained for all analytes, with intra- and inter-day precisions lower than 5.8% and 6.5% (RSD), respectively.

Analysis of Polyphenols by LC-MS Techniques

In this section, we discuss on the determination of caffeic acid in food matrices by LC-MS. As previously commented, caffeic acid is a secondary metabolite of plants widely distributed in the vegetal matrices such as cereals, seeds, nuts, fruits, oil, honey, tea and medicinal plants and their transformed products such as wine, beer and other beverages, chocolate, etc. [26, 27].

In all these samples, however, it is important to remark that other pholyphenols commonly occur together with caffeic acid. Hence, the vast majority of analytical methods reported in the literature are focused on the simultaneous determination of a broad range of polyphenolic compounds of interest [9, 10]. Natural polyphenols from plants are typically classified into four main families, namely: phenolic acids, flavonoids, stilbenes and lignans. Among them, phenolic acids and flavonoids are the most abundant, which account for approximately the 30 and 70% of polyphenols, respectively. In consequence, these two families of compounds represent the main source of dietary polyphenols [26, 27].

Caffeic acid is a phenolic acid belonging to the sub-family of hydroxycinnamic acids, which consist of compounds possessing the common phenyl-propyl (C6-C3) backbone structure. Other major hydroxycinnamic acids of interest in food analysis comprise coumaric, ferulic and sinapic acids which differ in the substitution of the phenyl moiety. As occurs with other polyphenols, caffeic acid can be found in plants as the single molecule or often conjugated with sugars (mainly hexoses, pentoses and their derivatives) (Figure 1). In some cases, conjugation takes place by esterification reaction among the carboxylic group of caffeic acid and one of the diverse hydroxyl groups of the sugar.

It should be noted that the number of potential combinations is certainly huge owing to the wide variety of possible saccharides and –OH groups of linkage. Anyway, the list of compounds actually remarkable due to qualitative and quantitative reasons is much more limited. Among them, caffeoylquinic and dicaffeoylquinic acids –containing one or two molecules of caffeic acid and one quinic (hexose) acid– have been described as the most relevant. Some of the principal reported derivatives include chlorogenic and neochlorogenic acids. For instance, the content of caffeic acid is at the level of ~1 g kg^{-1} in teas and coffee and ~20 mg L^{-1} in wines. Regarding derivatives, chlorogenic acid is found at ~20-40 mg kg^{-1} in peers, peaches, potatoes and coffee extracts; amounts of ~10 mg kg^{-1} of neochlorogenic acid are found in prunes. It should be noted that ester bonds caffeoyl derivatives are quite labile so the degradation of these substances may occur depending on the conditions of preservation of the samples. As a result, concentrations of caffeic acid in the samples may rise with storage, manipulation and transformation in detriment of a decay of related derivatives.

The determination of caffeic acid and other polyphenols is one of the priorities of food analysis because of their great implications in sensorial and functional properties of food products [28-32]. Hence, new analytical methods are welcome to satisfy the increasing demand of rapid, robust, precise and accurate quantifications of these compounds in diverse vegetal matrices and beverages [9, 10]. Regarding organoleptic concerns, it is well known that contents of some polyphenols strongly influence on color and taste attributes (e.g., bitterness and astringency) of samples [28, 30]. Although these features are also dependent on the contents of other polyphenols such as anthocyanins or tannins, it has also been described that caffeic acid and other phenolic acids play an important role on food quality, as they may have a subtle influence on some flavor and color characteristics [29]. In the case of health-promoting properties of foodstuffs, some of them have partly been attributed to the presence of caffeic acid and other polyphenols [31, 32]. First of all, it is well known that polyphenols are the principal source of antioxidants of the human diet and the moderate consumption of products such as green tea or wines is considered as beneficial because of this activity.

Also, anti-bacterial, anti-inflammatory, anti-allergic and anti-thrombotic properties have been related to compositional profiles of polyphenols. In the particular case of caffeic acid, its medical and pharmacological significance is attributed to hypolipidemic, immunoregulatory and anti-asthmatic activities. Also, the laxative effect of fruits such as prunes has tentatively been attributed to neochlorogenic acid. Complementarily, the role of these compounds as (bio)markers of plants, foods and beverages has recently been highlighted. The concept of biomarker refers to those compounds that may be specific (or highly characteristic) of some product varieties while they are absent or scare in others. In the case of caffeic acid and polyphenols, some of them are up- or down-expressed in some product varieties/classes in comparison to the others. Such differences can be statistically evaluated and those significant compounds are considered as product (discriminating) descriptors. As a consequence, polyphenolic profiles and fingerprints can be exploited to characterize, classify and authenticate some food products and beverages such as fruits, wines, etc. [33, 34].

In the last years, dozens of new analytical methods facing the determination of polyphenols, including caffeic acid, in food products have been developed. The vast majority of such methods are based on liquid chromatography (LC). Regarding detection, apart from the widespread UV-vis spectroscopy, fluorescence and MS have been introduced to increase the sensitivity and selectivity of the methods [9, 10, 35]. In lesser extension, alternative approaches based on electrochemical devices [36, 37], gas chromatography (GC) [38] and capillary electrophoresis (CE) [39-41] have been proposed to tackle the determination of these compounds.

Prior to LC analysis, sample treatments of diverse complexity may be required [42]. When dealing with simple matrices such as alcoholic and non-alcoholic beverages (e.g., juices, beer, wine and spirits) filtration through membranes of 0.45 μm of pore size may be sufficient to obtain clean solutions. Sample dilution with water or organic solvents may also be recommendable to avoid or minimize potential matrix effects. In the case of solid samples, solvent extraction with organic and hydro-organic mixtures allows to recover the analytes of interest. The wide range of physicochemical properties of polyphenols (e.g., molecular mass, solubility, polarity and acidity) entails important differences in the extraction conditions to be used. Complementarily, sonication and centrifugation are useful to facilitate the analyte transference to the extract as well as to obtain cleaner supernatant solutions. In some cases, solvent extraction is combined with solid-phase extraction (SPE) to perform further purification and preconcentration processes.

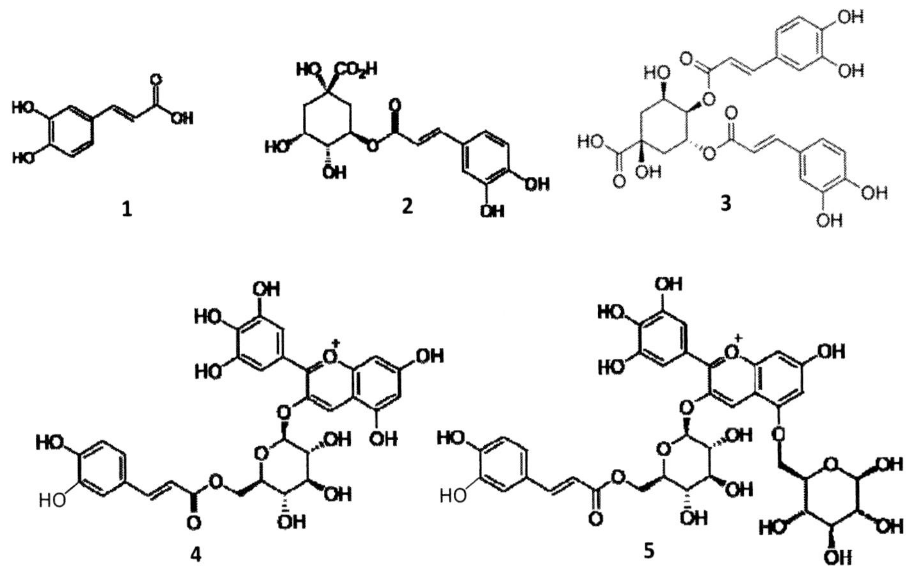

Figure 1. Examples of representative structures of some dietary caffeoyl derivatives. (1) caffeic acid; (2) caffeoyl hexoside (3-caffeoylquinic acid or chlorogenic acid); (3) dicaffeoyl hexoside (3,4-dicaffeoylquinic acid); (4) anthocyanin caffeoyl hexoside (delphinidin caffeoyl glucoside); (5) anthocyanin caffeoyl dihexoside (delphinidin caffeoyl diglucoside).

Reversed-phase C18 and anion-exchange cartridges are appropriate to retain the neutral (protonated) or basic (deprotonated) forms of phenolic acid species, respectively. Although scarcely applied to polyphenol determinations, solid-phase microextraction, accelerated solvent extraction, microwave assisted extraction and supercritical fluid extraction might be of interest to process food samples.

Pioneering methods for caffeic acid determination were developed in the seventieths combining liquid chromatography with UV-vis spectroscopic detention. Since then, hundreds of LC methods have been proposed to deal with the high diversity of polyphenols and derivatives as well as the great variety and complexity of some food matrices [9, 10, 33-35]. Briefly, typical mobile phases consisted of hydro-organic mixtures with MeOH or ACN as organic solvent and 10-100 mmol L^{-1} formic or acetic acid aqueous solutions to keep molecules protonated. Caffeic acid and related polyphenols can be separated by reversed-phase mode using C18 (or C8) analytical columns. Nowadays, this chromatographic mode continues being the most suitable for multi-family analysis. Throughout these four decades, the technology on column manufacturing has progressed tremendously and former HPLC columns packed with 5 μm particles are being replaced by those filled with sub-3 μm articles that allow more complex mixtures to be resolved. In this regard, the implementation of novel UHPLC has led to revisit and re-optimize the former HPLC methods. As a result, new methods exhibit excellent chromatographic performance, selectivity and reduced analysis time. Recently, the introduction of hydrophilic interaction chromatography (HILIC) as an alternative to reversed-phase mode has opened out great possibilities, especially for dealing with the most polar analytes.

This section is especially concerned with the main contributions of LC-MS to the determination of caffeic acid in food and beverages. Because of the great versatility and sensitivity of MS, it is not surprising that this detection technique is increasingly applied to the field of food analysis [35]. Of course, it is possible to use MS instruments alone, i.e., without being coupled to LC. In these cases, overall *m/z* spectra containing features corresponding to all analytes is obtained by direct sample infusion into the MS instrument. MS spectra can be used for target analysis or fingerprinting in different ways, namely: structural elucidation by MS^n experiments, analyte screening in complex food samples or compound quantification. In the absence of previous chromatographic separation, however, all compounds are simultaneously ionized so problems dealing with ion suppression, isobaric species and spectral interpretation may arise.

Coupling LC-MS is much more fruitful as the chromatographic separation of components over time enhances selectivity and reduces matrix effects. LC-MS provides signals consisting of ion abundances as a function of *m/z* and retention time. Compositional profiles or fingerprints by LC-MS have been exploited to establish comparisons among different types or classes of samples. Furthermore, the use of high resolution mass spectrometers (HRMS) coupled to the LC systems results in an exceptional tool for food analysis [35]. Obtaining accurate mass measurements of the compounds of interest is the major advantage of HRMS in order to provide unambiguous identifications and quantifications. This issue will be treated in detail in the following section.

Some representative applications of (U)HPLC-ESI-MS(/MS) dealing with instrumentation possibilities, quantification approaches, and identification and characterization of caffeic acid and its derivatives in a broad variety of food samples are described in the following paragraphs. Regarding ionization devices, ESI is the most generalized ion source while atmospheric pressure chemical ionization (APCI) has scarcely been used [43]. Owing to the intrinsic capacity of caffeic and other phenolic acids to easily form anion species, the analytes can be monitored satisfactorily in negative mode as $[M-H]^-$ ions. Only a few of compounds, such as anthocyanin derivatives, display higher sensitivities in positive mode. With respect to the analyzer, nowadays, low resolution instruments such as single quadrupoles are quite obsolete and more powerful detectors are preferred. Secondary ion monitoring (SIM) may be used to record the traces corresponding to each compound of interest. In general, the deprotonated molecular ion $[M-H]^-$ results in the most intense peak of spectra so it is commonly selected to carry out the detection of each component. In the case of caffeic acid and its caffeoyl and dicaffeoyl derivatives, SIM chromatograms can be monitored at *m/z* 171, 353 and 515, respectively. The occurrence of isobaric compounds of caffeoylquinic acids (e.g., chlorogenic, neochlorogenic and cryptochlorogenic acid) obviously requires to obtain acceptable chromatographic separations in order to avoid interferences. Similar considerations have to be taken into account in the case of dicaffeoylquinic isomers and other less common derivatives.

LC-MS methods based on triple quadrupole (QqQ) analyzers may also rely on SIM approaches as above. However, multiple reaction monitoring (MRM) of selected transitions established for each compound is preferred because of its superior performance regarding both selectivity and sensitivity. Typically, quantification and confirmation transitions are defined according to the most intense peak of the spectra of each component. In the case of caffeic acid, the main transition at low collision energy (CE) of 16 eV correspond to *m/z* 179 → 135, being $[M-H]^-$ and $[M-H-CO_2]^-$ the precursor and the product ions, respectively [44, 45]. When increasing CE up to 25 eV, another transition 179 → 134 is more suitable ($[M-H]^-$ → $[M-H-CHO_2]^{\cdot-}$). For chlorogenic acid, some of the main peaks of the spectrum are related to the breaking of the molecule into hexose (quinic) and caffeic acid moieties. Depending on the CE applied, various ions at the following *m/z* can be generated to be exploited for quantification and confirmation purposes: 191 (deprotonated molecular ion minus the caffeoyl moiety, $[M-H-C_9H_6O_3]^-$); 179 (deprotonated molecular of caffeic acid, i.e., ion minus the quinic acid moiety, $[M-H-C_7H_{11}O_5]^-$), 173, (deprotonated molecular ion minus the caffeoyl moiety and water, $[M-H-C_9H_8O_4]^-$); 85 $[M-H-C_{12}H_{12}O_7]^-$. For instance, in the study by Puigventós et al. [45], transitions 353 → 191 and 353 → 85 have been used for quantification and confirmation respectively. As indicated above, it should be noted that these ions are also characteristic of the other isomers (neochlorogenic and cryptochromogenic acids) as they only differ in the position of the OH group involved in the esterification. If so, the selected transitions are common to all these compounds and interferences may arise. Then, such analytes need to be separated chromatographically as MS discrimination is not possible. Many other applications reported in the literature for caffeic acid determination are based on QqQ analyzers. For instance, multiclass determination of polyphenols in vegetables such as tomato, broccoli, eggplant, etc., and olive oil were carried out by Flores et al. [46]. Most of the analytes, including caffeic and chlorogenic acids, were detected by negative ionization using the same MRM transitions given above. Similar approaches were considered in the study of major phenolic constituents of Salvia Officinalis L. [47], tomato and tomato-based products [48], and tea and dietary supplements [49]. In these examples, some LC-MS analyses were accomplished in less than 3 min. Regarding detection limits by MRM, values obtained were in the order of magnitude of 1 – 100 µg L^{-1}. Finally, although scarcely exploited, some authors have also proposed the detection of caffeic acid by positive ionization. In this case, the corresponding protonated molecular ion $[M+H]^+$ is found at *m/z* 181 and some relevant fragments appear at *m/z* 145 and 117. In these circumstances, chromatograms by MRM are recorded at 181 → 145 and 181 → 117 for quantification and confirmation, respectively [50].

Ion trap (IT) analyzers are especially suitable for the structural elucidation and investigation of complex polyphenol derivatives and oligomers. In this way, HPLC-MS/MS has provided excellent results in the characterization of condensed tannins consisting of dimmers, trimmers, tetramers, etc. of catechin and epicatechin as well as in the study polyphenolic glycosides such as anthocyanins and related derivatives than may involve a great diversity of sugar combinations. These compounds may differ, for instance, in the positions of the groups participating in the condensation and/or in the sugar molecule, being glucose, galactose, rhamnose and rutinose some of the most common residues of the glycosides. The study of ion fragmentations consists of the basis of a tentative elucidation of compounds. Apart from qualitative studies, of course, quantitative determinations relying on IT are also possible and several applications of this type have been reported in the literature. For example, a HPLC-DAD-ESI-IT-MS^n method, working with both positive and negative ionization modes, was established by Sentandreu et al. for a thorough evaluation of polyphenolic composition of pomegranate samples [51]. About 151 compounds were identified (65 of them not previously reported in these samples) by MS^2 to MS^4 experiments. Among other, several new compounds related to caffeic acid were tentatively assigned, such as various caffeoyl hexosides detected in negative mode at *m/z* 341. MS^2 fragmentations of this ion showed diverse ions with *m/z* values at 281, 251, 221, 179, 161 and 135. The most intense signal, corresponding to the deprotonated molecular anion of caffeic acid at 179 ($[M-H-C_7H_{11}O_5]^-$), was further fragmented and the resulting MS^3 spectrum displayed a peak at *m/z* 135 attributed to a decarboxylation process. Anthocyanins were analyzed in positive mode as they were positively charged. Various caffeoyl derivatives of anthocyanins were encountered, including several cyanidin, delphinidin and pelargonidin hexosides. For instance, cyanidin-3,5-caffeoyl hexosides were detected as $[M+H]^+$ at *m/z* 627, being their main ion fragmentation sequence 627 → 465 → 303. In another application by del Bubba et al. [52] polyphenolic fractions of various berry cultivars were evaluated by LC-MS/MS. The compositional profiles obtained were used to try to discriminate among the berry varieties.

Alternatively to conventional LC-MS, Cifuentes group established a new method relaying on comprehensive two-dimensional LC (LCxLC) with MS detection [53]. HILIC and reversed-phase separation modes were used in the first and second chromatographic dimensions, respectively. MS consisted of IT analyzer equipped with an ESI source. The overall polyphenolic profile, containing peaks of single and condensed molecules, was obtained in 50 min. Peaks of compounds such as dicaffeoylquinic derivatives, characterized by retention times in the HILIC and C18 domains (12.2 and 31 min) and *m/z* 515.4 and MS/MS fragments at 353 and 191, were detected in apples of different varieties. Comparative results of compositions of different kind of fruits were reported.

Beyond the qualitative and quantitative studies of caffeic acid and other polyphenols, an emerging trend relies on the analysis of compositional profiles and fingerprints as a source of information to be exploited for classification and authentication purposes [54-56]. In this working topic, the number of applications involving a chemometric data analysis has increased dramatically in the last years. LC-MS provides data of exceptional quality to be further analyzed by chemometric methods such as PCA and PLS-DA. Data to be analyzed comprise concentrations of polyphenols of interest (profiling approach) or instrumental signals consisting of intensity counts as a function of *m/z* and retention time (fingerprinting approach). Further data treatments have proved to be highly efficient to facilitate the extraction of relevant information on functional and descriptive characteristics of food products. As a first example, Pérez Ràfols et al. tackled the characterization and classification of beers belonging to Ale and Lager styles [57]. Concentrations of 13 selected phenolic acids and flavonoids were quantified and the resulting data was used to establish the training and prediction models. As an example, Figure 2 shows the plots of scores and loading of classifications, with circles and squares corresponding to lager and ale classes, respectively. The classification was entirely satisfactory in terms of selectivity and sensitivity and all the samples were correctly assigned to their actual classes.

Results indicated that caffeic, syringic and gentistic acids were typical of ale varieties while coumaric and ferulic acids were more abundant in lager ones. In another publication by Biasoto et al., MS spectra acquired were used for flavor characterization of wines [49]. In this study, it was found that caffeic and caffeoyl tartaric acids were not extensively distributed but they were only present in some wine varieties. Data was explored by PCA to assess the influence of grape varieties on the compositional profiles. Additionally, sensory attributes given by expert panelists were correlated with contents of some constituents in order to try to identify chemical descriptors of taste properties. Tentatively, caffeic acid and other organic acids were considered to contribute to wine sourness.

As another example by Puigventós et al., concentrations of selected compounds determined by HPLC-ESI-QqQ-MS were used to try to discriminate among cranberry- and grape-based products [45]. Although caffeic was not a noticeable descriptor, it was found that chlorogenic acid was more abundant in cranberry samples. In the application by Checa and coworkers, following a fingerprinting approach, Spanish wines from three protected designations of origin (PDO) were analyzed by HPLC-ESI-QqQ-MS [58].

The resulting chromatograms contained a lot of peaks of unknown compounds that could be selected and used as for classification purposes. Areas of representative peaks were considered as a source of analytical information.

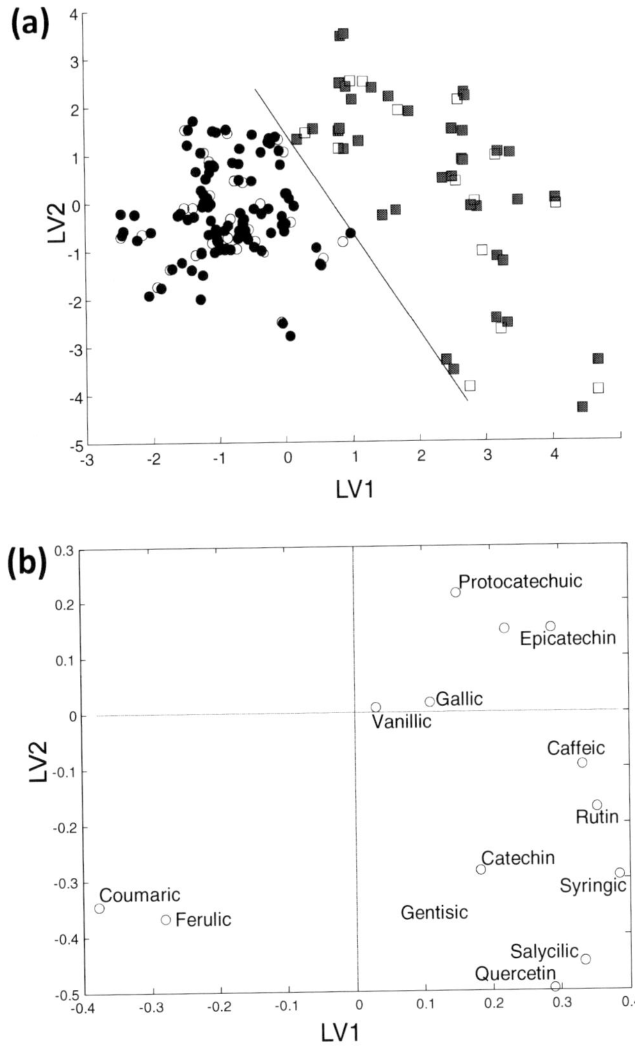

Figure 2. Classification of beers into lager and ale classes by PLS-DA. (a) Plot of scores; (b) Plot of loadings. Symbols: Circle = lager; square = ale; solid symbol = calibration sample; empty symbol = test sample.

Wine samples belonging to the different PDO were classified using PLS-DA. Classification models were highly satisfactory and the vast majority of samples were correctly assigned into their actual classes. The identification of chemical compound behind each of these contributions was based on MS and MS/MS analysis. Conclusions on characteristic components of each PDO were extracted and, for instance, gallic acid and trans-piceid were present in higher concentrations in Penedes wines, trans-caffeoyltartaric and trans-coumaroyltartaric were representative of Rioja, while quercetin-3-O-glucuronide and cis-piceid were descriptive of Ribera del Duero wines.

High Resolution Mass Spectrometry in the Analysis of Caffeic Acid

High-resolution mass spectrometry (HRMS) and accurate mass measurements have recently gained great popularity due to their capability to provide more comprehensive information concerning the exact molecular mass, elemental composition and detailed molecular structure of a compound [59]. Among the multiple benefits provided by HRMS over classical unit-mass-resolution tandem mass spectrometry we can find: *i*) differentiation of isobaric compounds (different compounds with the same nominal mass but different elemental composition); *ii*) simplification of sample-preparation procedures, thereby leading to faster methodologies requiring less sample manipulation; *iii*) information gathered by a single injection that can be used for quantification and screening purposes, including targeted, suspect and non-targeted analysis; *iv*) collection of full-scan spectra that can be stored and used in a later stage retrospective analysis, thus permitting the formulation of a posteriori hypotheses involving structural elucidation of unknown or suspected compounds [59, 60].

From the different HRMS instrumentation available [magnetic sector, time-of-flight (TOF), Orbitrap, and Fourier transform ion cyclotron (FTICR)] TOF and Orbitrap are the most-commonly used analyzers for the HPLC and/or UHPLC analysis of phenolic compounds. This is probably due to the recent advances in both LC-TOF-MS and Orbitrap mass analyzers that have significantly reduced instrument costs (especially when compared to FTICR) while maintaining high resolving powers of approximately 10,000-40,000 FWHM (full width at half maximum) and 10,000-140,000 FWHM for TOF and Orbitrap, respectively [61].

As already commented for LC-MS methods (previously section), the current trends in the analysis of caffeic acid by HRMS also focus on the simultaneous determination of a wide range of polyphenolic compounds of interest. LC-HRMS using a TOF analyzer has been successfully applied to the analysis of phenolic compounds (including caffeic acid) in Chinese traditional medicine [14], Lonicera species [62, 63], lemon verbena extract [64], buckwheat [65], and biological fluids [66]. For instance, Hossain et al. [24] examined polyphenols distributed in 4 major categories (hydroxycinnamic acid derivatives, hydroxybenzoic acid derivatives, flavonoids, and phenolic terpenes) in 5 Lamiaceae spices (rosemary, oregano, sage, basil, and thyme) by using liquid chromatography coupled to a quadrupole time-of-flight (Q-TOF) mass spectrometer. Phenolic compounds were identified on the basis of accurate mass measurement technique and collision-induced dissociation analysis.

Seven different polyphenols in the category of hydroxycinnamic acid derivatives, including caffeic acid, were found to occur in all of the spices examined. An example reported by the authors and that highlights the importance of the use of HRMS is the possibility to distinguish in the analyzed samples caffeic acid hexoside from dicaffeic acid. In fact, dicaffeic acid generated precursor and fragment ions similar to those of caffeic acid hexoside. As an example, Figure 3 shows an schematic diagram of the production of fragments from caffeic acid hexoside (*m/z* 341.09) during collision induced dissociation (CID) analysis. In this context, application of accurate mass measurement permitted to discriminate caffeic acid hexoside (calculated from $[M-H]^-$ = 341.0873) from dicaffeic acid (calculated from $[M-H]^-$ = 341.0660). In the same way, HRMS analysis allowed the authors to distinguish other compounds with the same nominal mass but dissimilar exact masses like epirosmannol (calculated *m/z* 345.1702, observed *m/z* 345.1702) from methyl carnosate (calculated *m/z* 345.2066, observed *m/z* 345.2054).

Chen et al. [14] performed a LC-TOF-MS analysis (mass range *m/z* 85–1100) in both positive and negative ion mode for the identification of multiple components in *Guanxinning* injection (GXNI), a Chinese medicine injections mainly composed of *Salvia miltiorrhiza* and *Ligusticum Chuanxiong* extracts. A total of 50 compounds from five compound classes, including saccharides, amino acids, organic acids, phenolic acids and phthalides, were identified or tentatively characterized on the basis of accurate mass measurements and subsequent TOF-MS product ions.

In this study, HRMS was extremely usefulness in the analysis of two groups of compounds, such as saccharides and phenolic acids, who shared some products ions at the same nominal mass such the ion at *m/z* 179.0, and also some neutral losses, such as 162 and 180 Da.

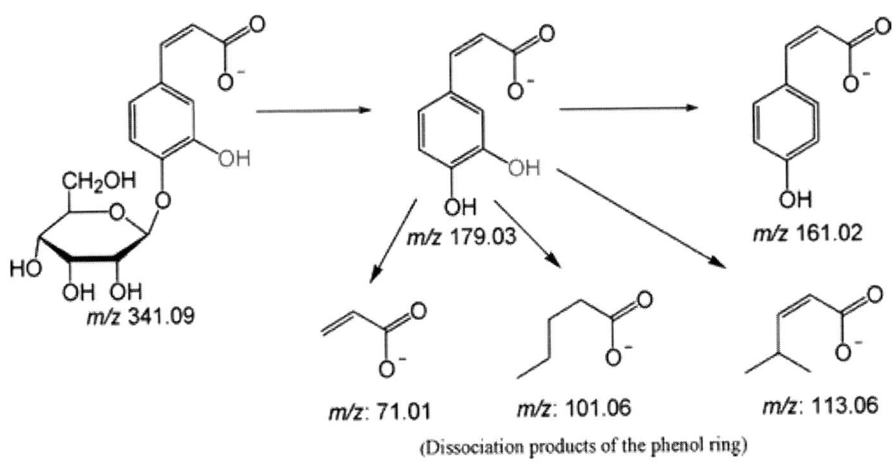

Figure 3. Schematic diagram of the production of fragments from caffeic acid hexoside (m/z 341.09) during CID analysis. Reprinted with permission from reference [24]. Copyright (2010) American Chemical Society.

In fact, thanks to exact mass measurement data provided by TOF-MS, the ion at *m/z* 179 have been attributed to deprotonated caffeic acid ($C_9H_7O_4^-$, *m/z* 179.0350) for phenolic acids and to deprotonated monosaccharide ($C_6H_{11}O_6^-$, *m/z* 179.0561) for saccharides. And it has also been possible to confirm the $C_6H_{10}O_5$ (162.0528 Da) and $C_6H_{12}O_6$ (180.0634 Da) fragments for saccharides, and $C_9H_6O_3$ (162.0317 Da) and $C_9H_8O_4$ (180.0423 Da) fragments for phenolic acids.

HRMS technique has also proven to be a reliable tool for the identification and structural elucidation of phenolic compounds in beer samples. A linear ion trap quadrupole-Orbitrap-mass spectrometry (LTQ-Orbitrap-MS) operating with resolution of 30,000 at *m/z* 400 has been recently used to identify 47 phenolic compounds, including simple phenolic acids, hydro-xycinnamoylquinics, flavanols, flavonols, flavones, alkylmethoxyphenols, alpha- and iso-alpha-acids, hydroxyphenylacetic acids and prenylflavonoids in four types of beer samples (ager, Pilsen, Märzenbier and nonalcoholic beer) [15].

Data-dependent scan experiments were very useful for the identification of unknown phenolic compounds since it provided high resolution and accurate mass product ion spectra from precursor ions that were unknown beforehand within a single analysis. As a result, 7 phenolic compounds, including caffeic acid-*O*-hexoside I and caffeic acid-*O*-hexoside II (*m/z* 341.0877, 0.43 mDa), have been reported in beer for the first time. An RP-HPLC-ESI-TOF-MS analytical method has also been used to separate and characterize a total of thirty phenolic compounds present in buckwheat flour [65]. Again, the use of HRMS allowed the tentative characterization, for the first time, of three compounds in buckwheat samples [i.e., the compound with $[M-H]^-$ 487.1463 *m/z* that was identified as 1-*O*-caffeoyl-6-*O*-alpha-rhamnopyranosyl-beta-glycopyranoside (swertiamacroside)].

LC-HRMS based metabolomics is another rapidly growing field, as reflected by the numerous articles which were published in the last few years in this area.

In fact, metabolite profiling and characterization has been greatly improved by the availability of modern mass spectrometers that are capable of providing high resolution and accurate mass. For instance, Ni et al. [13] described a UPLC-Q-TOF-MS method for the screening and identification of the constituents and their metabolites in rat plasma and urine after oral administration of *Glechoma longituba* extract. The results revealed that 21 parent components of *G. longituba* extract were absorbed into the blood circulation of the rats and a total of 80 metabolites of 9 parent compounds were tentatively detected in *vivo*. Ten metabolites deriving from caffeic acid (parent drug) have been identified as follows: $C_9H_8O_8S$ (*m/z* 274.9884), $C_9H_8O_6$ (*m/z* 211.0279), $C_{15}H_{16}O_{10}$ (*m/z* 355.0667), $C_9H_8O_7S$ (*m/z* 258.9891), $C_{14}H_{16}O_9$ (*m/z* 327.0707), $C_{12}H_{13}NO_5S$ (m/z 282.0479), $C_9H_8O_3$ (*m/z* 163.0417), $C_{10}H_{10}O_5$ (*m/z* 209.0458), $C_9H_{10}O_4$ (*m/z* 181.0543), $C_8H_7O_2$ (*m/z* 134.0382), and $C_{10}H_{10}O_4$ (*m/z* 193.0521).

Orbitrap mass analyzers are the other popular choice for labs looking for high-resolution mass spectrometers. Orbitrap-based mass spectrometer methodologies have been employed to analyze phenolic compounds in several matrices such as strawberry products [16], olive oil [67], barberry herb [68], as well as natural extracts [69, 70]. Recently, polyphenolic profiles of 44 unifloral Serbian honeys have been analyzed using UHPLC with hybrid mass spectrometer formed by Linear Trap Quadrupole (LTQ) and Orbitrap mass analyzers [71].

Caffeic acid (*Rt*=1.83 min) and caffeic acid phenylethyl ester (CAPE) (*Rt*=3.99 min) were among the identified compounds in the analyzed samples. The data obtained was then utilized by the authors to perform a PCA (principal component analysis) statistical analysis in order to select and establish floral markers of the botanical origin of Serbian honeys.

A LTQ-Orbitrap method has also been developed by Li et al. [69] for the identification of bioactive constituents in the whole plant of *Sarcandra glabra* and its related four preparations. The resolution of the hybrid mass spectrometer was set to 30,000 FWHM with a mass range of *m/z* 100-1000. According to the authors of the study, three monocaffeoylquinic acids were characterized as 3-*O*-caffeoylquinic acid, 5-*O*-caffeoylquinic acid and 4-*O*-caffeoylquinic acid, respectively, according to the elution order and the relative abundance of fragments. 4-*O*-Caffeoylquinic acid was easily distinguished from 3-*O*- and 5-*O*-caffeoylquinic acid by its different MS^2 base peak at *m/z* 173.0444.

On the other hand, although 3-*O*-caffeoylquinic acid and 5-*O*-caffeoylquinic acid generated an identical MS^2 base peak at *m/z* ~191.1, they have been distinguished by the relative intensity of the ion at *m/z* 179.0338 (deprotonated caffeic acid). Therefore, this study provides further evidence of the utility of HRMS in developing a high quality fingerprint for the comprehensive quality control of herbal remedies and conventional drugs.

Finally, Fourier transform ion cyclotron (FTICR) instruments have also demonstrated high resolution, high accuracy and broad dynamic range. However, they are very expensive and somewhat more complex to operate when compared to other MS instrumentations.

These aspects have probably reduced the number of applications reported in the literature through the years. An interesting and recent example of the application of FTICR in the analysis of phenolic acid compounds has been reported by Li et al. [72], who developed an electrospray ionization FTICR mass spectrometry method (ESI-FTICR-MS) operating in negative ion mode to determine the main components in crude extract of the Chinese herb Tanshen. Accurate mass analysis allowed to positively identify the elemental compositions of eleven phenolic acid compounds (mainly formed by the conjugates of danshensu and caffeic acid or caffeic acid dimmer).

CONCLUSION

Caffeic acid as well as other polyphenolic compounds are not essential nutrients but they are secondary metabolites synthesized by plants with important bioactive properties that may have an important effect on human health. It is for this reason that the determination of this kind of compounds in the characterization of plant extracts is increasing in the last years. Liquid chromatography coupled to mass spectrometry or to tandem mass spectrometry (LC-MS(/MS)) is one of the most frequently used techniques for that purpose. As many pholyphenols commonly occur together with caffeic acid, most of the analytical methods reported in the literature are focused on the simultaneous determination of a broad range of polyphenolic compounds of interest. LC-MS(/MS) chromatographic conditions, ionization sources, and MS and HRMS analyzers frequently used, as well as strategies for the structural characterization and the qualitative and quantitative analysis of caffeic acid, its active metabolites, and other phytochemicals have been discussed by means of relevant applications.

LC-MS/MS has been the technique of choice in the pharmacokinetic studies of caffeic acid as a biological active compound as well as its precursor and major metabolites (caffeic acid phenethyl ester, ferulic acid and isoferulic acid) in several biological fluids such as blood and plasma. Additionally, LC-MS/MS methods are frequently used for the determination of polyphenolic compounds if foodstuffs mainly using triple quadrupole or ion-trap mass analyzers in selected reaction monitoring acquisition mode for target analysis. However, beyond the qualitative and quantitative studies of caffeic acid and other polyphenols, an emerging trend relies on the analysis of compositional profiles and fingerprints as a source of information to be exploited for classification and authentication purposes. In this working topic, the number of applications involving a chemometric data analysis has increased dramatically in the last years. LC-MS provides data of exceptional quality to by further analyzed by chemometric methods such as PCA and PLS-DA. Data to be analyzed comprise concentrations of polyphenols of interest (profiling approach) or instrumental signals consisting of intensity counts as a function of *m/z* and retention time (fingerprinting approach).

Lately, high-resolution mass spectrometry and accurate mass measurements, mainly using time-of-flight and Orbitrap mass analyzers, have gained great popularity due to their capability to provide more comprehensive information concerning the exact molecular mass, elemental composition and detailed molecular structure of compounds, and the number of LC-HRMS dealing with the analysis of caffeic acid among other polyphenolic compounds is increasing. LC-HRMS based metabolomics is another rapidly growing field, as reflected by the numerous articles which were published in the last few years in this area. In fact, metabolite profiling and characterization has been greatly improved by the availability of modern mass spectrometers that are capable of providing high resolution and accurate mass measurements.

ACKNOWLEDGMENTS

This work has been funded by the Spanish Ministry of Economy and Competitiveness under the project CTQ2012-30836, and from the Agency for Administration of University and Research Grants (Generalitat de Catalunya, Spain) under the projects 2014 SGR-377 and 2014 SGR-539.

REFERENCES

[1] Scalbert, A., Manach, C., Morand, C., Remesy, C. & Jimenez, L. (2005). Dietary polyphenols and the prevention of diseases. *Crit. Rev. Food Sci. Nutr.*, 45, 287-306.

[2] Bach-Faig, A., Berry, E. M., Lairon, D., Reguant, J., Trichopoulou, A., Dernini, S., Medina, F. X., Battino, M., Belahsen, R., Miranda, G. & Serra-Majem, L. (2011). Mediterranean Diet Foundation Expert Group. Mediterranean diet pyramid today. Science and cultural updates. *Public Health Nutr.*, 14, 2274-2284.

[3] Middleton, E., Jr., Kandaswami, C. & Theoharides, T. C. (2000). The effects of plant flavonoids on mammalian cells: Implications for inflammation, heart disease, and cancer. *Pharmacol. Rev.*, 52, 673-751.

[4] Boerjan, W., Ralph, J. & Baucher, M. (2003). Lignin biosynthesis. *Annu. Rev. Plant Biol.*, 54, 519-546.

[5] Anwar, J., Spanevello, R. M., Pimentel, V. C., Gutierres, J., Thome, G., Cardoso, A., Zanini, D., Martins, C., Palma, H. E., Bagatini, M. D., Baldissarelli, J., Schmatz, R., Leal, C. A. M., da Costa, P., Morsch, V. M. & Schetinger, M. R. C. (2013). Caffeic acid treatment alters the extracellular adenine nucleotide hydrolysis in platelets and lymphocytes of adult rats. *Food Chem. Toxicol.*, 56, 459-466.

[6] Yamada, J. & Tomita, Y. (1996). Antimutagenic activity of caffeic acid and related compounds. *Biosci. , Biotechnol. , Biochem.*, 60, 328-329.

[7] Chung, T. W., Moon, S. K., Chang, Y. C., Ko, J. H., Lee, Y. C., Cho, G., Kim, S. H., Kim, J. G. & Kim, C. H. (2004). Novel and therapeutic effect of caffeic acid and caffeic acid phenethyl ester on hepatocarcinoma cells: complete regression of hepatoma growth and metastasis by dual mechanism. *FASEB J.*, 18, 1670-1681, 10.

[8] Piazzon, A., Vrhovsek, U., Masuero, D., Mattivi, F., Mandoj, F. & Nardini, M. (2012). Antioxidant Activity of Phenolic Acids and Their Metabolites: Synthesis and Antioxidant Properties of the Sulfate Derivatives of Ferulic and Caffeic Acids and of the Acyl Glucuronide of Ferulic Acid. *J. Agric. Food Chem.*, 60, 12312-12323.

[9] Ignat, I., Volf, I. & Popa, V. I. (2011). A critical review of methods for characterisation of polyphenolic compounds in fruits and vegetables. *Food Chem.*, 126, 1821-1835.

[10] Escarpa, A. & Gonzalez, M. C. (2001). An overview of analytical chemistry of phenolic compounds in foods. *Crit. Rev. Anal. Chem.*, 31, 57-139.

[11] Tang, C. & Sojinu, O. S. (2012). Simultaneous determination of caffeic acid phenethyl ester and its metabolite caffeic acid in dog plasma using liquid chromatography tandem mass spectrometry. *Talanta*, 94, 232-239.

[12] Su, S., Cui, W., Duan, J. A., Hua, Y., Guo, J., Shang, E., Liu, P. & Tang, Y. (2013). UHPLC-MS Simultaneous Determination and Pharmacokinetic Study of Three Aromatic Acids and One Monoterpene in Rat Plasma after Oral Administration of Shaofu Zhuyu Decoction. *Am. J. Chin. Med.*, 41, 697-715.

[13] Ni, S., Qian, D., Duan, J. A., Guo, J., Shang, E. x., Shu, Y. & Xue, C. (2010). UPLC-QTOF/MS-based screening and identification of the constituents and their metabolites in rat plasma and urine after oral administration of Glechoma longituba extract. *J. Chromatogr. B: Anal. Technol. Biomed. Life Sci.*, 878, 2741-2750.

[14] Chen, X., Lou, Z., Zhang, H., Tan, G., Liu, Z., Li, W., Zhu, Z. & Chai, Y. (2011). Identification of multiple components in Guanxinning injection using hydrophilic interaction liquid chromatography/time-of-flight mass spectrometry and reversed-phase liquid chromatography/time-of-flight mass spectrometry. *Rapid Commun. Mass Spectrom.*, 25, 1661-1674.

[15] Quifer-Rada, P., Vallverdu-Queralt, A., Martinez-Huelamo, M., Chiva-Blanch, G., Jauregui, O., Estruch, R. & Lamuela-Raventos, R. (2015). A comprehensive characterisation of beer polyphenols by high resolution mass spectrometry (LC-ESI-LTQ-Orbitrap-MS). *Food Chem.*, 169, 336-343.

[16] Álvarez-Fernández, M. A., Cerezo, A. B., Canete-Rodríguez, A. M., Troncoso, A. M. & Garcia-Canas, V. (2015). Composition of nonanthocyanin polyphenols in alcoholic-fermented strawberry products Using LC-MS (QTRAP), high-resolution MS (UHPLC-Orbitrap-MS), LC-DAD, and antioxidant activity. *J. Agric. Food Chem.*, 63, 2041-2051.

[17] Onori, P., DeMorrow, S., Gaudio, E., Franchitto, A., Mancinelli, R., Venter, J., Kopriva, S., Ueno, Y., Alvaro, D., Savage, J., Alpini, G. & Francis, H. (2009). Caffeic acid phenethyl ester decreases cholangiocarcinoma growth by inhibition of NF-ÎºB and induction of apoptosis. *Int. J. Cancer*, 125, 565-576.

[18] Abdel-Latif, M. M. M., Windle, H. J., El Homasany, B. S., Sabra, K. & Kelleher, D. (2005). Caffeic acid phenethyl ester modulates Helicobacter pylori-induced nuclear

factor-kappa B and activator protein-1 expression in gastric epithelial cells. *Br. J. Pharmacol.*, 146, 1139-1147.
[19] Son, S. & Lewis, B. A. (2002). Free Radical Scavenging and Antioxidative Activity of Caffeic Acid Amide and Ester Analogues: Structure-Activity Relationship. *J. Agric. Food Chem.*, 50, 468-472.
[20] Wang, X. Y., Stavchansky, S., Kerwin, S. M. & Bowman, P. D. (2010). Structure-activity relationships in the cytoprotective effect of caffeic acid phenethyl ester (CAPE) and fluorinated derivatives: Effects on heme oxygenase-1 induction and antioxidant activities. *Eur. J. Pharmacol.*, 635, 16-22.
[21] Celli, N., Dragani, L. K., Murzilli, S., Pagliani, T. & Poggi, A. (2007). In vitro and in vivo stability of caffeic acid phenethyl ester, a bioactive compound of propolis. *J. Agric. Food Chem.*, 55, 3398-3407.
[22] Gülçin, I. (2010). Antioxidant activity of caffeic acid (3,4-dihydroxycinnamic acid). *Toxicology*, 217, 213-220.
[23] Rubio, L., Serra, A., Macia, A., Borras, X., Romero, M. P. & Motilva, M. J. (2012). Validation of determination of plasma metabolites derived from thyme bioactive compounds by improved liquid chromatography coupled to tandem mass spectrometry. *J. Chromatogr. B: Anal. Technol. Biomed. Life Sci.*, 905, 75-84.
[24] Hossain, M. B., Rai, D. K., Brunton, N. P., Martin-Diana, A. B. & Barry-Ryan, C. (2010). Characterization of Phenolic Composition in Lamiaceae Spices by LC-ESI-MS/MS. *J. Agric. Food Chem.*, 58, 10576-10581.
[25] Wang, X., Li, W., Ma, X., Chu, Y., Li, S., Guo, J., Jia, Y., Zhou, S., Zhu, Y. & Liu, C. (2014). Simultaneous determination of caffeic acid and its major pharmacologically active metabolites in rat plasma by LC-MS/MS and its application in pharmacokinetic study. *Biomed. Chromatogr.*, Ahead-
[26] Manach, C., Scalbert, A., Morand, C., Remesy, C. & Jimenez, L. (2004). Polyphenols: Food sources and bioavailability. *Am. J. Clin. Nutr.*, 79, 727-747.
[27] Bueno, J. M., Ramos-Escudero, F., Saez-Plaza, P., Munoz, A. M., Jose Navas, M. & Asuero, A. G. (2012). Analysis and Antioxidant Capacity of Anthocyanin Pigments. Part I: General Considerations Concerning Polyphenols and Flavonoids. *Crit. Rev. Anal. Chem.*, 42, 102-125.
[28] Bueno, J. M., Saez-Plaza, P., Ramos-Escudero, F., Jimenez, A. M., Fett, R. & Asuero, A. G. (2012). Analysis and Antioxidant Capacity of Anthocyanin Pigments. Part II: Chemical Structure, Color, and Intake of Anthocyanins. *Crit. Rev. Anal. Chem.*, 42, 126-151.
[29] Boulton, R. (2001). The copigmentation of anthocyanins and its role in the color of red wine: A critical review. *Am. J. Enol. Vitic.*, 52, 67-87.
[30] Lesschaeve, I. & Noble, A. C. (2005). Polyphenols: Factors influencing their sensory properties and their effects on food and beverage preferences. *Am. J. Clin. Nutr.*, 81, 330S-335S.
[31] Nijveldt, R. J., Van Nood, E., Van Hoorn, D. E. C., Boelens, P. G., Van Norren, K. & Van Leeuwen, P. A. M. (2001). Flavonoids: a review of probable mechanisms of action and potential applications. *Am. J. Clin. Nutr.*, 74, 418-425.
[32] Cabrera, C., Artacho, R. & Gimenez, R. (2006). Beneficial effects of green tea - a review. *J. Am. Coll. Nutr.*, 25, 79-99.

[33] Saurina, J. (2010). Characterization of wines using compositional profiles and chemometrics. *TrAC, Trends Anal. Chem.*, 29, 234-245.

[34] Saurina, J. & Sentellas, S. (2015). Determination of Phenolic Compounds in Food Matrices: Applications to Characterization and Authentication, Chapter 13, in: Núñez, O., Gallart-Ayala, H., Martins, C.P.B.., & Lucci, P. (Eds), Fast Liquid Chromatography-Mass Spectrometry Methods in Food and Environmental Analysis, Imperial College Press, London, p. 517-547.

[35] Motilva, M. J., Serra, A. & Macia, A. (2013). Analysis of food polyphenols by ultra high-performance liquid chromatography coupled to mass spectrometry: An overview. *J. Chromatogr. A*, 1292, 66-82.

[36] Arribas, A. S., Martinez-Fernandez, M. & Chicharro, M. (2012). The role of electroanalytical techniques in analysis of polyphenols in wine. *TrAC, Trends Anal. Chem.*, 34, 78-96.

[37] Dobes, J., Zitka, O., Sochor, J., Ruttkay-Nedecky, B., Babula, P., Beklova, M., Kynicky, J., Hubalek, J., Klejdus, B., Kizek, R. & Adam, V. (2013). Electrochemical tools for determination of phenolic compounds in plants. A review. *Int. J. Electrochem. Sci.*, 8, 4520-4542.

[38] Viñas, P., Campillo, N., Martinez-Castillo, N. & Hernandez-Cordoba, M. (2009). Solid-phase microextraction on-fiber derivatization for the analysis of some polyphenols in wine and grapes using gas chromatography-mass spectrometry. *J. Chromatogr. A*, 1216, 1279-1284.

[39] Kartsova, L. A. & Alekseeva, A. V. (2008). Chromatographic and electrophoretic methods for determining polyphenol compounds. *J. Anal. Chem.*, 63, 1024-1033.

[40] Jac, P., Polasek, M. & Pospisilova, M. (2006). Recent trends in the determination of polyphenols by electromigration methods. *J. Pharm. Biomed. Anal.*, 40, 805-814.

[41] García-Canas, V. & Cifuentes, A. (2008). Recent advances in the application of capillary electromigration methods for food analysis. *Electrophoresis*, 29, 294-309.

[42] Naczk, M. & Shahidi, F. (2004). Extraction and analysis of phenolics in food. *J. Chromatogr. A*, 1054, 95-111.

[43] Nelson, B. C. & Sharpless, K. E. (2003). Quantification of the Predominant Monomeric Catechins in Baking Chocolate Standard Reference Material by LC/APCI-MS. *J. Agric. Food Chem.*, 51, 531-537.

[44] Franquet-Griell, H., Checa, A., Núñez, O., Saurina, J., Hernández-Cassou, S. & Puignou, L. (2012). Determination of Polyphenols in Spanish Wines by Capillary Zone Electrophoresis. Application to Wine Characterization by Using Chemometrics. *J. Agric. Food Chem.*, 60, 8340-8349.

[45] Puigventós, L., Navarro, M., Alechaga, E., Núñez, O., Saurina, J., Hernández-Cassou, S. & Puignou, L. (2014). Determination of polyphenolic profiles by liquid chromatography-electrospray-tandem mass spectrometry for the authentication of fruit extracts. *Anal. Bioanal. Chem.*, Ahead-

[46] Alarcón-Flores, M. I., Romero-González, R., Vidal, J. L. M. & Frenich, A. G. (2013). Multiclass determination of phytochemicals in vegetables and fruits by ultra high performance liquid chromatography coupled to tandem mass spectrometry. *Food Chem.*, 141, 1120-1129.

[47] Zimmermann, B. F., Walch, S. G., Tinzoh, L. N., Stuehlinger, W. & Lachenmeier, D. W. (2011). Rapid UHPLC determination of polyphenols in aqueous infusions of Salvia

officinalis L. (sage tea). *J. Chromatogr. B: Anal. Technol. Biomed. Life Sci.*, 879, 2459-2464.

[48] Di Lecce, G., Martinez-Huelamo, M., Tulipani, S., Vallverdu-Queralt, A. & Lamuela-Raventos, R. M. (2013). Setup of a UHPLC-QqQ-MS Method for the Analysis of Phenolic Compounds in Cherry Tomatoes, Tomato Sauce, and Tomato Juice. *J. Agric. Food Chem.*, 61, 8373-8380.

[49] Biasoto, A. C. T., Catharino, R. R., Sanvido, G. B., Eberlin, M. N. & da Silva, M. A. A. P. (2010). Flavour characterization of red wines by descriptive analysis and ESI mass spectrometry. *Food Qual. Prefer.*, 21, 755-762.

[50] Vrhovsek, U., Masuero, D., Gasperotti, M., Franceschi, P., Caputi, L., Viola, R. & Mattivi, F. (2012). A Versatile Targeted Metabolomics Method for the Rapid Quantification of Multiple Classes of Phenolics in Fruits and Beverages. *J. Agric. Food Chem.*, 60, 8831-8840.

[51] Sentandreu, E., Cerdan-Calero, M. & Sendra, J. M. (2013). Phenolic profile characterization of pomegranate (Punica granatum) juice by high-performance liquid chromatography with diode array detection coupled to an electrospray ion trap mass analyzer. *J. Food Compos. Anal.*, 30, 32-40.

[52] Del Bubba, M., Checchini, L., Chiuminatto, U., Doumett, S., Fibbi, D. & Giordani, E. (2012). Liquid chromatographic/electrospray ionization tandem mass spectrometric study of polyphenolic composition of four cultivars of Fragaria vesca L. berries and their comparative evaluation. *J. Mass Spectrom.*, 47, 1207-1220.

[53] Montero, L., Herrero, M., Ibanez, E. & Cifuentes, A. (2013). Profiling of phenolic compounds from different apple varieties using comprehensive two-dimensional liquid chromatography. *J. Chromatogr. A*, 1313, 275-283.

[54] Antignac, J. P., Courant, F., Pinel, G., Bichon, E., Monteau, F., Elliott, C. & Le Bizec, B. (2011). Mass spectrometry-based metabolomics applied to the chemical safety of food. *TrAC, Trends Anal. Chem.*, 30, 292-301.

[55] Ibáñez, C., Garcia-Canas, V., Valdes, A. & Simo, C. (2013). Novel MS-based approaches and applications in food metabolomics. *TrAC, Trends Anal. Chem.*, 52, 100-111.

[56] Hu, C. & Xu, G. (2013). Mass-spectrometry-based metabolomics analysis for foodomics. *TrAC, Trends Anal. Chem.*, 52, 36-46.

[57] Pérez-Ràfols, C., Vinas, D., Hernández-Cassou, S. & Saurina, J. (2015). Experimental design for the determination of polyphenols by liquid chromatography: application to the chemometric characterization and classification of beers. *Anal. Methods*, 7, 3283-3290.

[58] Serrano-Lourido, D., Saurina, J., Hernandez-Cassou, S. & Checa, A. (2012). Classification and characterisation of Spanish red wines according to their appellation of origin based on chromatographic profiles and chemometric data analysis. *Food Chem.*, 135, 1425-1431.

[59] Lucci, P. & Martins, C.P.B. (2015) Liquid Chromatography High Resolution Mass Spectrometry in Environmental and Food Analysis, in: Núñez O., Gallart-Ayala, H., Martins, C.P.B. & Lucci, P. (Editors), Fast Liquid Chromatography-Mass Spectrometry Methods in Food and Environmental Analysis, Imperial College Press, London, UK, pp. 33-56.

[60] Kaufmann, A. (2012). The current role of high-resolution mass spectrometry in food analysis. *Anal. Bioanal. Chem.*, 403, 1233-1249.

[61] Lucci, P., Busquets, R., & Núñez, O. (2014) UHPLC–MS(/MS) Analysis of Pesticides in Food, in Mu Naushad, M., Rizwan Khan M. (Editors), Ultra Performance Liquid Chromatography Mass Spectrometry: Evaluation and Applications in Food Analysis, CRC Press- Taylor & Francis Group, London, UK, pp. 17-50.

[62] Qi, L. W., Chen, C. Y. & Li, P. (2009). Structural characterization and identification of iridoid glycosides, saponins, phenolic acids and flavonoids in Flos Lonicerae Japonicae by a fast liquid chromatography method with diode-array detection and time-of-flight mass spectrometry. *Rapid Commun. Mass Spectrom.*, 23, 3227-3242.

[63] Ren, M. T., Chen, J., Song, Y., Sheng, L. S., Li, P. & Qi, L. W. (2008). Identification and quantification of 32 bioactive compounds in Lonicera species by high performance liquid chromatography coupled with time-of-flight mass spectrometry. *J. Pharm. Biomed. Anal.*, 48, 1351-1360.

[64] Quirantes-Piné, R., Funes, L., Micol, V., Segura-Carretero, A. & Fernández-Gutierrez, A. (2009). High-performance liquid chromatography with diode array detection coupled to electrospray time-of-flight and ion-trap tandem mass spectrometry to identify phenolic compounds from a lemon verbena extract. *J. Chromatogr. A*, 1216, 5391-5397.

[65] Verardo, V., Arráez-Román, D., Segura-Carretero, A., Marconi, E., Fernández-Gutiérrez, A. & Caboni, M. F. (2010). Identification of buckwheat phenolic compounds by reverse phase high performance liquid chromatography-electrospray ionization-time of flight-mass spectrometry (RP-HPLC-ESI-TOF-MS). *J. Cereal Sci.*, 52, 170-176.

[66] Wang, Y., Hao, H., Wang, G., Tu, P., Jiang, Y., Liang, Y., Dai, L., Yang, H., Lai, L., Zheng, C., Wang, Q., Cui, N. & Liu, Y. (2009). An approach to identifying sequential metabolites of a typical phenylethanoid glycoside, echinacoside, based on liquid chromatography-ion trap-time of flight mass spectrometry analysis. *Talanta*, 80, 572-580.

[67] Kanakis, P., Termentzi, A., Michel, T., Gikas, E., Halabalaki, M. & Skaltsounis, A. L. (2013). From olive drupes to olive oil. An HPLC-orbitrap-based qualitative and quantitative exploration of olive key metabolites. *Planta Med.*, 79, 1576-1587.

[68] Kukula-Koch, W., Aligiannis, N., Halabalaki, M., Skaltsounis, A. L., Glowniak, K. & Kalpoutzakis, E. (2013). Influence of extraction procedures on phenolic content and antioxidant activity of Cretan barberry herb. *Food Chem.*, 138, 406-413.

[69] Li, X., Zhang, Y., Zeng, X., Yang, L. & Deng, Y. (2011). Chemical profiling of bioactive constituents in Sarcandra glabra and its preparations using ultra-high-pressure liquid chromatography coupled with LTQ Orbitrap mass spectrometry. *Rapid Commun. Mass Spectrom.*, 25, 2439-2447.

[70] Navarro, M., Núñez, O., Saurina, J., Hernández-Cassou, S. & Puignou, L. (2014). Characterization of fruit products by capillary zone electrophoresis and liquid chromatography using the compositional profiles of polyphenols: Application to authentication of natural extracts. *J. Agric. Food Chem.*, 62, 1038-1046.

[71] Keckes, S., Gasic, U., Velickovic, T. C., Milojkovic-Opsenica, D., Natic, M. & Tesic, Z. (2013). The determination of phenolic profiles of Serbian unifloral honeys using ultra-high-performance liquid chromatography/high resolution accurate mass spectrometry. *Food Chem.*, 138, 32-40.

[72] Li, H., Song, F., Zheng, Z., Liu, Z. & Liu, S. (2008). Characterization of saccharides and phenolic acids in the Chinese herb Tanshen by ESI-FT-ICR-MS and HPLC. *J. Mass Spectrom.*, 43, 1545-1552.

INDEX

#

A

D

F

G

H

I

J

K

L

M

Q

R

S

T

U

V

W

X

Y

Z